Flow and Heat or Mass Transfer in the Chemical Process Industry

Flow and Heat or Mass Transfer in the Chemical Process Industry

Special Issue Editors

Dimitrios V. Papavassiliou
Quoc T. Nguyen

MDPI • Basel • Beijing • Wuhan • Barcelona • Belgrade

MDPI

Special Issue Editors
Dimitrios V. Papavassiliou
The University of Oklahoma
USA

Quoc T. Nguyen
The University of Oklahoma
USA

Editorial Office
MDPI
St. Alban-Anlage 66
Basel, Switzerland

This is a reprint of articles from the Special Issue published online in the open access journal *Fluids* (ISSN 2311-5521) from 2017 to 2018 (available at: http://www.mdpi.com/journal/fluids/special_issues/flow_heat_mass_transfer)

For citation purposes, cite each article independently as indicated on the article page online and as indicated below:

LastName, A.A.; LastName, B.B.; LastName, C.C. Article Title. *Journal Name* **Year**, *Article Number, Page Range.*

ISBN 978-3-03897-238-9 (Pbk)
ISBN 978-3-03897-239-6 (PDF)

Contents

About the Special Issue Editors

Dimitrios V. Papavassiliou, C.M. Sliepcevich Professor, received a BS degree from the Aristotle University of Thessaloniki, and MS and Ph.D. degrees from the University of Illinois at Urbana-Champaign. He joined the University of Oklahoma after working at Mobil's Upstream Strategic Research Center in Dallas, Texas. His research contributions are in the area of computations and numerical methods for turbulent flows and flows in porous media, in the area of micro- and nano-fluidics, and in the area of biologically relevant flows. He is the author or co-author of over 115 archival journal publications. His research work has been funded by federal institutions, private consortia of companies and private foundations. Dr. Papavassiliou has served as the Program Director for Fluid Dynamics at the National Science Foundation between 2013 and 2016. He is currently a member of the Consulting Editors Board of the AIChE Journal, and he is a Fellow of the AIChE.

Quoc T. Nguyen, Postdoctoral Research Fellow in the University of Oklahoma, obtained a BE degree from the Ho Chi Minh University of Technology, and Ph.D. degree in Chemical Engineering from the University of Oklahoma. He has been working in the area of computations and numerical modeling of turbulent flows, with applications found in chemical, mechanical and biomedical engineering. He has published several peer-reviewed articles and has been a reviewer of multiple scientific journals. His work has been recognized and funded by federal organizations, private companies and private foundations. He is currently a member of the American Institute of Chemical Engineers (AIChE) and American Physical Society (APS).

fluids

MDPI

Editorial

Flow and Heat or Mass Transfer in the Chemical Process Industry

Dimitrios V. Papavassiliou * and Quoc Nguyen

School of Chemical, Biological and Materials Engineering, The University of Oklahoma, Norman, OK 73019, USA; quocnguyen@ou.edu
* Correspondence: dvpapava@ou.edu; Tel.: +1-405-325-5811

Received: 24 August 2018; Accepted: 28 August 2018; Published: 28 August 2018

Keywords: convection; diffusion; reactive flows; two-phase flow; computational fluid mechanics

Flow through processing equipment in a chemical or manufacturing plant (e.g., heat exchangers, reactors, separation units, pumps, pipes, etc.) is coupled with heat and/or mass transfer. Rigorous investigation of this coupling is important for equipment design. Generalizations and empiricisms served practical needs in prior decades; however, such empiricisms can now be revised or altogether replaced by understanding the interplay between flow and transfer. Currently available experimental and computational techniques can make this possible. Typical examples of the importance of flow in enhancing the transfer of heat and mass is the contribution of coherent flow structures in turbulent boundary layers, which are responsible for turbulent transfer and mixing in a heat exchanger, and the contribution of swirling and vortex flows in mixing. Furthermore, flow patterns that are a function of the configuration of a porous medium are responsible for transfer in a fixed-bed reactor or a fluid-bed regenerator unit.

The goal of this special issue is to provide a forum for recent developments in theory, state-of-the-art experiments, and computations on the interaction between flow and transfer in single and multi-phase flow, and from small scales to large scales, as they are important for the design of industrial processes. It includes papers that cover applications in biological fluid mechanics, microfluidics, membranes, turbulent flows, and gas–liquid flows.

In microfluidics, Kanaris and Mouza [1] proposed a new design for a micromixer based on the insertion of helical structures into a straight tube. Mixing is induced by creating a swirling flow. This design is different from current devices that are based on microtubes, which are themselves helical or twisting. A detailed computational fluid-dynamics study of the proposed design led to the development of model equations for the prediction of mixing efficiency and pressure drop in the microfluidic mixer.

In bio-applications, Williams et al. [2] explored the use of perfusion bioreactors for bone-tissue engineering. In order to monitor the process of cell proliferation for stem cells seeded on polymeric scaffolds, they combined experiments with simulations. Intermittent samples were taken over a period of 16 days from the bioreactor and imaged with tomographic techniques. Flow simulations conducted on flow domains generated by exactly these images followed, allowing the calculation of the flow-induced stresses on the scaffolds. This process led to the generation of data on the average shear stress vs. reaction time—data that are critical for the thoughtful design of tissue engineering scaffolds. In the second bio-application paper, Passos et al. [3] investigated the delivery of drugs through dentinal tissue. After validation with sophisticated experiments that use micro-laser-induced fluorescence (μ-LIF), computational studies followed to characterize the details of the diffusion of the drugs through the tissue, and to develop a model with appropriate design parameters. Both the Williams et al. [2] and the Passos et al. [3] papers deal with flows and transport in microfluidic environments.

In large-scale applications with industrial interest, Sobhansrbandi et al. [4] investigated methane reformation with cold plasma. The conversion of methane to syngas (hydrogen and carbon monoxide) was found to be economically feasible, taking advantage of the vortex flow created by a smart design of the reformer. Computational fluid dynamics revealed that the vortex flow enhances the mixing of the reactant gases and the rate of hydrogen production. In the same area of thermal fluids, Duong et al. [5] presented the process of designing eco-friendly cellulose fibers from paper waste. Recycling paper, as described in this work, produced aerogels for thermal insulation that performed as well or better than commercially available insulation materials. The third paper with direct industrial relevance is a detailed analysis of the behavior of the spiral-wound membrane—a separation method used in reverse osmosis and desalination processes for water treatment. Koutsou et al. [6] used three-dimensional (3D) direct numerical simulations to show how flow and mass transfer are affected by the membrane fouling process. Correlations between the friction factor and the mass transfer coefficient obtained through this study have practical use and can also be employed to develop realistic dynamic models for the operation of spiral-wound membranes in water treatment plants. Staying in porous media, Dixon and Madeiros [7] investigated radial dispersion in fixed-bed columns. They used computations to simulate the radial velocity and concentration profile for several different computer-generated tubes packed with spheres. The flow of air and methane was simulated, and the effects of the diameter of the packed column relative to the packing particle diameter were explored. It was found that the concentration exhibited a sharp decrease close to the column wall, resulting in a two- or three-parameter model being required to accurately predict this process, while models that assumed a single dispersion coefficient across the column were inadequate.

Two-phase flow phenomena in bench-scale bubble columns were studied in the work of Mohagheghian and Elbing [8]. Experiments were used to image the full distribution of air bubble diameters as air was injected into a water column, revealing the length scale of the most frequent bubble size. In addition, it was seen that the higher-order statistics of the bubble size distribution (i.e., skewness and kurtosis) affected the flow field in the column and could be used to indicate flow regime transitions. The wakes of bubbles in bubble-column reactors were studied by Ruttinger et al. [9], with emphasis on how these wakes were influenced by flow structures such as vortex streets. High-speed particle image velocimetry (PIV) was used to illuminate the flow structure in the wake of a single bubble. It was found that the vortex street interacted with the wake behind the bubble, enhancing transfer of momentum and mixing of mass. The paper of Cortes Garcia et al. [10] also focused on gas–liquid phases, but examined the determination of the gas–liquid mass transfer coefficient by critically examining Danckwert's method commonly used to determine this coefficient. It was found that, in cases where the mass transfer between the gas–liquid phases was intense, or in cases where a lot of liquid is present, there were significant errors in the calculated mass transfer coefficient. Cortes Garcie et al. [10] suggested the use of computations and experiments to fit the theoretical reaction and diffusion equations as best practice, while Mohagheghian and Elbing [8], Ruttinger et al. [9], and Cortes Garcia et al. [10] focused on adiabatic gas–liquid flow. Liao and Lucas [11] examined flash boiling flows, where the generation of bubbles was the result of a vaporization process as heat was added to the system. Available theories and correlations were evaluated with comparisons to computational fluid-dynamics simulation results. It was found that both conduction and convection were important for the bubble growth rate and that the existing correlations are applicable within specific ranges of the Jakob and Reynolds numbers.

Turbulent flow effects on mixing were examined by Nguyen and Papavassiliou [12]. Qualitative measures of mixing effectiveness and mixing quality were suggested, and Lagrangian computations and direct numerical simulations of turbulent flow allowed the evaluation of these measures for mixing particles with different Schmidt numbers. It was found that molecular diffusion and turbulence were both important in mixing when the turbulent flow was anisotropic. Dharmarathne et al. [13] also used direct numerical simulations of turbulent flow to investigate the formation of hot and cold spots when jets of fluid with lower temperature were injected into a channel flow. Details of the development of coherent structures and their role in the transfer of heat in anisotropic turbulence were documented.

Fluids **2018**, *3*, 61

It was found that the coherent flow structures were modified by the injections, and these modifications were critical for the movement of hot fluid from the wall to the outer region. Engineering applications where cooling is necessary should, thus, ensure the generation of such coherent structures.

Finally, we thank the contributors to this special issue for sharing their research, and the reviewers for generously donating their time to select and improve the manuscripts.

Conflicts of Interest: The authors declare no conflict of interest.

References

1. Kanaris, A.; Mouza, A. Design of a Novel μ-Mixer. *Fluids* **2018**, *3*, 10. [CrossRef]
2. Williams, C.; Kadri, O.; Voronov, R.; Sikavitsas, V. Time-Dependent Shear Stress Distributions during Extended Flow Perfusion Culture of Bone Tissue Engineered Constructs. *Fluids* **2018**, *3*, 25. [CrossRef]
3. Passos, A.; Tziafas, D.; Mouza, A.; Paras, S. Computational Modelling for Efficient Transdentinal Drug Delivery. *Fluids* **2018**, *3*, 4. [CrossRef]
4. Sobhansarbandi, S.; Maharjan, L.; Fahimi, B.; Hassanipour, F. Thermal Fluid Analysis of Cold Plasma Methane Reformer. *Fluids* **2018**, *3*, 31. [CrossRef]
5. Duong, H.; Xie, Z.; Wei, K.; Nian, N.; Tan, K.; Lim, H.; Li, A.; Chung, K.-S.; Lim, W. Thermal Jacket Design Using Cellulose Aerogels for Heat Insulation Application of Water Bottles. *Fluids* **2017**, *2*, 64. [CrossRef]
6. Koutsou, C.; Karabelas, A.; Kostoglou, M. Fluid Dynamics and Mass Transfer in Spacer-Filled Membrane Channels: Effect of Uniform Channel-Gap Reduction Due to Fouling. *Fluids* **2018**, *3*, 12. [CrossRef]
7. Dixon, A.G.; Medeiros, N.J. Computational Fluid Dynamics Simulations of Gas-Phase Radial Dispersion in Fixed Beds with Wall Effects. *Fluids* **2017**, *2*, 56. [CrossRef]
8. Mohagheghian, S.; Elbing, B. Characterization of Bubble Size Distributions within a Bubble Column. *Fluids* **2018**, *3*, 13. [CrossRef]
9. Rüttinger, S.; Hoffmann, M.; Schlüter, M. Experimental Analysis of a Bubble Wake Influenced by a Vortex Street. *Fluids* **2018**, *3*, 8. [CrossRef]
10. Cortes Garcia, G.; van Eeten, K.; de Beer, M.; Schouten, J.; van der Schaaf, J. On the Bias in the Danckwerts' Plot Method for the Determination of the Gas–Liquid Mass-Transfer Coefficient and Interfacial Area. *Fluids* **2018**, *3*, 18. [CrossRef]
11. Liao, Y.; Lucas, D. Evaluation of Interfacial Heat Transfer Models for Flashing Flow with Two-Fluid CFD. *Fluids* **2018**, *3*, 38. [CrossRef]
12. Nguyen, Q.; Papavassiliou, D. Quality Measures of Mixing in Turbulent Flow and Effects of Molecular Diffusivity. *Fluids* **2018**, *3*, 53. [CrossRef]
13. Dharmarathne, S.; Pulletikurthi, V.; Castillo, L. Coherent Vortical Structures and Their Relation to Hot/Cold Spots in a Thermal Turbulent Channel Flow. *Fluids* **2018**, *3*, 14. [CrossRef]

fluids

MDPI

Article

Design of a Novel μ-Mixer

Athanasios G. Kanaris [1] **and Aikaterini A. Mouza** [2,*]

1   Scientific Computing Department, STFC, Rutherford Appleton Laboratory, Didcot OX11 0QX, UK;
    agkanaris@gmail.com
2   Department of Chemical Engineering, Aristotle University of Thessaloniki, 54124 Thessaloniki, Greece
*   Correspondence: mouza@auth.gr; Tel.: +30-231-099-4161

Received: 15 December 2017; Accepted: 24 January 2018; Published: 28 January 2018

Abstract: In this work, the efficiency of a new μ-mixer design is investigated. As in this type of devices the Reynolds number is low, mixing is diffusion dominated and it can be enhanced by creating secondary flows. In this study, we propose the introduction of helical inserts into a straight tube to create swirling flow. The influence of the insert's geometrical parameters (pitch and length of the propeller blades) and of the Reynolds number on the mixing efficiency and on the pressure drop are numerically investigated. The mixing efficiency of the device is assessed by calculating a number—i.e., the index of mixing efficiency—that quantifies the uniformity of concentration at the outlet of the device. The influence of the design parameters on the mixing efficiency is assessed by performing a series of 'computational' experiments, in which the values of the parameter are selected using design of experiments (DOE) methodology. Finally using the numerical data, appropriate design equations are formulated, which, for given values of the design parameters, can estimate with reasonable accuracy both the mixing efficiency and the pressure drop of the proposed mixing device.

Keywords: mixing; computational fluid dynamics (CFD); microfluidics; chaotic advection; helical insert

1. Introduction

The increasing demand for more economical and, at the same time, more environmentally friendly production methods has led to the design of new process equipment. By the term micro-device (μ-device), we refer to devices with at least one characteristic dimension of the order of a few millimeters. Key benefits of μ-devices are the development of energy friendly, productive, and cost-effective performance processes which at the same time provide greater security. For example, μ-reactors—i.e., devices with characteristic dimensions in the submillimeter range—offer significant advantages over conventional reactors, such as increased safety and reliability, as well as better process control and scalability. Due to the small characteristic dimension of the conduit, the flow in μ-reactors is laminar. As the extent of the chemical reactions is governed by the slow diffusive mass transfer, which in turn is proportional to the interfacial area between the reacting phases, the μ-reactor design turns out to be a μ-mixer design problem. That is why mixing in small devices has been studied extensively in recent years.

In general, mixing is achieved by stirring, where, due to the turbulent nature of flow, mixing is accomplished by advection followed by diffusion. In μ-mixers, the flow is laminar and in this case depending on the way mixing is enhanced, μ-mixers are distinguished in passive and active ones. In active μ-mixers, an external source of energy is used [1]. Active micromixers are unfortunately difficult to integrate and, in general, they have a higher implementation cost. In passive μ-mixers, whose key advantage is the low operating cost, mixing efficiency is enhanced by incorporating parts that promote secondary flows (e.g., curved sections, backward or forward-facing steps). Several comprehensive reviews of μ-mixing devices and their principles can be found in the relevant literature [1–6]. As it has been reported [1,7–9] in the micro-scale, mixing can be improved by "chaotic advection", which involves breaking, stretching, and folding of liquid streams leading to an increase of the interfacial contact area of the fluids.

In our laboratory, the mixing efficiency of several types of passive μ-mixers was investigated both experimentally and numerically [5–7,10]. Figure 1 shows a typical passive μ-mixer (i.e., Dean-type) whose functional characteristics were studied in our previous works [7,10]. Static mixers that create swirl flow are common in the macro scale and recently the application of swirl inducing configurations has been also studied in the μ-scale [11]. It is also known [12] that static helical mixers are widely used in the chemical industry for in-line blending of liquids under laminar flow conditions and also that the geometric modification of their elements can significantly improve their mixing performance. Moreover, it is suggested [13] that the addition of obstructions as part of the channel geometry can be also beneficial to micromixer efficiency.

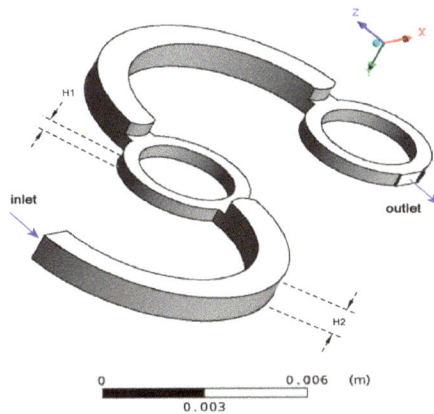

Figure 1. Typical passive μ-mixer (Dean-type) [7]. Reproduced with permission from [7].

Motivated by the above, the purpose of this study is to investigate the possibility of using a helical insert (Figure 2) that enhances mixing by generating swirl flow. More precisely, our aim is to numerically assess the effect of the geometric parameters of the helical insert (pitch and length of the insert blades) (Figure 3) as well as the physical properties of the fluids to be mixed on both the mixing efficiency of the mixing device and the resulting pressure drop.

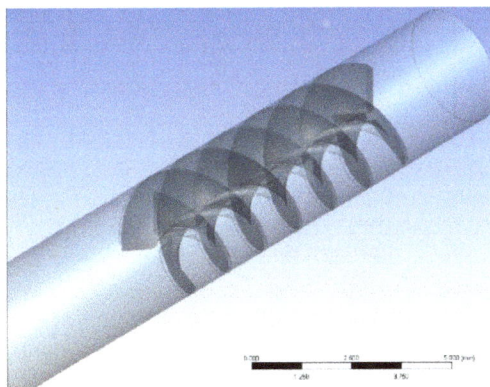

Figure 2. Schematic of the μ-mixer.

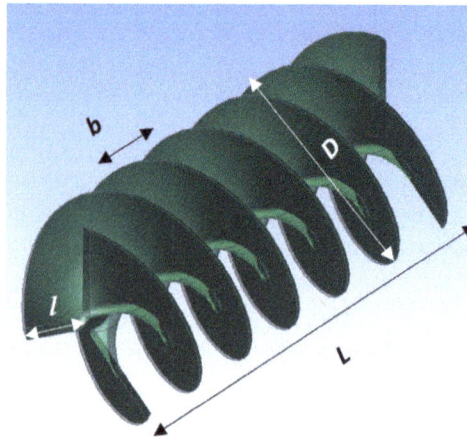

Figure 3. Geometry of the swirl generator insert.

2. Numerical Methodology

The velocity field was visualized using a computational fluid dynamics (CFD) code (ANSYS CFX 18.1, ANSYS, Inc., Karnosboro, PA, USA) while the computational geometry and the mesh were designed using the parametric features of ANSYS Workbench package. The μ-channel was modeled as a 3D computational domain. The simulations are performed using the ANSYS-CFX code, which includes the usual parts of a standard CFD code. The flow domain, constructed using the geometry section of the code, is presented in Figure 4. A grid dependency study was performed for choosing the optimum grid density. Detail of the insert is shown in Figure 4b, while the geometrical characteristics of the apparatus appear in Table 1.

Table 1. Geometrical parameters of the mixing device

Parameter	Value
Total length of the mixer, L_m	52 mm
Internal diameter of the mixer, D_m	3.1 mm
Diameter of the insert, D	3.0 mm
Position of the insert (distance from mixer inlet), e	3.0 mm
Length of the insert, L	7.5 mm
Angle between blades, φ	120°
Blade length, l	0.60–1.74 mm
Blade pitch, b	0.625–2.75 mm

In the present calculations, the computational fluid dynamics code uses the laminar flow model of ANSYS CFX and the high-resolution advection scheme for the discretization of the momentum equations. A pressure boundary condition of atmospheric pressure is set on the outlet port, while the convergence criterion is the mass-balance residual value is less than 10^{-9}.

As the numerical diffusion in the CFD calculations can influence the accuracy of the calculations, a thorough grid dependency study was performed to ensure that the solution is independent of the grid density. Pressure drop between inlet and outlet of the device, as well as the water mass fraction profile at the outlet, were considered as metrics for the evaluation of the dependence of the grid density on the solution. The final grid parameters (minimum/maximum element size, number of divisions for the sweep mesh areas, etc.) for a representative screening run were used to define the meshing parameters for the rest of the simulation runs. Number of elements for all cases varied between 850,000 and 3,000,000 elements, approximately. Due to the small characteristic dimension of the conduit the flow is laminar ($Re = 9$–100) and hence the laminar flow model was selected. The boundary conditions imposed are:

- velocities of the two fluids, on each of the two semicircles comprising the inlet,
- non-slip boundary condition at all walls,
- zero relative pressure at the outlet.

Moreover, and for the sake of simplicity, it was assumed that one of the fluids, the base fluid, has the properties of water, while the diffusion coefficient between the two fluids equals the self-diffusion coefficient of water (1×10^{-9} m^2/s).

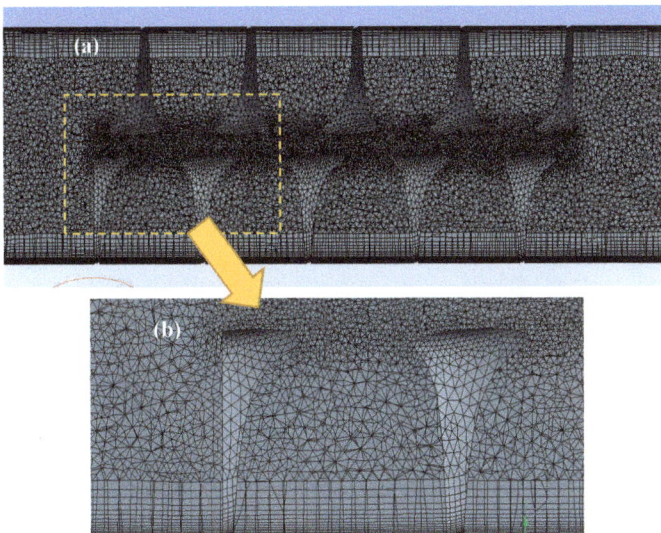

Figure 4. (a) The flow domain and the grid and (b) grid detail.

Design parameters are declared as a mix of geometrical parameters and physical properties. Two geometrical design parameters, namely the number of turns, n, of the blade within the limits of the length of the insert, and the ratio, l/D, are used. The other variables applied are the ratios of density and dynamic viscosity for the two different mixing fluids as well as the ratio of their velocities at the entrance of the conduit. The finite volume method, a fully coupled solver for the pressure and velocity coupling, and the "high order" method to distinguish the momentum equations, all provided by ANSYS CFX, are used in the solver definition section [14]. Simulations were performed for a number of iterations that ensure a satisfactory reduction in residual mass and momentum (10^{-12}).

The mixing efficiency over a cross-section A was quantified by calculating the Index of Mixing Efficiency, I_{ME}, proposed by Kanaris et al. [5], which is based on the standard deviation of mass fraction from the mean concentration over a cross section

$$I_{ME} = 1 - \frac{\sqrt{\int_A (c - \bar{c})^2 dA}}{\sqrt{\int_A (\bar{c})^2 dA}},$$

(1)

where c is the mass fraction of the base fluid, in our case the water, in each cross-section unit (i.e., each grid element) and $\bar{c}$ is the mean concentration of the same fluid over the whole cross-sectional area, A, of the device. Complete mixing is achieved when the volume fraction of the base fluid in each cross-section unit equals its mean mass fraction, $\bar{c}$, which must be calculated for each individual fluid pair by also taking into account the density of the mixing fluids:

$$\bar{c} = \frac{1}{1 + \frac{\rho}{\rho_w}},$$ (2)

A value of $I_{ME} = 1$ denotes perfect mixing, i.e., the variance from the optimal mass fraction is zero.

3. Results

3.1. Screening Experiments

To assess the effect of geometrical parameters on mixing efficiency, preliminary simulations—i.e., screening experiments—were performed. For the sake of simplicity, in these simulations both fluids had the properties of water. Initially only one helical insert was used and the simulations revealed that the length of the fluid path after the exit of the insert has only marginal contribution to the overall efficiency of the μ-mixer. Thus, to further improve mixing efficiency, a second insert was placed downstream and adjacent to the first one (Figure 5). The geometrical characteristics of the two inserts are identical, except that the blades of the two insert drive the fluid to opposite directions. A pipe extension, without any mixing promotion features, is placed downstream of the second insert, to allow the dissipation of vortices and a developed flow profile at the measuring cross section. The length-to-diameter ratio of this extension is selected to be around 10. As stated above, additional simulations ran for a length-to-diameter ratio of 20, showing borderline improvement (around +3%) of mixing efficiency and, therefore, it is considered that the originally selected ratio is adequate for the purpose of mixing characterization in this study. Also, as part of the initial screening simulations, a design with a single helical insert of up to twice the length of the proposed insert has been tested; results have shown that the addition of a counter-clockwise insert generally improves the mixing efficiency, with the magnitude being dependent on the geometry of the helical blades, as expected.

One of the initial screening runs, where the mixing fluids have identical properties and inlet velocities, while $n = 3$ and $l/D = 0.2$, is used to generate the typical results presented in Figures 5–8. In Figure 5, the flow pattern (streamlines) along the proposed device with the two inserts is presented. It is evident that the addition of the second insert improves mixing. The effect of this addition is more clearly illustrated in Figure 6, where the mass fraction distribution of the base fluid is presented at three cross sections of the device, more precisely at the exit of the first and the second insert as well as at the outlet of the device.

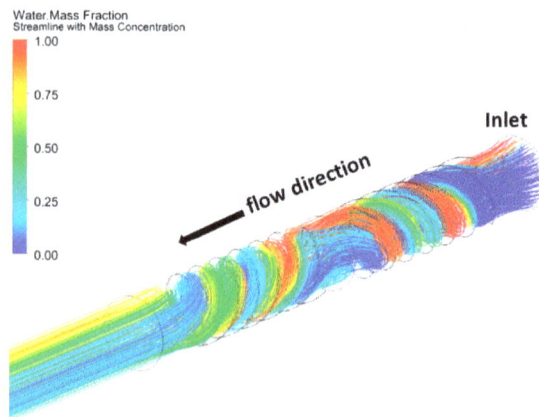

Figure 5. Typical mass concentration along streamlines ($Re = Re_w = 10$).

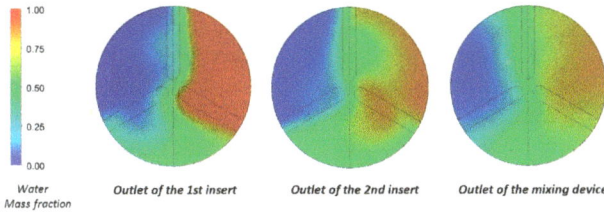

Figure 6. Mass fraction distribution at three cross sections of the device ($Re = Re_w = 10$).

In Figure 7, the performance of the proposed device—i.e., the one with the two helical inserts—is compared with the one that contains a single helical insert with the same length as well as with that of a straight pipe, whose cross section and length are the same as that of the proposed device. For a certain set of the design parameters and for the same Re number ($Re_w = Re = 10$) the use of the mixing device configuration almost doubles the value of the I_{ME}. The increase of the fluid velocity considerably affects the mixing efficiency, or equally the uniformity of the mass fraction distribution at the exit of the device (Figure 8). It is evident that, as it is expected, the mixing efficiency is mainly influenced by the value of Re, or equally the velocity of the fluid, which leads to more intense swirling flow, despite the fact that at lower velocities the contact time of the fluids is longer.

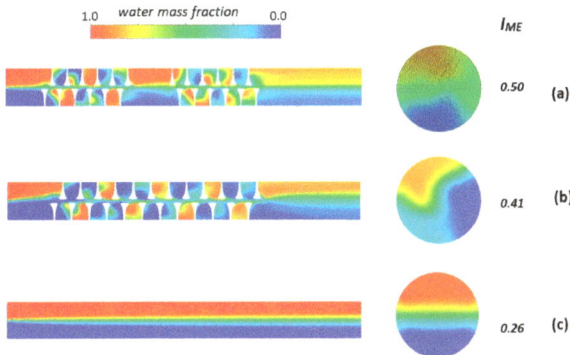

Figure 7. Comparison of the mixing performance, in terms of mass fraction distribution at a plane along the axis and at the cross section at the outlet, as well as expressed as I_{ME} at the exit of the device: (**a**) device with two inserts; (**b**) device with one insert of double length; and (**c**) straight pipe ($Re = Re_w = 10$).

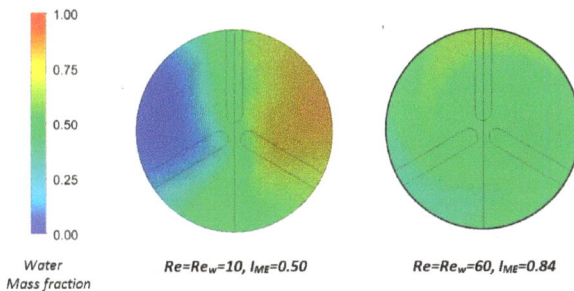

Figure 8. Effect of fluid velocity to the mixing efficiency at the outlet of the screening run device.

3.2. Parametric Study

The efficacy of the proposed mixing device was assessed by conducting a parametric study. However, to reduce the complexity of the problem, some variables—namely the length and the inside diameter of the mixing device as well as the position, the length, the radius and the angle between the blades of the insert—were kept constant (as presented in Table 1), while the two inserts are considered adjacent to each other. Also, for the sake of simplicity, the physical properties of one of the mixing fluids were assumed to be those of water, while the properties of the second fluid are variable and correspond to those of various types of water-based inks. For the same reason only two inserts were used, although it is evident that mixing will be further enhanced by inserting a series of inserts that would alternate the direction of flow. Table 2 presents the independent, dimensionless variables involved in the parametric study as discussed above together with their upper and lower bound values. Based on the values used, Re/Re_w varies between 0.05 and 1.60.

The effect of the design parameters on the efficiency of the μ-device is investigated by performing a series of simulations for certain values of the design parameters chosen by employing the Box–Behnken method, i.e., an established design-of-experiments (DOE) technique that allows the designer to extract as much information as possible from a limited number of test cases [15]. For the present study, the number of design points dictated by the Box–Behnken method is 42 and presented in Table 3.

Table 2. Constraints of the design variables

Parameter	Lower Bound	Upper Bound
Number of turns, n	1	3
Blockage ratio, l/D	0.20	0.58
Dynamic viscosity ratio, μ/μ_w	1.0	20.0
Density ratio, ρ/ρ_w	0.6	1.0
Inlet velocity ratio, u/u_w	1.0	3.0

Table 3. Design points for the simulation runs based on Box–Behnken design-of-experiments (DOE) methodology.

Run#	Turns, n	l/D	μ/μ_w	ρ/ρ_w	u/u_w	Run#	Turns, n	l/D	μ/μ_w	ρ/ρ_w	u/u_w
DP01	3	0.20	10.5	0.8	2	DP22	2	0.39	10.5	0.8	2
DP02	1	0.20	10.5	0.8	2	DP23	2	0.39	20.0	0.8	1
DP03	1	0.39	10.5	0.6	2	DP24	2	0.39	20.0	0.8	3
DP04	1	0.39	1.0	0.8	2	DP25	2	0.39	1.0	1.0	2
DP05	1	0.39	10.5	0.8	1	DP26	2	0.39	10.5	1.0	1
DP06	1	0.39	10.5	0.8	3	DP27	2	0.39	10.5	1.0	3
DP07	1	0.39	20.0	0.8	2	DP28	2	0.39	20.0	1.0	2
DP08	1	0.39	10.5	1.0	2	DP29	2	0.58	10.5	0.6	2
DP09	1	0.58	10.5	0.8	2	DP30	2	0.58	1.0	0.8	2
DP10	2	0.20	10.5	0.6	2	DP31	2	0.58	10.5	0.8	1
DP11	2	0.20	1.0	0.8	2	DP32	2	0.58	10.5	0.8	3
DP12	2	0.20	10.5	0.8	1	DP33	2	0.58	20.0	0.8	2
DP13	2	0.20	10.5	0.8	3	DP34	2	0.58	10.5	1.0	2
DP14	2	0.20	20.0	0.8	2	DP35	3	0.39	10.5	0.6	2
DP15	2	0.20	10.5	1.0	2	DP36	3	0.39	1.0	0.8	2
DP16	2	0.39	1.0	0.6	2	DP37	3	0.39	10.5	0.8	1
DP17	2	0.39	10.5	0.6	1	DP38	3	0.39	10.5	0.8	3
DP18	2	0.39	10.5	0.6	3	DP39	3	0.39	20.0	0.8	2
DP19	2	0.39	20.0	0.6	2	DP40	3	0.39	10.5	1.0	2
DP20	2	0.39	1.0	0.8	1	DP41	3	0.58	10.5	0.8	2
DP21	2	0.39	1.0	0.8	3	DP42	2	0.20	10.5	0.8	2

The generic form of a fully quadratic model with two design parameters, x_1 and x_2, is

$$Y = a_{11}x_1{}^2 + a_{22}x_2{}^2 + a_1x_1 + a_2x_2 + a_{12}x_1x_2, \tag{3}$$

The model includes the quadratic, the linear terms, and their interaction. To avoid overfitting, it is important that the researcher addresses the importance of each factor of the model. Additionally, it is also significant to take into account any possible insight regarding the potential form of the final equation and its non-linearity. For this reason, a different approach is followed in this case: the fitting approach uses the natural logarithms of the design parameters and responses. Based on the outcome, it is safe, within a certain degree of acceptable error, to ignore the quadratic terms and fit a model with only the linear terms and some of their interactions.

The following equation represents the proposed model

$$\ln(Y) = \alpha \ln(n) + \beta \ln\left(\tfrac{l}{D}\right) + \gamma \ln\left(\tfrac{\mu}{\mu_w}\right) + \delta \ln\left(\tfrac{\rho}{\rho_w}\right) + \varepsilon \ln\left(\tfrac{u}{u_w}\right) + \zeta \ln\left(\tfrac{l}{D}\right) \ln\left(\tfrac{l}{D}\right) + \\ + \eta \ln(n) \ln\left(\tfrac{\mu}{\mu_w}\right) + \theta \ln\left(\tfrac{l}{D}\right) \ln\left(\tfrac{\mu}{\mu_w}\right) + \iota \ln\left(\tfrac{u}{u_w}\right) \ln\left(\tfrac{\mu}{\mu_w}\right) + cst, \tag{4}$$

whose parameters can be calculated with regression based on response surface methodology; Y is the response, i.e., ΔP or I_{ME}, and cst is a constant. The annexed form of the final model would then be

$$Y = n^\alpha \left(\frac{l}{D}\right)^{\beta + \zeta \ln\left(\frac{l}{D}\right)} \left(\frac{\mu}{\mu_w}\right)^{\gamma + \eta \ln(n) + \theta \ln\left(\frac{l}{D}\right) + \iota \ln\left(\frac{u}{u_w}\right)} \left(\frac{\rho}{\rho_w}\right)^\delta \left(\frac{u}{u_w}\right)^\varepsilon e^{cst}, \tag{5}$$

A transform of the above model to include the Re ratio instead of the velocity ratio would be

$$Y = n^\alpha \left(\frac{l}{D}\right)^{\beta + \zeta \ln\left(\frac{l}{D}\right)} \left(\frac{\mu}{\mu_w}\right)^{\gamma + \eta \ln(n) + \theta \ln\left(\frac{l}{D}\right) + \iota \ln\left(\frac{Re}{Re_w} \frac{\mu}{\mu_w} \frac{\rho_w}{\rho}\right) + \varepsilon} \left(\frac{\rho}{\rho_w}\right)^{\delta - \varepsilon} \left(\frac{Re}{Re_w}\right)^\varepsilon e^{cst}, \tag{6}$$

Table 4 contains the parameters of the fitting models for ΔP and I_{ME}, respectively.

Table 4. Response surface model parameters for (a) ΔP & (b) I_{ME}.

Parameter	ΔP	I_{ME}
α	0.817014	0.026161
β	5.456162	−1.48235
γ	0.594749	−0.34662
δ	0.312127	−0.18884
ε	0.72966	−0.42494
ζ	1.472507	−0.99062
η	−0.02689	0.097432
θ	0.014815	−0.16046
ι	0.126574	0.092324
cst	5.91789	−0.66962

From Figure 9, where the values calculated using the proposed equations are compared with the CFD results, it is evident that they can predict with $\pm10\%$ accuracy both ΔP (Figure 9a) and I_{ME} (Figure 9b) values. The validity of the proposed correlations is further examined by comparing them with CFD the results generated using the six 'verification points' presented in Table 5 and are also in agreement (Figure 9).

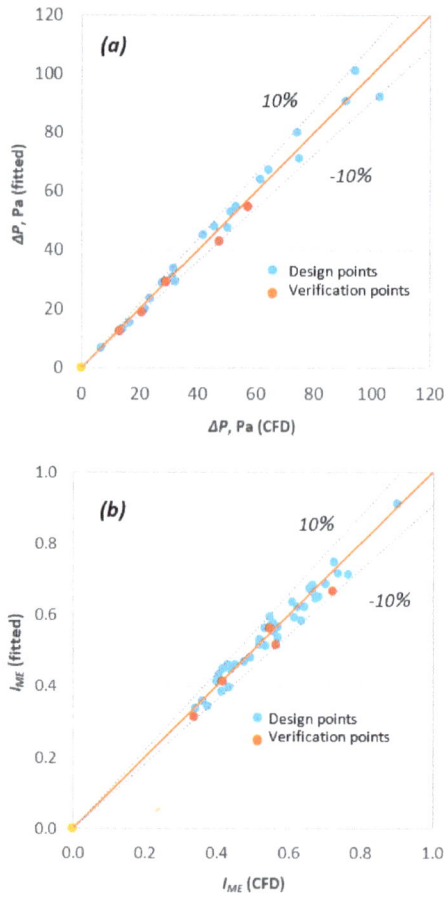

Figure 9. Comparison between CFD and fitted values for (a) ΔP and (b) I_{ME}.

Table 5. Verification points

Run#	Turns, n	l/D	μ/μ_w	ρ/ρ_w	u/u_w
VP01	3	0.20	10.5	0.8	2
VP02	1	0.20	10.5	0.8	2
VP03	1	0.39	10.5	0.6	2
VP04	1	0.39	1.0	0.8	2
VP05	1	0.39	10.5	0.8	1
VP06	1	0.39	10.5	0.8	3

4. Conclusions

In this study, we have numerically investigated the mixing efficiency as well as the corresponding ΔP inside a novel type of μ-mixer. The device comprises two successive helical inserts that propel the fluid to opposite directions and induces mixing by generating swirling flow. Screening experiments reveal that the addition of the second insert improves mixing considerably. It was also found that for the range of Re numbers investigated, the resulting pressure drop is maintained at low levels (<150 Pa).

Appropriate 'computational experiments' were then conducted to investigate the effect of the various geometrical parameters of the novel μ-mixer, i.e., the one with the two inserts, and the parameters of the mixing fluids (physical properties and flow rate) on the mixing performance of the proposed device. Both the number of the required 'computational experiments' and the values of the design parameters were selecting using a DOE methodology. Using the data obtained from the 'computational experiments' correlations, which are able to predict the mixing efficiency and the associated pressure drop with ±10% accuracy, have been formulated.

In the next stage of the study, experiments will be performed using a prototype helical insert that has been already constructed by 3D printing (Figure 10). The experimental data acquired using the device will be used for evaluating the CFD code. The aim is to provide a means of constructing the type of helical insert that would be more suitable for a given application.

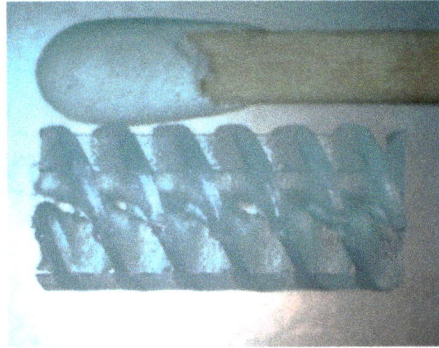

Figure 10. The helical insert constructed by 3D printing.

Acknowledgments: The authors would like to thank Spiros V. Paras for his helpful comments.

Author Contributions: Aikaterini A. Mouza had the initial conception of this work; Athanasios G. Kanaris and Aikaterini A. Mouza designed the CFD experiments, acquired and analyzed the data, and interpreted the results; Athanasios G. Kanaris has provided insights on the improvement of computational performance for the required simulations; Aikaterini A. Mouza drafted the work and revised the manuscript. This work is not affiliated with the Science and Technology Facilities Council (STFC).

Conflicts of Interest: The authors declare no conflict of interest.

Nomenclature

A	cross-section of the device	mm^2
b	blade pitch	mm
c	mass fraction in each cell of a cross-section A	dimensionless
$\bar{c}$	mean concentration over a cross-section A	dimensionless
D	diameter of the insert	mm
D_m	inside diameter of the device	mm
I_{ME}	index of mixing efficiency	dimensionless
l	blade length	mm
L	length of the insert	mm
L_m	total length of mixer	mm
n	number of turns of the insert	dimensionless
Re	Reynolds number of each liquid based on D	dimensionless
ΔP	pressure drop	Pa
μ	liquid viscosity	Pa s
ρ	liquid density	kg/m^3
φ	angle between blade	deg

References

1. Hessel, V.; Löwe, H.; Schönfeld, F. Micromixers—A review on passive and active mixing principles. *Chem. Eng. Sci.* **2005**, *60*, 2479–2501. [CrossRef]
2. Nguyen, N.-T. Micromixers—A review. *J. Micromech. Microeng.* **2005**, *15*, R1. [CrossRef]
3. Hardt, S.; Drese, K.S. Passive micromixers for applications in the microreactor and μTAS fields. *Microfluid. Nanofluid.* **2005**, *1*, 108–118. [CrossRef]
4. Mansur, E.A.; Ye, M. A state-of-the-art review of mixing in microfluidic mixers. *Chin. J. Chem. Eng.* **2008**, *16*, 503–516. [CrossRef]
5. Kanaris, A.G.; Stogiannis, I.A.; Mouza, A.A.; Kandlikar, S.G. Comparing the mixing performance of common types of chaotic micromixers: A numerical study. *Heat Transf. Eng.* **2015**, *36*, 1122–1131. [CrossRef]
6. Aref, H.; Blake, J.R. Frontiers of chaotic advection. *Rev. Mod. Phys.* **2017**, *89*, 025007. [CrossRef]
7. Kanaris, A.G.; Mouza, A.A. Numerical investigation of the effect of geometrical parameters on the performance of a micro-reactor. *Chem. Eng. Sci.* **2011**, *66*, 5366–5373. [CrossRef]
8. Yuan, F.; Isaac, K.M. A study of MHD-based chaotic advection to enhance mixing in microfluidics using transient three dimensional CFD simulations. *Sens. Actuators B Chem.* **2017**, *238*, 226–238. [CrossRef]
9. Aref, H. The development of chaotic advection. *Phys. Fluids* **2002**, *14*, 1315–1325. [CrossRef]
10. Mouza, A.A.; Patsa, C.M.; Schönfeld, F. Mixing performance of a chaotic micro-mixer. *Chem. Eng. Res. Des.* **2008**, *86*, 1128–1134. [CrossRef]
11. Huang, S.-W.; Wu, C.-Y. Fluid mixing in a swirl-inducing microchannel with square and T-shaped cross-sections. *Microsyst. Technol.* **2017**, *23*, 1971–1981. [CrossRef]
12. Pahl, M.H.; Muschelknautz, E. Static mixers and their applications. *Int. Chem. Eng.* **1982**, *92*, 205–228.
13. Rawool, A.S.; Mitra, S.K.; Kandlikar, S.G. Numerical simulation of flow through microchannels with designed roughness. *Microfluid. Nanofluid.* **2006**, *2*, 215–221. [CrossRef]
14. Versteeg, H.K.; Malasekera, W. *Computational Fluid Dynamics*; Longman Press: London, UK, 1995.
15. Box, G.E.P.; Hunter, J.S.; Hunter, W.G. *Statistics for Experimenters: Design, Innovation and Discovery*, 2nd ed.; J. Wiley and Sons, Inc.: Hoboken, NJ, USA, 2005.

fluids

MDPI

Article

Time-Dependent Shear Stress Distributions during Extended Flow Perfusion Culture of Bone Tissue Engineered Constructs

Cortes Williams III [1], Olufemi E. Kadri [2], Roman S. Voronov [2] and Vassilios I. Sikavitsas [1,*]

[1]   Stephenson School of Biomedical Engineering, University of Oklahoma, Norman, OK 73019, USA; cwilliams@ou.edu

[2]   Otto H. York Department of Chemical and Materials Engineering, New Jersey Institute of Technology, Newark, NJ 07102, USA; ok26@njit.edu (O.E.K.); rvoronov@njit.edu (R.S.V.)

*   Correspondence: vis@ou.edu; Tel.: +1-405-325-1511

Received: 9 January 2018; Accepted: 29 March 2018; Published: 3 April 2018

Abstract: Flow perfusion bioreactors have been extensively investigated as a promising culture method for bone tissue engineering, due to improved nutrient delivery and shear force-mediated osteoblastic differentiation. However, a major drawback impeding the transition to clinically-relevant tissue regeneration is the inability to non-destructively monitor constructs during culture. To alleviate this shortcoming, we investigated the distribution of fluid shear forces in scaffolds cultured in flow perfusion bioreactors using computational fluid dynamic techniques, analyzed the effects of scaffold architecture on the shear forces and monitored tissue mineralization throughout the culture period using microcomputed tomography. For this study, we dynamically seeded one million adult rat mesenchymal stem cells (MSCs) on 85% porous poly(L-lactic acid) (PLLA) polymeric spunbonded scaffolds. After taking intermittent samples over 16 days, the constructs were imaged and reconstructed using microcomputed tomography. Fluid dynamic simulations were performed using a custom in-house lattice Boltzmann program. By taking samples at different time points during culture, we are able to monitor the mineralization and resulting changes in flow-induced shear distributions in the porous scaffolds as the constructs mature into bone tissue engineered constructs, which has not been investigated previously in the literature. From the work conducted in this study, we proved that the average shear stress per construct consistently increases as a function of culture time, resulting in an increase at Day 16 of 113%.

Keywords: tissue engineering; microcomputed tomography; computational fluid dynamics; shear stress distribution; flow perfusion

1. Introduction

Every year in the United States, there are more than 500,000 bone graft surgeries [1]. In most cases, bone will regenerate after fracture with minimal complications; however, when there is a critically-sized defect or fracture healing is impaired, bone grafts must be used in order to regain proper bone function. Furthermore, bone diseases such as osteoporosis, infection, skeletal defects and bone cancer may also cause a need for bone grafts. Bone tissue engineering is a possible solution to the problems plaguing the current bone graft therapies. Because tissue engineered bone would be made using the patient's own cells, immune rejection would be eliminated. For this to work, four components are needed for tissue growth: cells that can be differentiated into bone cells, osteoconductive scaffolds acting as a matrix while the tissue grows, growth factors and other chemical stimulation and mechanical stimulation to encourage osteogenic differentiation. Mechanical stimulation, in particular, is implemented through the use of bioreactors.

Previous studies have given the indication that the shear stresses bone cells experience inside the body are between 8 and 30 dynes/cm^2 [2]. In vitro culture studies combined with computational fluid dynamic simulation results have shown that shear stresses below 15 dynes/cm^2 are conducive to increased matrix production and osteoblastic differentiation. However, if the shear rates are too high, detachment or cell death can occur [3]. In addition, whenever inhomogeneous cell seeding distributions occur, especially in the case of cell aggregates, even modest shear rates that are otherwise beneficial may result in cell detachment [4–7]. Due to this, it is important to properly model and evaluate the flow profile inside cell-seeded scaffolds [8–11]. Ideally, the localized shear rates should be anticipated in order to give proper fluid control. However, the largest barrier to this goal is the continual deposition of mineralized tissue during the culture period. After the stem cells differentiate into mature osteoblasts, both soft and hard extracellular matrices grow into the pores of the construct. This effectively alters the flow field, due to the porosity of the scaffold decreasing, and renders simulations performed on empty scaffolds invalid after the start of culture.

To combat this issue, we aimed to evaluate the localized fluid shear distributions throughout the culture period, giving an indication of the effects of tissue growth on the flow-induced stress fields, which has been extensively investigated in the literature [12–16]. Using spunbonded poly(L-lactic acid) scaffolds and a custom flow perfusion bioreactor, we cultured rat mesenchymal stem cells for 16 days under shear-induced differentiation flow ranges. The resulting constructs were imaged utilizing microcomputed tomography (μCT), segmented and reconstructed following previously published techniques [2,17,18]. This flow path allows for subsequent computational fluid dynamic (CFD) simulations on the cultured constructs.

In this manuscript, we hypothesized that the levels of fluid shear present at the walls of a scaffold, where the cells are located, will increase as a function of culture time. Previous studies have assumed that (1) the shear field predicted using a non-cultured scaffold is representative of cultured constructs and (2) that the average wall shear experienced by the cells is constant throughout a culture period [18,19]. The intention of this study is to use CFD simulations in conjunction with microcomputed tomography of mature constructs and biochemical assays to bring to light the relationship between the localized shear field and culture time, which would give researchers the ability to predict the time-dependent shear distribution in conjunction with the growing extracellular matrix within three dimensional scaffolds exposed to flow perfusion.

2. Results

2.1. Construct Cellularity

In order to validate the presence of cells in the constructs, we conducted a destructive dsDNA quantification assay. As shown in Figure 1, there is a slight decrease in scaffold cellularity between Day 1 and Day 4, and a statistically steady cellularity through the end of culture. The vertical dotted line between Day 1 and Day 4 indicates the switch in flow rate from 0.15 mL/min–0.5 mL/min. Hence the decrease between these two days represents a loss of cells either due to cell detachment or as a result of cell death [20,21]. This loss is a common occurrence, as MSCs display weak adherence to poly(L-lactic acid) (PLLA). The horizontal dashed line represents the amount of cells initially seeded on the constructs. The ratio between this line and Day 1 is known as the seeding efficiency, which in this case is 40%.

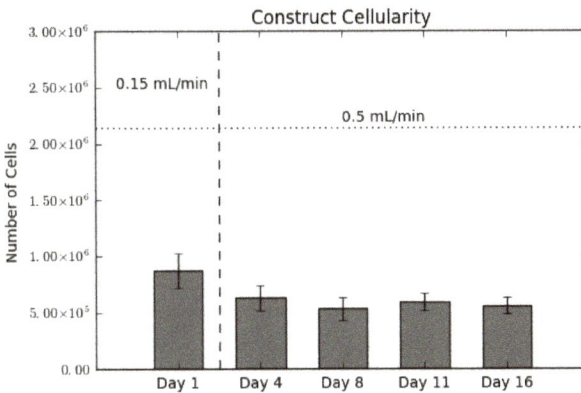

Figure 1. Construct cellularity for each construct over the culture period. The horizontal dashed line indicates the initial amount of cells seeded. The vertical dotted line indicates the switch in flow rates, from 0.15 mL/min during seeding to 0.5 mL/min for culture. Values are given as the mean ± the standard error of the mean ($n = 4$).

2.2. Calcium Deposition

Calcium deposition was measured using a calcium assay at each sacrificial time point, with results shown in Figure 2. As seen in the graph, there is a sharp increase in calcium deposition around Day 8. In conjunction with the calcium assay, we rendered a 3D representation of the constructs (Figures 3 and 4), which were imaged using μCT. Extensive mineralized tissue can be seen in samples sacrificed after Day 8, while samples prior to that time displayed only minor mineralized tissue. It can be noted that the largest standard deviation in mineralization was observed in samples sacrificed on Day 8, a time point that matches the onset of extensive mineralization (Figure 3). A similar spike in mineralized tissue can be seen around Day 11.

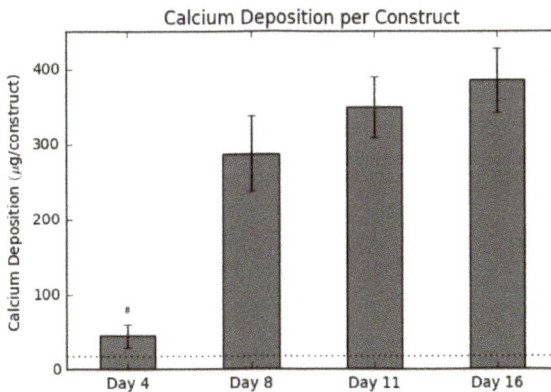

Figure 2. Calcium levels present within each construct over the culture period. The horizontal dotted line represents the background signal for an empty construct. Values are given as the mean ± the standard error of the mean ($n = 4$). The # signifies the significantly lowest value ($p < 0.01$).

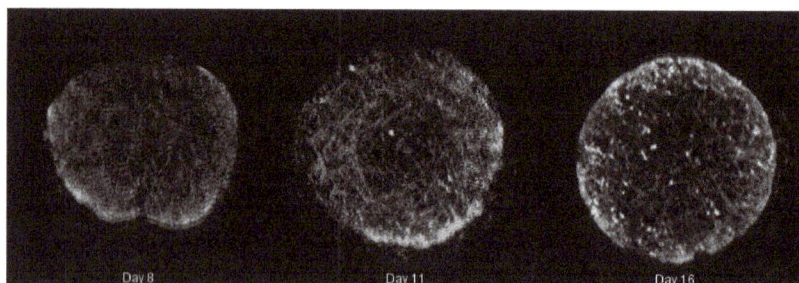

Figure 3. Summary of mineralized tissue (hard extracellular matrix (ECM)) deposited in cultured constructs rendered during microcomputed tomography (μCT) with Simple Viewer for samples taken on Days 8, 11 and 16, respectively.

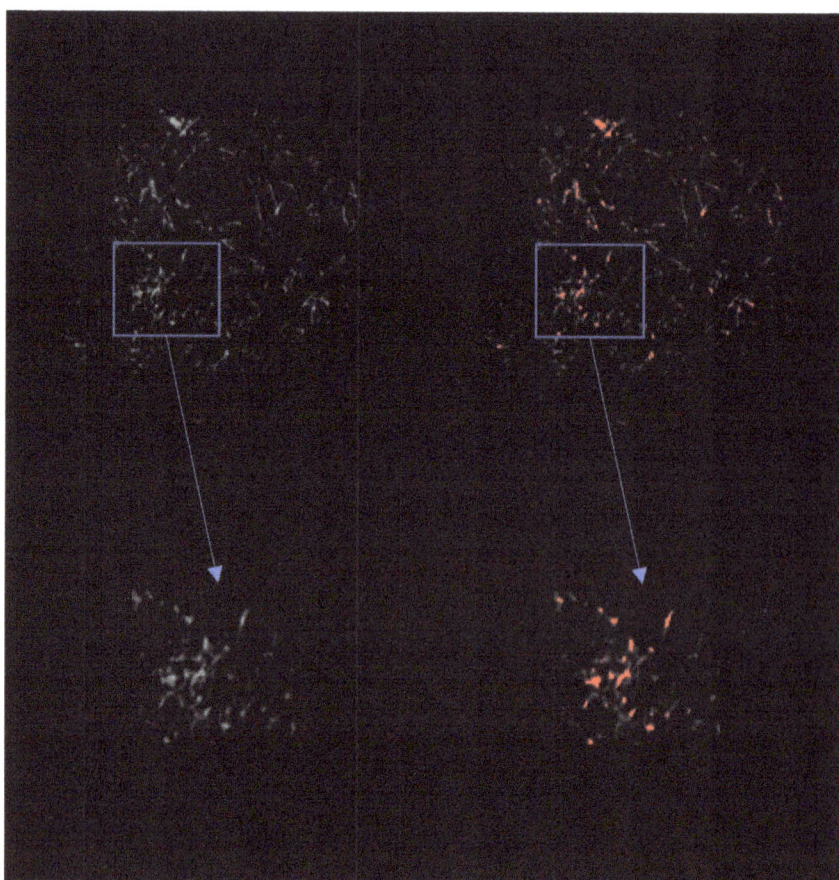

Figure 4. (Left) 2D grayscale view of the scaffold after μCT imaging. The top row is a view of the entire scaffold, while the bottom row is a magnified view of the indicated area of interest. (Right) 2D view of the scaffold after μCT imaging with ECM indicated in red.

2.3. Shear Stress Distributions over Time

The localized fluid shear stress distributions for the reconstructions at each intermittent time point are shown in Figure 5, where light blue is Day 1, green is Day 4, red is Day 8 and dark blue is Day 11. The data presented show a pronounced increase in shear stress levels with an increase in culture time, which is evident by the rightwards shift in distributions shown in the graph.

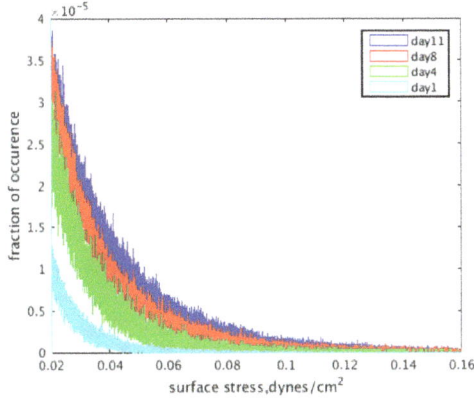

Figure 5. Wall shear stress distributions based on the day a construct was removed from culture and imaged.

2.4. Effects of Calcium Deposition on Localized Shear Fields

Figure 6 shows isometric sections of reconstructed scaffolds, with wall shear stress heat maps overlaid on top. It is evident that there are higher levels of shear stress present during the later time points, which is supported by the distributions shown in Figure 5. As pore size decreases, bottlenecks occur resulting in increased fluid velocity and, subsequently, increased shear stress (yellow and red colors in the figure). Furthermore, these increased levels of shear stress are more widely distributed as culture time increases.

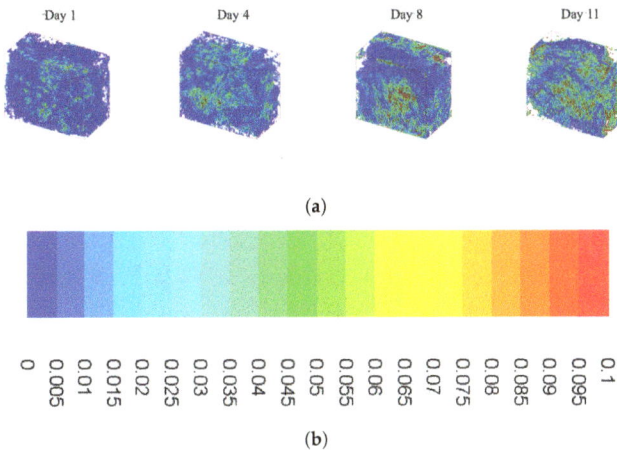

Figure 6. Summary of wall shear stress heat maps for constructs cultured under osteoconductive conditions. (**a**) Wall shear heat maps for Day 1 (far left) to Day 11 (far right) obtained using the custom lattice Boltzmann method code. (**b**) Scale bar: values given in $\frac{g}{cm \cdot s^2}$.

2.5. Average Wall Shear Stress

Figure 7 displays the average wall shear stress calculated from simulations following construct culture, resection, imaging and reconstruction. Indeed, the results do show a continuous increase in shear as culture time increases, which is consistent with the results presented in Figure 5. The most significant takeaway from this graph is the large jump in average wall shear between Day 4 and Day 8, due to matrix clogging the pores, consistent with the calcium deposition results (Figure 2).

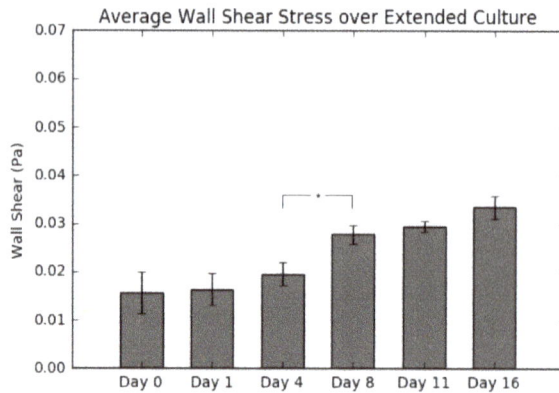

Figure 7. Summary of average shear stress per layer for a 0.5 mL/min flow rate. Values are given as the mean ± the standard error of the mean (n = 3). Significance calculated via analysis of variance (ANOVA) with the Tukey honest significant difference (HSD) post-hoc analysis.

3. Discussion

Following destructive analysis of the constructs, we evaluated the average wall shear stress per construct as a function of culture time. As seen in Figure 7, the average shear stress remains statistically the same throughout the first four days of culture; however, after Day 8, there is a continual increase in the value to the end of the culture period (113% by Day 16). This finding is in agreement with theoretical estimations of shear forces in scaffolds having similar levels of tissue formation [13]. Furthermore, it also confirms our hypothesis about bone tissue engineered cultures; that the shear stress experienced by the cells will increase significantly during culture, if the circulation fluid flow remains constant. We attribute this to extracellular matrix deposition resulting in a clogging of the construct pores. By holding the circulation fluid flow constant and simultaneously decreasing the pore sizes, we effectively are increasing the fluid velocity within the construct interior and, along with it, the wall shear stress. Considering the influence of shear rate on osteoblastic differentiation, it is evident that a continuous increased shear rate exposure will potentially accelerate the differentiation towards an osteoblastic lineage. Obviously, if the initial shear rates are near the range of flow-induced detachment, the observed increase could lead to undesirable detachment, implying the need for further investigating the shear levels at which detachment occurs.

After evaluating the construct reconstructions, we can see a clear correlation to the amount of mineralized tissue and an overall increase in the magnitude of wall shear experienced within the pores of the construct. Indeed, this relationship is obvious in both Figure 5, showing the frequency distributions, and Figure 6, which shows the wall shear heat maps. For the former, the distributions show an increased frequency of elevated shear stress as culture time increases, supporting the aforementioned increase in average wall shear stress. This finding is consistent when evaluating the heat maps. As culture time increases, there is a higher density of elevated shear seen within the constructs. This increase is most pronounced after Days 8–11, which, according to the reconstructions, are when large amounts of mineralized tissue starts depositing.

Additionally, the scaffold cellularity and the levels of mineralized tissue deposited were evaluated. As seen in Figure 1, a seeding efficiency of 40% was achieved, higher than most perfusion-based seeding methods, and can be directly attributed to the oscillatory seeding protocol we established in previous studies. Following the complete seeding process, the seeding efficiency drops further to 28% at Day 4 with the addition of a higher unidirectional fluid flow established at the end of Day 1. We believe that the decrease in cellularity seen between Day 1 and Day 4 is due to the change from basic maintenance flow employed immediately after seeding until Day 1 (0.15 mL/min) to culture flow beyond Day 1 (0.5 mL/min), and potentially causing a portion of cells to detach from the scaffold, especially those that form aggregates that are loosely bound to the surface. However, it must be noted that this drop can also be attributed to cell death or apoptosis, as seen in previous studies [4,5].

In Figure 2, a spike in calcium production is seen between Day 4 and Day 8. The point at which the spike occurs remains a critical parameter in bone tissue engineering research and better understanding of the factors influencing it will result in more efficient scheduling of culturing bone tissue engineering constructs in vitro. Recent studies identifying the initiation of extensive osteoblastic differentiation using nondestructive metabolic monitoring may allow us in future studies to better predict the exact timing of mineralization and thus allow for accurate prediction of the end of the culture period [5].

Figures 3 and 4 demonstrate qualitatively that the mineralized tissue begins developing from the outer layer of the top surface of the scaffold directly exposed to the fluid flow. This result contradicts our previous assumption of a homogeneous cell distribution and cannot be justified by any localized fluid shear forces. It is potentially an artifact due to the scaffold radius being slightly larger than the bioreactor cassette, which causes both a more snug fit, but also a decreased porosity at the edges in contact with the wall; thusly, this presents an environment where cells can attach in greater numbers due the larger surface area available for attachment. Such a phenomenon is not expected to occur in rigid three-dimensional scaffold, but it will persist whenever deformable meshes are utilized, as is in this case.

In addition, Figure 3 shows an increase in mineralized tissue found utilizing μCT between Day 8 and Day 11. We believe this lag time is due to the cells beginning to deposit calcium that is not dense enough to be picked up during imaging segmentation around Day 8. Along with this, Figure 4 shows the 2D images obtained from μCT with the growing tissue highlighted in red. These images illustrate the state of the construct at the end of culture and give insight into the density of mineralization that would occur if culture continued. It is to be expected that the soft tissue visualized in the image will eventually transition into fully-mineralized tissue.

4. Materials and Methods

4.1. Scaffold Manufacturing

Poly(L-lactic acid) (PLLA; Grade 6251D; 1.4% D enantiomer; 108,500 MW; 1.87 PDI; NatureWorks LLC) non-woven fiber mesh scaffolds were produced via spunbonding, as previously indicated [22]. Scaffolds were cut from an 8 mm-thick PLLA mat, resulting in a porosity of 88% and a radius of 3.5 mm. A Nikon HFX-II microscope (Nikon Corporation, Tokyo, Japan) was used to evaluate fiber diameter, found to be 24.5 μm, and was confirmed by scanning electron microscopy, shown in Figure 8.

4.2. Cell Expansion, Seeding and Culture

Adult mesenchymal stem cells were extracted from the tibias and femurs of male Wistar rats (Harlan Laboratories, Inc., Indianapolis, IN, USA) using methods identified in previous publications [1,4]. Cells were cultured at 37 °C and 5% CO_2 in standard minimum essential medium eagle alpha modification (α-MEM) (Invitrogen, Thermo Fisher Scientific corporation, Waltham, MA, USA) supplemented with 10% fetal bovine serum (Atlanta Biologicals, Flowery Branch, GA, USA) and

1% antibiotic-antimycotic (Invitrogen). Passage 2 cells were used for this study at a density of two million cells/mL for scaffold seeding.

We prepped the scaffolds for cell seeding using an established pre-wetting technique [5]. Vacuum air removal of scaffolds was conducted in 75% ethanol. Pre-wet scaffolds were placed in cassettes within a flow perfusion bioreactor for one hour in α-MEM to remove any remaining ethanol [23,24]. Schematics of the perfusion system used for this study may be found in Figure 9. Following the removal of ethanol, two million MSCs/150 μL of osteogenic α-MEM were pipetted in each scaffold chamber. The seeding mixture was dynamically perfused at 0.15 mL/min, forwards and backwards, in five-minute intervals for two hours. Osteogenic media consist of standard α-MEM supplemented with dexamethasone, beta-glycerophosphate and ascorbic acid, which have been shown to induce osteogenic differentiation [25]. After dynamic seeding, the bioreactor was allowed to rest for two hours, without flow, to facilitate cell attachment. Finally, osteogenic α-MEM was continuously perfused at a rate of 0.5 mL/min for the remainder of the culture period. Scaffolds were collected for analysis at Days 1, 4, 8, 11 and 16.

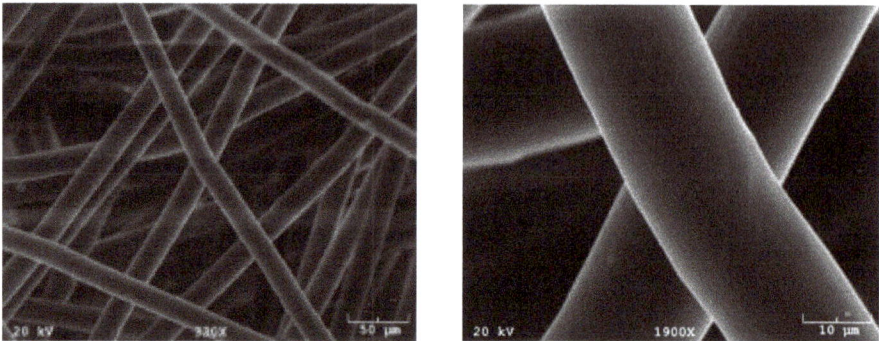

Figure 8. Common synthetic polymeric scaffolds used for tissue engineering. Scaffolds manufactured using spunbonding and imaged using scanning electron microscopy (SEM).

Figure 9. Schematic of the custom in-house perfusion bioreactor system. The right image shows the combination of the bioreactor body, scaffold cassettes and stand.

4.3. Construct Cellularity

The number of cells present in each construct was evaluated using the fluorescent PicoGreen® dsDNA assay (Invitrogen). At each sacrificial time point, the construct was removed from the cassette and rinsed in PBS to remove any cells not adhered to the scaffold. Subsequently, the scaffolds were cut into eight pieces, placed in 1 mL of deionized (DI) H_2O and stored at $-20\,°C$. Each construct underwent three freeze/thaw cycles to lyse the cells. Fluorescent analysis was conducted on a Synergy HT Multi-Mode Microplate Reader (BioTek Instruments, Inc., Winooski, VT, USA) at an excitation wavelength of 480 nm and an emission wavelength of 520 nm. All samples and standards were run in triplicate. Resulting values were then divided by the previously-determined dsDNA content per cell.

4.4. Construct Calcium Deposition

Calcium deposition was measured utilizing the scaffolds following the freeze/thaw cycles of the previous section. The solution was measured with a calcium colorimetric assay (Sigma-Aldrich Corporation, St. Louis, MO, USA, Cat. # MAK022). Samples were read on a Synergy HT Multi-Mode Microplate Reader (Bio-Tek) at an absorbance of 575 nm. All samples and standards were run in triplicate.

4.5. Imaging and Reconstruction

Micro-computed tomography was used to non-destructively scan the scaffolds at a resolution of 20 μm and 45 kV energy (Quantum FX, Perkin Elmer, Waltham, MA, USA; L10101, Hamamatsu Photonics, Hamamatsu, Japan; PaxScan 1313, Varian Medical Systems, Palo Alto, CA, USA). The resulting 2D image slices were filtered, thresholded and stacked using the open-source visualization software 3D Slicer (slicer.org). Following reconstruction, the porosity of the digital scaffold was measured and compared to the actual spunbonded scaffolds in order to assure a proper segmentation. Using methods described in a previous publication, we also identified in each construct three different materials (polymer scaffold, soft tissue and mineralized tissue). This was done using a segmentation method that allowed us to distinguish between the aforementioned materials, based on their attenuation to X-rays and structural differences. Details for the methodology appear in a previous study [26]. For this study, we investigated a total of six time points (three constructs per time point).

4.6. CFD Simulations

Simulations were performed via the lattice Boltzmann method implemented using a previously-validated custom in-house lattice Boltzmann code, which has been extensively utilized for computing surface shear stresses on μCT reconstructions [2,17,27].

Lattice Boltzmann Simulations

The lattice Boltzmann method (LBM) is a numerical technique for simulating fluid flow that consists of solving the discrete Boltzmann equation [28,29]. In addition to its computational advantages such as inherent parallelizability on high-end parallel computers [30,31] and relative ease of implementation, LBM techniques have been used in a wide spectrum of applications including turbulence [32], non-Newtonian flows [33–35] and multiphase flows [36]. More importantly, for the present application, LBM is especially appropriate for modeling pore-scale flow through porous media (such as scaffolds) due to the simplicity with which it handles complicated boundaries [2,12,18].

LBM is based on the discrete Boltzmann equation, which is an evolution equation for a particle distribution function, calculated as a function of space and time [37] as follows:

$$f_i(x + e_i\Delta t, t + \Delta t) = f_i(x,t) + \Omega_i(x,t) \pm ff_i \tag{1}$$

where x represents the particle position, t is time, Δt is the evolution time step, e is the microscopic velocity, Ω is the collision operator, f_i is the particle distribution function, f_i^{eq} is the equilibrium particle distribution function and ff_i is the forcing factor. The terms on the right-hand side of Equation (1) describe the three steps of the LBM algorithm: streaming, collision and forcing steps. In the streaming stage, the particle distribution function f_i at position x and time t moves in the direction of the velocity to a new position on the lattice at time $t + \Delta t$. The collision stage subsequently computes the effects of collisions that have occurred during movements in the streaming step, which is considered a relaxation towards equilibrium. The collision term can be computed for single-relaxation time (SRT) or multiple-relaxation time LBM models. The SRT approximation given by Bhatnagar, Gross and Krook (BGK) [38] is common. It is expressed as:

$$\Omega_i(x,t) = \frac{-1}{\tau}(f_i - f_i^{eq}) \tag{2}$$

The particle equilibrium distribution function f_i^{eq} is given as:

$$f_i^{eq}(x) = w_i\rho(x)\left[1 + 3\frac{e_i U_i}{c^2} + 9\frac{(e_i U_i)^2}{c^4} - \frac{3}{2}\frac{U_i^2}{c^2}\right] \tag{3}$$

where c is the lattice speed and defined as $c = \frac{\Delta x}{\Delta t}$, Δx is the lattice constant, iis an index that selects between possible discrete velocity directions (e.g., $i = 0, 1, 2, \ldots, 14$ for the D3Q15 lattice), w_i is a lattice dependent weighting factor, ρ is the fluid density and U is the macroscopic fluid velocity. The time τ is the time scale by which the particle distribution function relaxes to equilibrium and is related to kinematic fluid viscosity using:

$$v = \frac{1}{3}\left(\tau - \frac{1}{2}\right) \tag{4}$$

During the forcing step of the algorithm, a pressure drop is specified by adding a forcing factor ff_i to the fluid particle distribution components moving in the positive stream-wise direction and by subtracting from those in the negative direction.

The final stage of the algorithm involves computing the macroscopic fluid density ρ and velocity U at any instant using the conservation equations of mass and momentum given as:

$$\rho = \sum_{i=0}^{n} f_i \tag{5}$$

$$\rho U_i = \sum_{i=0}^{n} f_i e_i \tag{6}$$

A custom-written, in-house code was developed for this work, and a detailed description may be found in previous publications [2,17,39]. The three-dimensional, 15 lattice (D3Q15) for LBM [40], in conjunction with the single-relaxation time approximation of the collision term given by Bhatnagar, Gross and Krook [38], was used to perform simulations. LBM results have been validated for several flow cases for which analytical solutions are available: forced flow in a slit, flow in a pipe and flow through an infinite array of spheres [2].

Due to the computationally-intensive nature of fluid flow simulations in scaffolds, representative portions cut from whole scaffolds are used. In this work, a single cuboid portion was extracted from the center of the 3D scaffold reconstruction to avoid end effects in flow simulations. The resulting simulation domain size was 153 µm × 277 µm × 221 µm, with the center of the domain located at the center of the scaffold. Periodic boundary conditions were applied in all three directions, in order to approximate the whole scaffold. In addition, it was assumed that the whole extracellular matrix (ECM) was a rigid, non-permeable domain.

A no-slip boundary condition was applied at wall faces using the "bounce-back" technique [29]. To take advantage of the inherent LBM parallelizability, the domain was decomposed using message passing interface [17]. Simulation convergence was defined as when the minimum, average and highest velocities computed for the simulation domain vary by at least 0.001% for two consecutive time steps.

In order to estimate the mechanical stimulation of the cells by the flow of the culture media, the fluid-induced shear stresses on the surface of the scaffold were calculated following a scheme suggested by Porter et al. [18]. Here, the total stress stress tensor σ is represented by a summation of the hydrostatic pressure p and the viscous (also known as deviatoric) stress tensor τ:

$$\sigma_{ij} = -p\delta_{ij} + \tau_{ij} \tag{7}$$

where δ_{ij} is the unit tensor, such that δ_{ij} is 1 if $i = j$ and 0 if $i \neq j$.

While the former is assumed to be negligible, the latter is the component of the total stress tensor that is of interest since as it is responsible for shearing the cells. Because it describes how momentum is transported across the fluid layers due to velocity shear, it must be related to the deformation tensor:

$$D_{ij} = \frac{1}{2}(\nabla U + \nabla U^T) + \frac{1}{2}(\nabla U - \nabla U^T) \tag{8}$$

This tensor can be decomposed into a symmetric (rate of strain) and an anti-symmetric (vorticity) contributions. Furthermore, we assume that the constitutive relationship between the viscous stress and the rate of strain is linear: in other words, the fluid is Newtonian, with a constant viscosity. Consequently, the viscous stress tensor depends only on the symmetric component of the deformation tensor due to this assumption. Hence, the shear stress at every location within the scaffold was calculated using the following equation, where σ is the shear stress tensor and U is local velocity vector:

$$\underline{\underline{\sigma}} = \mu \frac{1}{2}(\nabla U + \nabla U^T) \tag{9}$$

The derivatives for the velocity field were approximated numerically using the centered finite difference method (Equations (10)–(12)), where lu is the length of one side of an element in the LBM model. The same was done for the partials of U_y and U_z. Following this, the symmetric strain matrices were found by adding the 3×3 partials matrix for each field location to its own transpose.

$$\frac{dU_x}{dx}(i,j,k) = \frac{U_x(i+lu,j,k) - U_x(i-lu,j,k)}{2 \times lu} \tag{10}$$

$$\frac{dU_x}{dy}(i,j,k) = \frac{U_x(i,j+lu,k) - U_x(i,j-lu,k)}{2 \times lu} \tag{11}$$

$$\frac{dU_x}{dz}(i,j,k) = \frac{U_x(i,j,k+lu) - U_x(i,j,k-lu)}{2 \times lu} \tag{12}$$

Finally, eigenvalues of the symmetric matrix were obtained using the Jacobi method, and the largest absolute-value eigenvalue (i.e., largest principal component of the tensor) for each fluid voxel was used to determine the stresses experienced by the cells.

The fluid dynamic viscosity was 0.01 $\frac{g}{cm \cdot s}$, which is close to that of α-MEM supplemented with 10% FBS typically used in cell culturing experiments [41]. Velocity vectors used in calculations were derived from a flow rate of 0.1 mL/min. Computed shear stress values are the largest eigenvalues of $\underline{\underline{\sigma}}$. Stress maps generated using Tecplot 360 EX 2016 (Tecplot Inc., Bellevue, WA, USA) were used to visualize computed shear stresses. Additionally, we modeled the ECM as an impermeable wall without elasticity (static mesh) and did not distinguish between hard and soft ECM during the computations.

4.7. Statistical Analysis

A one-way analysis of variance (ANOVA) was used to compare the mean $\pm$ the standard deviation of pore measurements, in which Tukey's honestly significant difference (HSD) test was performed to identify significant differences (p-value < 0.05). One-way ANOVA and Tukey's HSD were used for the rest of the results. All statistical analysis was performed using a custom Python code utilizing the open source Numpy, matplotlib and SciPy libraries.

5. Conclusions

In the presented manuscript, we hypothesized that the distribution of fluid shear present at the walls of a construct cultured under osteoconductive conditions will exhibit higher magnitudes as culture increases. In order to accomplish this, rat mesenchymal stem cells were dynamically seeded on 85% porous spunbonded poly(L-lactic acid) and cultured with osteogenic media for up to 16 days in a flow perfusion bioreactor. Following culture, these constructs were either destructively evaluated with assays for cellularity and calcium deposition or imaged using μCT and reconstructed to allow for CFD simulations to be performed. Average shear stress values and shear stress frequency distributions obtained from simulations were compared with the assays and confirmed our original hypothesis. In terms of the calcium quantification assay, a spike is seen around Day 8. This finding is supported by the reconstructions, where imaging identified an increase in mineralized tissue between Days 8 and 11. Additionally, the shear distribution heat maps show elevated magnitudes of shear stress in the same time period. Finally, we proved that both the shear stress distributions and the average shear stress per construct consistently increase as a function of culture time. This is due to mineralization occurring within the pores of the scaffold, decreasing pore diameter and effectively increasing velocity within the pores. In future studies, a correlation or algorithm may be developed that will give users the ability to predict, for the culture period, fluid shear distributions in bone tissue engineered cultures using μCT images of empty scaffolds, fluid dynamic simulations on the reconstructions and nondestructive metabolic monitoring that allows for the identification of the point of sharp increase of mineralized deposition.

Acknowledgments: The authors acknowledge support from Ghani Muhammed and Hong Liu for imaging assistance. This work was funded in part by the Gustavus and Louise Pfeiffer Research Foundation and the Oklahoma Center for the Advancement of Science and Technology (HR13-214). C.W. gratefully acknowledges fellowship funding from the Oklahoma Louis Stokes Alliance for Minority Participation (OK-LSAMP) Bridge to the Doctorate, Cohort 6 (EHR/HRD (Education & Human Resources/Human Resource Development) #1249206). Any opinions, findings, conclusions or recommendations expressed in this material are those of the authors and do not necessarily reflect the views of the Gustavus and Louise Pfeiffer Research Foundation, Oklahoma Center for the Advancement of Science and Technology, or the National Science Foundation.

Author Contributions: C.W. and V.I.S. conceived of and designed the experiments. C.W. performed cell culture, performed bioreactor experiments, performed image analysis, performed reconstructions, performed data analysis and wrote the manuscript. O.E.K. performed LBM simulations, performed data analysis and wrote the manuscript. R.S.V. and V.I.S. performed data analysis and wrote the manuscript.

Conflicts of Interest: The authors declare no conflict of interest. The founding sponsors had no role in the design of the study; in the collection, analyses or interpretation of data; in the writing of the manuscript; nor in the decision to publish the results.

References

1. VanGordon, S.B. Three-Dimensional Bone Tissue Engineering Strategies Using Polymeric Scaffolds. Ph.D. Thesis, The University of Oklahoma, Norman, OK, USA, 2012.
2. Voronov, R.; VanGordon, S.; Sikavitsas, V.I.; Papavassiliou, D.V. Computational modeling of flow-induced shear stresses within 3D salt-leached porous scaffolds imaged via micro-CT. *J. Biomech.* **2010**, *43*, 1279–1286.
3. Melchels, F.P.; Tonnarelli, B.; Olivares, A.L.; Martin, I.; Lacroix, D.; Feijen, J.; Wendt, D.J.; Grijpma, D.W. The influence of the scaffold design on the distribution of adhering cells after perfusion cell seeding. *Biomaterials* **2011**, *32*, 2878–2884.
4. Alvarez-Barreto, J.F.; Linehan, S.M.; Shambaugh, R.L.; Sikavitsas, V.I. Flow perfusion improves seeding of tissue engineering scaffolds with different architectures. *Ann. Biomed. Eng.* **2007**, *35*, 429–442.
5. Alvarez-Barreto, J.F.; Sikavitsas, V.I. Improved mesenchymal stem cell seeding on RGD-modified poly(L-lactic acid) scaffolds using flow perfusion. *Macromol. Biosci.* **2007**, *7*, 579–588.
6. Du, D.J.; Furukawa KS, U.T. 3D culture of osteoblast-like cells by unidirectional or oscillatory flow for bone tissue engineering. *Biotechnol. Bioeng.* **2009**, *102*, 1670–1678.

7. Papantoniou, I.; Guyot, Y.; Sonnaert, M.; Kerckhofs, G.; Luyten, F.P.; Geris, L.; Schrooten, J. Spatial optimization in perfusion bioreactors improves bone tissue-engineered construct quality attributes. *Biotechnol. Bioeng.* **2014**, *111*, 2560–2570.

8. Campolo, M.; Curcio, F.; Soldati, A. Minimal perfusion flow for osteogenic growth of mesenchymal stem cells on lattice scaffolds. *AIChE J.* **2013**, *59*, 3131–3144.

9. Maes, F.; Van Ransbeeck, P.; Van Oosterwyck, H.; Verdonck, P. Modeling fluid flow through irregular scaffolds for perfusion bioreactors. *Biotechnol. Bioeng.* **2009**, *103*, 621–630.

10. Cioffi, M.; Boschetti, F.; Raimondi, M.T.; Dubini, G. Modeling evaluation of the fluid-dynamic microenvironment in tissue-engineered constructs: A micro-CT based model. *Biotechnol. Bioeng.* **2006**, *93*, 500–510.

11. Sandino, C.; Checa, S.; Prendergast, P.J.; Lacroix, D. Simulation of angiogenesis and cell differentiation in a CaP scaffold subjected to compressive strains using a lattice modeling approach. *Biomaterials* **2010**, *31*, 2446–2452.

12. Hossain, S.; Bergstrom, D.J.; Chen, X.B. A mathematical model and computational framework for three-dimensional chondrocyte cell growth in a porous tissue scaffold placed inside a bi-directional flow perfusion bioreactor. *Biotechnol. Bioeng.* **2015**, *112*, 2601–2610.

13. Guyot, Y.; Luyten, F.P.; Schrooten, J.; Geris, L. A three—Dimensional computational fluid dynamics model of shear stress distribution during neotissue growth in a perfusion bioreactor. *Biotechnol. Bioeng.* **2015**, *112*, 2591–2600.

14. Guyot, Y.; Papantoniou, I.; Luyten, F.P.; Geris, L. Coupling curvature-dependent and shear stress-stimulated neotissue growth in dynamic bioreactor cultures: A 3D computational model of a complete scaffold. *Biomech. Model. Mechanobiol.* **2015**, *15*, 169–180.

15. Nava, M.M.; Raimondi, M.T.; Pietrabissa, R. A multiphysics 3D model of tissue growth under interstitial perfusion in a tissue-engineering bioreactor. *Biomech. Model. Mechanobiol.* **2013**, *12*, 1169–1179.

16. Checa, S.; Prendergast, P.J. A Mechanobiological Model for Tissue Differentiation that Includes Angiogenesis: A Lattice-Based Modeling Approach. *Ann. Biomed. Eng.* **2009**, *37*, 129–145.

17. Voronov, R.S.; VanGordon, S.B.; Sikavitsas, V.I.; Papavassiliou, D.V. Distribution of flow-induced stresses in highly porous media. *Appl. Phys. Lett.* **2010**, *97*, 024101.

18. Porter, B.; Zauel, R.; Stockman, H.; Guldberg, R.; Fyhrie, D. 3-D computational modeling of media flow through scaffolds in a perfusion bioreactor. *J. Biomech.* **2005**, *38*, 543–549.

19. Childers, E.P.; Wang, M.O.; Becker, M.L.; Fisher, J.P.; Dean, D. 3D printing of resorbable poly(propylene fumarate) tissue engineering scaffolds. *MRS Bull.* **2015**, *40*, 119–126.

20. McCoy, R.J.; Jungreuthmayer, C.; O'Brien, F.J. Influence of flow rate and scaffold pore size on cell behavior during mechanical stimulation in a flow perfusion bioreactor. *Biotechnol. Bioeng.* **2012**, *109*, 1583–1594.

21. Wendt, D.; Marsano, A.; Jakob, M.; Heberer, M.; Martin, I. Oscillating perfusion of cell suspensions through three-dimensional scaffolds enhances cell seeding efficiency and uniformity. *Biotechnol. Bioeng.* **2003**, *84*, 205–214.

22. VanGordon, S.B.; Voronov, R.S.; Blue, T.B.; Shambaugh, R.L.; Papavassiliou, D.V.; Sikavitsas, V.I. Effects of scaffold architecture on preosteoblastic cultures under continuous fluid shear. *Ind. Eng. Chem. Res.* **2011**, *50*, 620–629.

23. Kasper, F.K.; Liao, J.; Kretlow, J.D.; Sikavitsas, V.I.; Mikos, A.G. *Flow Perfusion Culture of Mesenchymal Stem Cells for Bone Tissue Engineering*; StemBook, Harvard Stem Cell Institute: Cambridge, MA, USA, 2008.

24. Bancroft, G.N.; Sikavitsas, V.I.; Van Den Dolder, J.; Sheffield, T.L.; Ambrose, C.G.; Jansen, J.A.; Mikos, A.G. Fluid flow increases mineralized matrix deposition in 3D perfusion culture of marrow stromal osteoblasts in a dose-dependent manner. *Proc. Natl. Acad. Sci. USA* **2002**, *99*, 12600–12605.

25. Porter, J.R.; Ruckh, T.T.; Popat, K.C. Bone tissue engineering: A review in bone biomimetics and drug delivery strategies. *Biotechnol. Prog.* **2009**, *25*, 1539–1560.

26. Voronov, R.S.; VanGordon, S.B.; Shambaugh, R.L.; Papavassiliou, D.V.; Sikavitsas, V.I. 3D tissue-engineered construct analysis via conventional high-resolution microcomputed tomography without X-ray contrast. *Tissue Eng. Part C Methods* **2013**, *19*, 327–335.

27. Pham, N.H.; Voronov, R.S.; VanGordon, S.B.; Sikavitsas, V.I.; Papavassiliou, D.V. Predicting the stress distribution within scaffolds with ordered architecture. *Biorheology* **2012**, *49*, 235–247.

28. Chen, S.; Doolen, G.D. Lattice Boltzmann method for fluid flows. *Annu. Rev. Fluid Mech.* **1998**, *30*, 329–364.

29. Sukop, M.C.; Thorne, D.T., Jr. *Lattice Boltzmann Modeling an Introduction for Geoscientists and Engineers*; Springer: Berlin, Germany; New York, NY, USA, 2006; 172p.

30. Kandhai, D.; Koponen, A.; Hoekstra, A.G.; Kataja, M.; Timonen, J.; Sloot, P.M.A. Lattice-Boltzmann hydrodynamics on parallel systems. *Comput. Phys. Commun.* **1998**, *111*, 14–26.

31. Wang, J.Y.; Zhang, X.X.; Bengough, A.G.; Crawford, J.W. Domain-decomposition method for parallel lattice Boltzmann simulation of incompressible flow in porous media. *Phys. Rev. E* **2005**, *72*, 016706.

32. Cosgrove, J.A.; Buick, J.M.; Tonge, S.J.; Munro, C.G.; Greated, C.A.; Campbell, D.M. Application of the lattice Boltzmann method to transition in oscillatory channel flow. *J. Phys. Math. Gen.* **2003**, *36*, 2609–2620.

33. Gabbanelli, S.; Drazer, G.; Koplik, J. Lattice Boltzmann method for non-Newtonian (power-law) fluids. *Phys. Rev. E* **2005**, *72*, 046312.

34. Boyd, J.; Buick, J.; Green, S. A second-order accurate lattice Boltzmann non-Newtonian flow model. *J. Phys. Math. Gen.* **2006**, *39*, 14241–14247.

35. Yoshino, A.; Hotta, Y.; Hirozane, T.; Endo, M. A numerical method for incompressible non-Newtonian fluid flows based on the lattice Boltzmann method. *J. Non-Newton. Fluid Mech.* **2007**, *147*, 69–78.

36. Swift, M.R.; Orlandini, E.; Osborn, W.R.; Yeomans, J.M. Lattice Boltzmann simulations of liquid-gas and binary fluid systems. *Phys. Rev. E* **1996**, *54*, 5041–5052.

37. McNamara, G.; Zanetti, G. Use of the Boltzmann equation to simulate Lattice-gas automata. *Phys. Rev. Lett.* **1988**, *61*, 2332–2335.

38. Bhatnagar, P.L.; Gross, E.P.; Krook, M. A Model for Collision Processes in Gases. I. Small amplitude processes in charged and neutral one-component systems. *Phys. Rev.* **1954**, *94*, 511–525.

39. Voronov, R.S.; VanGordon, S.B.; Sikavitsas, V.I.; Papavassiliou, D.V. Efficient Lagrangian scalar tracking method for reactive local mass transport simulation through porous media. *Int. J. Numer. Methods Fluids* **2011**, *67*, 501–517.

40. Qian, Y.H.; Dhumieres, D.; Lallemand, P. Lattice Bgk models for Navier-Stokes equation. *Europhys. Lett.* **1992**, *17*, 479–484.

41. Lakhotia, S.; Papoutsakis, E.T. Agitation induced cell injury in microcarrier cultures—Protective effect of viscosity is agitation intensity dependent—Experiments and modeling. *Biotechnol. Bioeng.* **1992**, *39*, 95–107.

![fluids logo] *fluids*

MDPI

Article

Computational Modelling for Efficient Transdentinal Drug Delivery

Agathoklis D. Passos [1], Dimitris Tziafas [2], Aikaterini A. Mouza [1] and Spiros V. Paras [1,*]

[1]   Department of Chemical Engineering, Aristotle University of Thessaloniki, 541 24 Thessaloniki, Greece;
      passos@auth.gr (A.D.P.); mouza@auth.gr (A.A.M.)
[2]   Hamdan Bin Mohamed College of Dental Medicine, DHCC Dubai, Dubai, UAE; dtziaf@dent.auth.gr
*   Correspondence: paras@auth.gr; Tel.: +30-231-099-6174

Received: 14 November 2017; Accepted: 25 December 2017; Published: 27 December 2017

Abstract: This work deals with the numerical investigation of the delivery of potential therapeutic agents through dentinal discs (i.e., a cylindrical segment of the dentinal tissue) towards the dentin–pulp junction. The aim is to assess the main key features (i.e., molecular size, initial concentration, consumption rate, disc porosity and thickness) that affect the delivery of therapeutic substances to the dental pulp and consequently to define the necessary quantitative and qualitative issues related to a specific agent before its potential application in clinical practice. The computational fluid dynamics (CFD) code used for the numerical study is validated with relevant experimental data obtained using micro Laser Induced Fluorescence (μ-LIF) a non-intrusive optical measuring technique. As the phenomenon is diffusion dominated and strongly dependent on the molecular size, the time needed for the concentration of released molecules to attain a required value can be controlled by their initial concentration. Finally, a model is proposed which, given the maximum acceptable time for the drug concentration to attain a required value at the pulpal side of the tissue along with the aforementioned key design parameters, is able to estimate the initial concentration to be imposed and vice versa.

Keywords: drug delivery; dentine; diffusion; bio-active molecules; CFD; μ-LIF; microfluidics

---

1. Introduction

During the last few decades, clinical practice has been oriented towards the design and development of modern dental treatment techniques that would ensure the long-term maintenance of vitality and function of the dentine–pulp complex [1]. The traditional treatment strategies in vital pulp therapy have been mainly focused on protection of dental pulp from possible irritation presented by components released from dental materials and induction of replacement of diseased dentin by a layer of tertiary dentine [2]. Furthermore, the fact that the pulp–dentin complex can be repaired, and, as a part of reparative events, tertiary dentin is formed, has recently led the scientific community towards the development of novel optimal procedures and production of potential therapeutic agents for use in regenerative pulp therapies. There is a growing weight of evidence that bioactive molecules diffused through the dentinal tubules can regulate the biosynthetic activity of pulpal cells [3–5]. The idea of using the dentinal tubules, i.e., structures running in parallel to the whole thickness of dentin, for drug delivery in pulp, was originally proposed by Pashley [6]. Later, Pashley [7] suggested that the therapeutic agents must consist of small molecules or ions that exhibit relatively high diffusion coefficients.

Although it is well known that a remaining dentin zone (Remaining Dentin Thickness, *RDT*) is necessary for maintaining the dental pulp function [8], much still needs to be learned about the effect of potent toxic components that are released from traditional restorative materials or bioactive signaling molecules, a knowledge that will form the basis for novel regenerative therapeutic procedures. As it is expected, the size of the *RDT* plays an important role in the level of penetration of exogenous compounds to the pulp. This is in accordance with Tak and Usumez [9], who concluded that the amount of HEMA that diffused through the dentin to the pulp is increased in the case of decreasing *RDT*. It is reasonable to suggest that any toxic interruption of pulp functions or therapeutic upregulation of the responsible for the biosynthetic activity of teeth odontoblasts will probably reflect the molecular concentration of these constituents at the dentin–pulp interface area [10]. Thus, a number of delivery considerations have to be initially addressed. However, due to the minute dimensions of the dentinal tubules, the experimental study of the drug delivery through the dentinal tubules under the same conditions as in clinical therapy is practically unachievable using the current experimental techniques. Thus, it is necessary to develop novel techniques that would permit the prediction of the transport rate of selected therapeutic molecules.

Several in vitro experimental studies reported in the literature concern the inward diffusion of substances in dentinal slabs that are placed in custom made split-chamber devices under the absence/existence of simulated pulpal pressure. Pashley and Matthews [11] conducted experiments aiming to evaluate the influence of outward forced convective flow on the inward diffusion of radioactive iodide. In addition Hanks and Wataha [12] investigated the diffusion of various biological and synthetic molecules through dentine, including resin components and dyes. However, due to the limitations imposed by the experimental techniques, these studies fail to fully represent the actual tooth case and can only provide information on the permeability of dentinal discs with respect to several substances. Due to the non-uniformity of the cross-section of dentinal tubules, the theoretical equations, which are derived from Fick's Second Law and are used for estimating the flow through a tubulus, are very complicated to be applied. Consequently, computational fluid dynamics (CFD) appear to be the most feasible method for studying the problem under consideration. Several studies confirmed the suitability of CFD for dental pulp study e.g., [13–16]. In a previous study [17], we have numerically investigated the delivery of bioactive therapeutic agents to the dentin-enamel junction (DEJ) through a typical dentin tubule that has no obstructions or alterations of any type. The results of this initial approach led to the formulation of a simplified model that is able to estimate the initial drug concentration that must be imposed, so as a given type of therapeutic molecules to be effectively transported through a single and enclosed dentinal tubule.

The present work extends our previous study by numerically investigating the delivery of potential therapeutic agents through dentinal discs (i.e., a cylindrical segment of the dentinal tissue) towards the dentin–pulp junction in enclosed cavities. New key design parameters (i.e., Remaining Dentin Thickness, Porosity and Consumption Rate) have been addressed in a way that fully represents the actual tooth case. The aim is to quantify the effect of these parameters on the transdentinal diffusion characteristics. The study will also include the effect of the number of the active dentinal tubules expressed by a porosity value as well as the effect of the dentin–pulp barrier expressed as a consumption rate at the pulp on the drug delivery characteristics.

2. Transport of Therapeutic Compounds through the Dentinal Tissue

Dentin is a mineralized matrix that is porous and yellow-hued. It is made up of 70% inorganic materials (mainly hydroxyapatite and some non-crystalline amorphous calcium phosphate), 20% organic materials (90% of which is collagen type 1 and the remaining 10% ground substance, which includes dentine-specific proteins), and 10% water [18]. The dentin tissue consists of microscopic channels, the dentinal tubules, which radiate outwards through the dentin from the pulp to the exterior cementum or enamel border (Figure 1a). In the tooth crown, the dentinal tubules extend from the dentino-enamel junction (DEJ) to the pulp periphery following an S-shaped path. Tapering from the inner to the outermost surface, they have an average diameter of 2.5 μm near the pulp, 1.2 μm in the middle of the dentin, and 0.9 μm at the DEJ, while the total tissue width is estimated to be about 2.5 mm [19]. Most of the dentinal tubules contain non-myelinated terminal nerves and odontoblasts that are placed in an environment filled with dentinal fluid [20–22].

The number of dentinal tubules per unit area depends on the age and gender of each individual, while their density and diameter define the surface area that they occupy. For example, the area of dentin occupied by tubules at the dentino-enamel junction is only 1% and increases to 45% at the pulp chamber [18], while their density near the pulp varies between 59,000 to 76,000 tubules/mm^2. In addition, the dentinal tubules are connected to each other by branching canalicular systems.

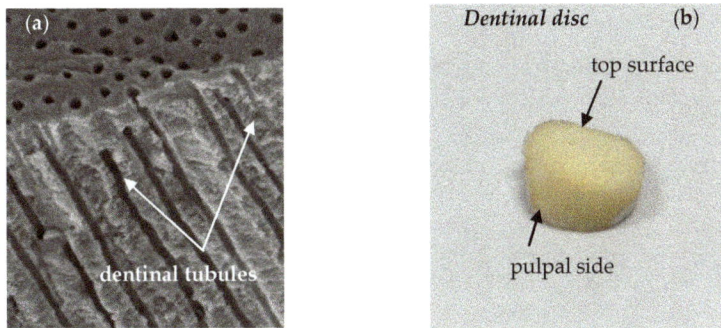

Figure 1. (a) dentinal tubules running perpendicularly from pulpal wall towards dentino-enamel junction (DEJ) scanning electron microscopy (SEM), reproduced with permission from [17] and (b) dentinal disc of a human tooth.

In clinical practice, when the dentin pulp complex is affected by caries or trauma, biological compounds are placed in enclosed cavities and then diffuse through the dentinal fluid to the pulp. In this case, the surface of the traumatized dentin tissue must be covered with suitable dental materials after the addition of the therapeutic agent. In the case of open cavities, compounds/bacteria and consequently also therapeutic agents placed in the vicinity of the cavity may never enter the dentinal tubules because of the fluid tendency to flow outwards.

In the framework of the present computational study, the dentinal tissue is modeled as a porous disc with tortuosity value $\tau = 1$. The transport of substances in a cylindrical section of the tissue is investigated under the effect of the geometrical characteristics of the tissue, the size and the initial concentration of the applied molecules as well as their consumption rate at the pulpal side of the tooth.

3. Materials and Methods

3.1. Code Validation

Prior to proceeding with the computational procedure, the CFD simulations were validated using experimental data acquired by performing experiments using dentinal porous discs. The dentinal discs employed were 0.85 ± 0.05 mm thick (Figure 1b) and were cut perpendicular to the vertical axis of exported human teeth, just above the level of the pulp horns, by a low speed saw ISOMET (Buehlel, Lake Bluff, IL, USA) under constant water flow. The dentine discs were hand-sanded under tap water, by a 600-grit silicon carbide paper to achieve a uniform flat surface with smear layer, acid-etched on both sides, with 35% phosphoric acid (Ultra etch, Ultradent, South Jordan, UT, USA) for 15 s and then thoroughly rinsed with water spray in order to clean the disc and remove the smear layer from the dentine. The disc was placed between two transparent glass tubes with a standard central hole (32 mm^2) and it was attached with a minimum quantity of aquarium marine silicone (Top sil, Mercola, Athens, Greece) carefully placed on the enamel margins of the dentine disc. After the silicone was set, all of the joints were reinforced externally with sticky wax (Kem-den, Purton, UK) and the whole system was filled with Ringer's solution (Vioser, Iraklio, Greece) to be checked for leakages.

The experimental conduit, which is schematically presented in Figure 2, is a closed system that was constructed at the Dental School of Aristotle University of Thessaloniki.

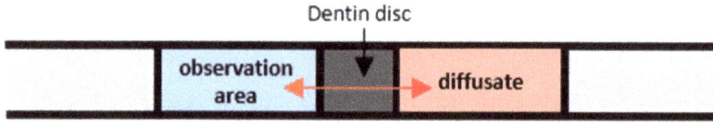

Figure 2. Schematic of the experimental conduit. Reproduced with permission from [17].

The μ-LIF experimental setup, available in our Lab, is shown in Figure 3. The measuring section was illuminated by a double cavity Nd:YAG Laser emitting at 532 nm. The fluid flux was measured by a high sensitivity charge-coupled device (CCD) camera (Hisense MkII, Qingdao, China), connected to a Nikon (Eclipse LV150, Tokyo, Japan) microscope, which moves along the vertical axis with an accuracy of one micron. A 10X air immersion objective with NA = 0.20 was used. For each measurement, at least 20 images were acquired at a sampling rate of 5 Hz.

Experiments were conducted using two liquids, namely distilled water as reference and distilled water marked with the fluorescent dye Rhodamine B, completely dissolved in water. The diffusion of Rhodamine B in the aqueous solution (diffusion coefficient in water 4.5×10^{-10} m^2/s [23]) through the dentinal disc to the opposite area (i.e., the observation area) is measured using the non-intrusive experimental technique μ-LIF (μ-Laser Induced Fluorescence). Two concentration groups of the aqueous solution were used, namely G1 for $M_0 = 0.1$ mg/L and G2 for $M_0 = 0.05$ mg/L.

As reported by Bindhu and Harilal [24], the Rhodamine B quantum yield (up to $\varphi = 0.97$ at low concentrations) depends on the solvent and on the type of excitation (i.e., continuous or pulsed), while it weakly decreases with increasing concentration. Rhodamine B, dissolved in water, is most effectively excited by green light and emits red light with the maximum intensity in the range 575–585 nm [24,25]. Image processing and concentration calculations were performed using appropriate software (Flow Manager by Dantec Dynamics, Skovlunde, Denmark).

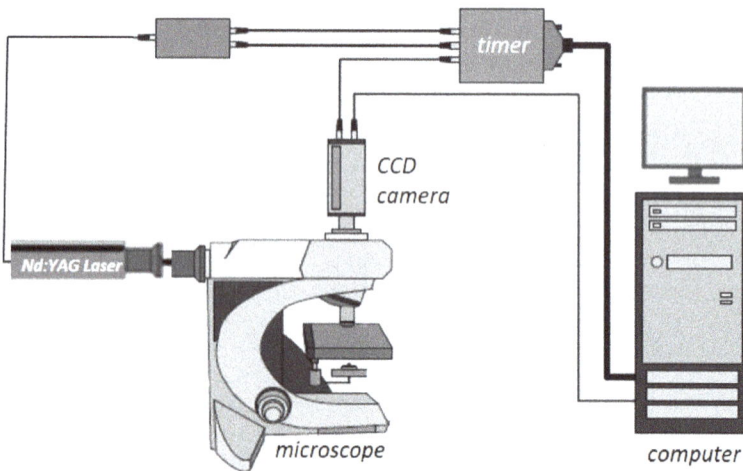

Figure 3. Schematic representation of the μ-LIF experimental setup. Reproduced with permission from [17].

A typical procedure for μ-LIF measurements is as follows [17]:

- Prior to the actual measurements, the system is calibrated using suitable solutions. In our case aqueous Rhodamine B solutions with known concentrations, namely, $M = 0.05, 0.025$ and 0.00 mg/L were employed, while, for each concentration C_i, an image $C_{(x,y)}$ is taken.
- To reduce noise, image masking of the acquired images is performed before defining an appropriate Region of Interest (ROI), at which the fluorescence intensity is measured.
- The relationship between the measured fluorescence intensity field $I_{(x,t)}$ and the concentration field $C_{(x,y)}$ is determined.
- A set of 20 images is acquired and the mean image is calculated.
- Each mean image is compared with the previously defined μ-LIF calibration curve.

During our experiments, the lens was focused on the middle plane of the capillary, while the optical system was set to have an ROI of 300×580 µm. To minimize the uncertainty of the experimental procedure, care was taken for the experimental conditions to be identical to those of the calibration procedure. The uncertainty of the method is estimated to be less than $\pm 5\%$ [26]. The concentration measurements were performed at the observation area (Figure 2), namely at a distance of 1 mm from the pulpal side of the dentinal disc. As there is no lateral diffusion, i.e., the process is 1D, the concentration can be considered constant along the diameter of the tube, meaning that the μ-LIF measurement corresponds to the mean concentration at a cross-section of the conduit.

The porosity of the dentin disc was defined by employing Scanning Electron Microscopy (SEM). More precisely, using the acquired images, the porosity was estimated by measuring the area occupied by dentinal tubules. Apart from the experimental measurements, we also performed computational simulations in a computational domain that fully represents the experimental setup (i.e., dentinal disc dimensions and initial Rhodamine B concentrations). The obtained numerical results are in very good agreement with the corresponding experimental ones. In Figure 4, the measured temporal molar concentration (C_L) values of Rhodamine B at the observation area of a dentinal disc with porosity $\varphi = 10\%$ are compared with the corresponding CFD simulation results.

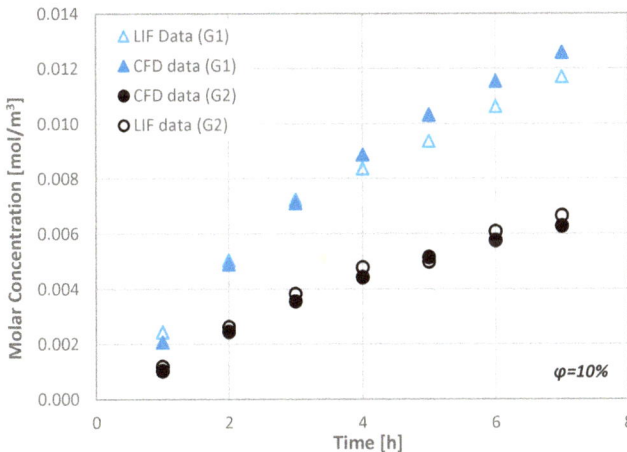

Figure 4. Typical comparison of the molar concentration values at the observation area of the μ-LIF technique with the corresponding computational results ($\varphi = 10\%$).

Since the comparison of the CFD data with the available experimental results indicate a very good agreement, i.e., better than $\pm 15\%$, the CFD code can be considered suitable for performing additional simulations concerning the transport of substances through the porous disc.

3.2. Numerical Procedure

In this study, we aim to investigate the delivery of potential therapeutic substances, which are placed at the top surface of the dentinal tissue, to the pulpal side by molecular diffusion through the dentinal fluid that occupies the micro-pores of the tissue. In order to accomplish this, we assume that dentine is a porous tissue with a porosity value related directly to the volume of dentinal fluid per unit volume of the tissue. Since the porosity of the dentin tissue is generally unknown and its thickness, especially in cases of traumatized teeth, is variable, we perform a parametric study that incorporates the effect of these two parameters on the diffusion process characteristics. In addition, the present parametric study includes the effect of the consumption of the diffusates at the pulpal side of the dentinal tissue, where the odontoblasts are located. The consumption at the pulpal side of the dentinal tissue is modeled by employing a constant consumption rate of the diffusate using an appropriate built-in feature of the CFD code. Since the consumption rate of the agents is considered to be unknown, we employed a range of consumption rates between 0 (i.e., no consumption is assumed) and 10^{-10} kg/(m^2·s). Finally, a parametric study is performed that includes the effect of:

- the porosity of the tissue (φ),
- the thickness of the tissue (Remaining Dentinal Thickness, RDT),
- the initial concentration (M_0) of the substances to be diffused,
- the molecular size of the substances to be diffused, i.e., their Diffusion Coefficient and
- the consumption rate (R) of the diffusate at the pulpal side

on the transport characteristics through an enclosed dentinal disc.

The porous disc is considered to be initially full with dentinal fluid, i.e., a fluid that has the thermophysical properties of water. In Figure 5, the computational domain is schematically presented. The disc is cylindrical with a cross-section area of 32 mm^2, while the diffusate occupies a volume of 32 mm^3.

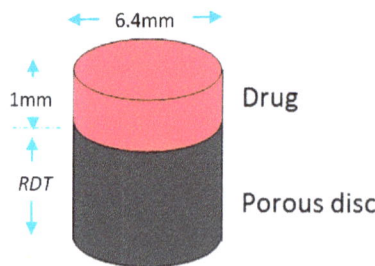

Figure 5. Computational domain for the porous disc model. (RDT = 0.5–2.5 mm).

The permeability of porous media is usually expressed as a function of certain physical properties of the interconnected system, such as porosity and tortuosity. Although it is natural to assume that permeability values depend on porosity, an appropriate relationship is not simple to be determined, since this would require a detailed knowledge of size distribution and spatial arrangement of the pore channels in the porous medium. For instance, two porous systems can have the same porosities but different permeabilities. One of the most widely accepted and simplest models for the permeability–porosity relationship is the Kozeny–Carman (KC) model, which provides a link between media properties and flow resistance in pore channels. In this study, we apply the general Poiseuille equation in a straight channel with a generic cross-sectional shape [27]:

$$U_{channel} = -\frac{A}{a}\frac{1}{n}\frac{dp}{dx} \tag{1}$$

where A is the generic cross-sectional area of the micro-channel and a is a dimensionless geometric factor. Classical derivations of the KC model [28,29] consider the case of a circular cross-section as a starting point in Equation (1), i.e., $A/a = R^2/8$. They extend and adapt Equation (1) to the equivalent channel model of the porous body by taking into consideration the effects of (i) the effective tortuous path, (ii) the effective volume and (iii) the concept of effective radius R:

$$U = -c\frac{R^2}{8}\frac{\varphi}{\tau}\frac{1}{n}\frac{dp}{dx} \qquad (2)$$

where c (=2 for cylindrical micro-pores) was introduced as an empirical geometrical parameter, φ is the porosity, τ the tortuosity defined as the square of the ratio between the effective channel length L_e due to the tortuous path and the length L of the porous body ($\tau \equiv L_e/L$). Comparison of Equation (2) with Darcy's law gives the following expression for defining the permeability (k) (Equation (3)):

$$k = c\frac{R^2}{8}\frac{\varphi}{\tau} \qquad (3)$$

To study the diffusion process, we employ substances, whose molecular sizes are the same as those of proteins or therapeutic agents used in dental clinic practice. The radius R_{min} and the diffusion coefficient D of a protein can be estimated by Equations (4) and (5) [30]:

$$R_{min} = 0.066 M_w^{1/3} \qquad (4)$$

where R_{min} is given in nanometers and the molecular weight M_w in Daltons (Da)

$$D = \frac{kT}{6\pi\mu R_s} \qquad (5)$$

where $k = 1.38 \times 10^{-16}$ g·cm^2·s^{-2}·K^{-1} is the Boltzmann constant and T the absolute temperature. k is given here in centimeter–gram–second units because D is expressed in centimeter–gram–second, while μ is the solute viscosity in g/(cm·s). In our case, it is assumed to be that of water, as its thermophysical properties are very close to those of the actual dentinal fluid [31]. R_s, called Stokes radius, represents the radius of a smooth sphere that would have the same frictional coefficient f with a protein and is expressed in centimeters in this equation. Assuming that f equals f_{min}, i.e., the minimal frictional coefficient that a protein of a given mass would obtain if the protein were a smooth sphere of radius R_{min}, in Equation (5), R_s can be replaced by R_{min}.

In this study, three different diffusates, which cover a wide range of molecular sizes that correspond to the size of actual bioactive molecules, are considered. The radius R_s of the molecules tested is in the range of 2.2–22.0 nm and consequently the corresponding diffusion coefficient, D, in water at 37 °C, is in the range of 1.36–0.14 × 10^{-10} m^2/s. For example, bone morphogenetic protein (BMP-7) is a bioactive protein used in the dental clinic practice with approximately 50 kDa molecular weight and R_s of 2.2 nm. To conduct a parametric study, three initial mass concentration values for each diffusing substance are employed, i.e., 0.10, 0.05 and 0.01 mg/mL following our previous study [17].

The parametric study was performed using three porosity values (φ), namely 5%, 10% and 20%, which can be regarded to adequately represent the porosity values of the dentin tissue, as it can be seen from Section 2. Employing the Design Exploration features of the ANSYS Workbench® package (18, ANSYS, Canonsburg, PA, USA) and following the Design of Experiments (DOE) methodology, a set of "computational experiments" was initially designed. The DOE methods allow the designer to extract as much information as possible from a limited number of test cases and makes the method ideal for CFD models that are significantly time-consuming [32]. Eventually, the temporal and spatial concentration at the pulpal side of the disc will be calculated using the computational procedure. A summary of the design variables range is given in Table 1.

Table 1. Constraints of the design variables.

Parameter	Lower Bound	Upper Bound
Porosity (φ), %	5	20
Remaining Dentinal Thickness (*RDT*), mm	0.5	2.5
Initial Concentration (M_0), mg/mL	0.01	0.10
Molecular Size (R_s), nm	2.2	22.0
Consumption Rate (*R*), kg/(m^2·s)	0	10^{-10}

A grid dependence study was also performed for the case involving the diffusate with the highest diffusion coefficient value. This leads to the construction of an unstructured mesh of 1×10^6 of tetrahedral elements, while the porosity model is used to account for the space occupied by dentinal tubules per unit area in the disc. All simulations were run in transient mode, while the total simulation time for each run varied in the range of 10–25 h, depending on the consumption rate (*R*) applied. A time-step dependence study was also performed (i.e., time steps in the range of 5–10 s). Zero flux was inposed as boundary conditions at all the walls of the computational domain, while, when a consumption rate is employed, negative flux is imposed at the pulpal side. For the discretization scheme, the High Resolution one was employed and as for the transient scheme, a Second Order Backward Euler scheme was used. A high-performance unit for parallel computing, which runs on a Gentoo Linux distribution, was used.

4. Results

Since the behavior observed among the numerous computational results of the present study is similar, only typical examples of the effect of each key design parameter on the molar concentration at the pulpal side of the dentinal discs (C_L) are presented. The general case, i.e., when a consumption rate is imposed at the pulp, is presented first. For the simulations concerning the effect of the other parameters, no consumption is assumed, i.e., *R* = 0.

4.1. Effect of the Agent Consumption Rate at the Pulpal Side

Figure 6 shows the concentration at the pulpal side for the various consumption rates applied, when an agent with R_s = 2.2 nm and M_0 = 0.10 mg/mL is employed at the top of the dentinal disc of φ = 10%. As it can be seen, the concentration value at the pulpal side is lower than the initial one regardless of the consumption rate applied. The concentration value may exhibit 100% decrease when the lowest and the highest consumption rates are compared. During the first time steps of the agent application on the disc, the concentration gradient along the disc thickness is high and the concentration at the pulpal side C_L increases until t^* = 1 (i.e., dashed line), if the agents are consumed, and, after this critical time, equilibrium cannot be reached. On the contrary, as time passes, the agent is being continuously consumed, which means that the concentration at the pulp will decrease until the agent disappears.

It is expected that, after a few hours, the tissue will be free of the agent presence. The total estimated time for this to happen depends on the consumption rate, the porosity of the dentinal disc and the initial concentration. For a consumption rate of 10^{-10} kg/(m^2·s), the total estimated time for an agent of R_s = 2.2 nm to be fully consumed is approximately 18 h in a porous disc of φ = 10%.

From Figure 6, it becomes evident that the consumption rate of the therapeutic agent does not affect the form of the curve but only the magnitude of the C_L values. In order to facilitate the evaluation of the effect of the other parameters, a zero consumption rate at the pulp end is assumed for the remaining CFD simulations.

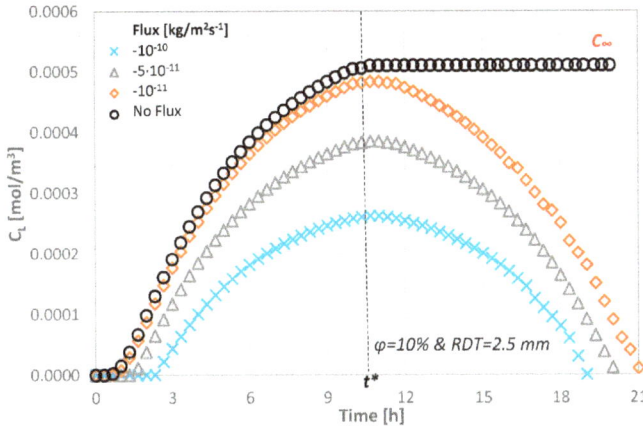

Figure 6. Molar concentration at the pulpal side of the dentinal discs as a function of different consumption rates for φ = 10% and RDT = 2.5 mm.

4.2. Effect of Remaining Dentin Thickness (RDT)

The effect of the RDT on the concentration at the pulpal side (C_L) is shown in Figure 7, when a substance of R_s = 4.4 nm at M_0 = 0.10 mg/mL is diffused through a dentinal disc of φ = 5%. As it is expected, since the process is considered one-dimensional, it is dominated by the diffusion length. For discs with significant low thickness, the delivery of substances to the pulp is rapid, and a steady state condition can be quickly achieved. For example, when a dentinal disc of 0.5 mm is employed, the concentration at the pulpal side reaches its final value after two hours of application. The time barrier for steady state condition can be easily estimated from the dimensionless time variable $t = D/L^2$.

In order to gain a greater insight into the transport process through dentinal discs, the temporal molar flux is calculated and presented in Figure 8. The magnitude of the diffusional flux across the dentinal tissue is inversely proportional to the dentine thickness, an observation that agrees with the relevant literature [11,33].

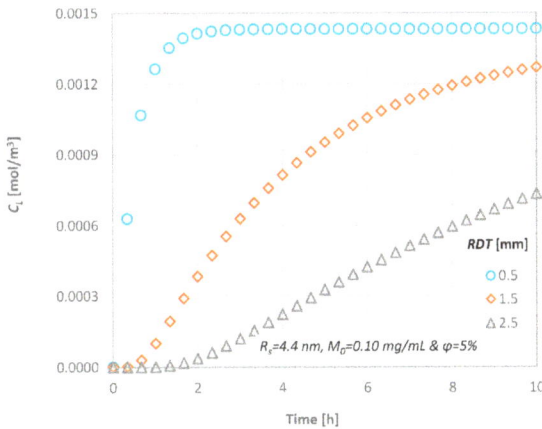

Figure 7. Effect of the Remaining Dentin Thickness (RDT) on the diffusion characteristics; C_L versus time.

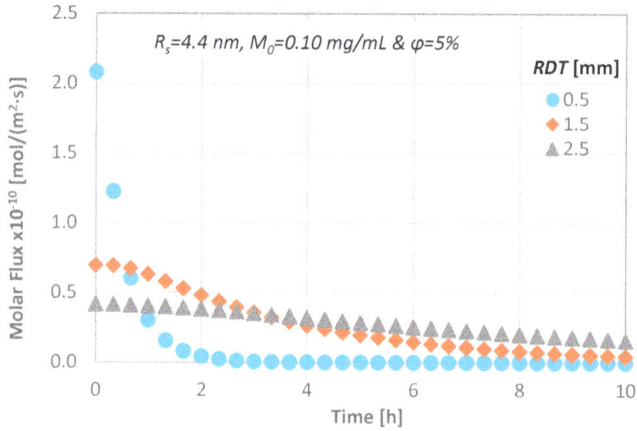

Figure 8. Effect of the *RDT* on the diffusion characteristics; molar flux versus time.

4.3. Effect of the Molecular Size

For a given dentinal disc (i.e., given *RDT* and porosity), the diffusion is controlled by the diffusion coefficient of each diffusing agent, while the estimated final value will be the same in all cases. The concentration variation with time at the bottom end area is presented in Figure 9 for three substances with different molecular sizes. As it is expected, substances/proteins with larger molecules penetrate at a lower rate inside the disc because their diffusion coefficient is a function of the size of the molecules as described in Section 3.2.

For the delivery of substances with different molecular sizes, the concentration gradient inside the dentinal tissue is presented in Figure 10. The molar flux is greater during the middle steps of the process, as the substance penetrates at a greater rate inside the tissue. Obviously, the substances with greater molecular sizes diffuse at a lower rate through media, and, as it can also be seen from Figure 9, it needs more than 10 h for the concentration C_L of the substance with the greatest molecular size to reach steady state conditions (i.e., the concentration gradient to be zero).

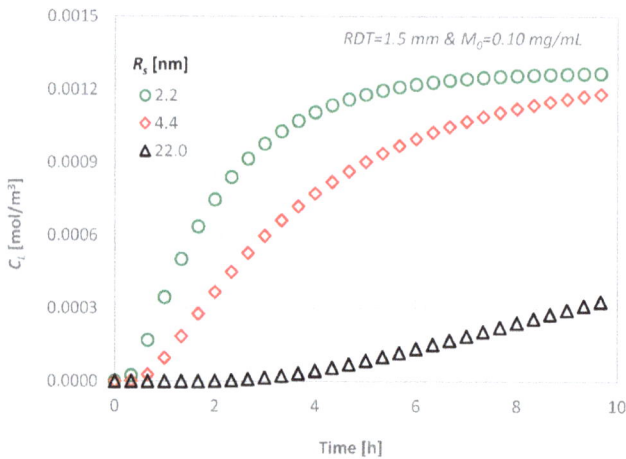

Figure 9. Effect of the molecular size on the diffusion characteristics; C_L versus time.

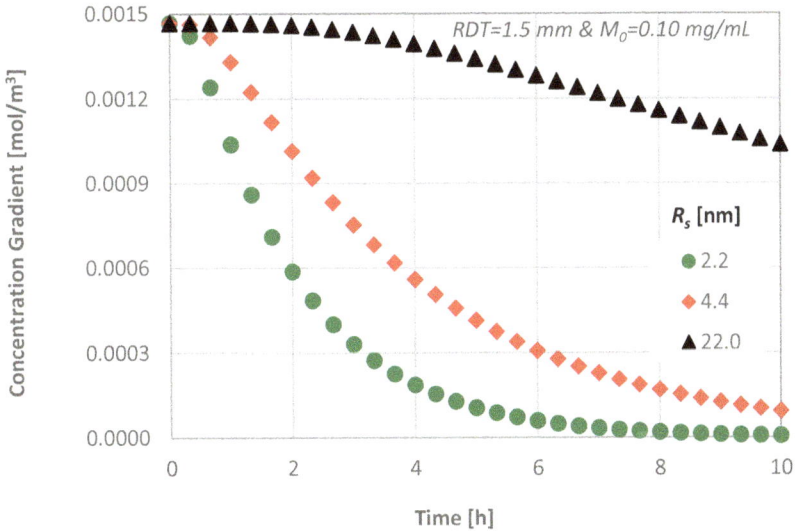

Figure 10. Effect of the molecular size on the diffusion characteristics; concentration gradient versus time.

4.4. Effect of Dentine Porosity

Figure 11 presents the calculated temporal concentration values of the diffusing agent at the pulpal side of the dentinal discs as a function of the porosity of the tissue. As the porosity of the tissue decreases, higher values of concentration are observed at the pulpal side because the available area for the substance to be diffused is reduced. However, the porosity of the discs does not affect the time needed for the steady state condition to be reached.

The variation in concentration at the pulpal side depends strongly on both the porosity and the dentin thickness. The results indicate that there is 5% difference between the minimum and the maximum *RDT* of the dentinal discs investigated in this study for low porosity values, i.e., 5%. As the total pore volume increases, this difference also increases and reaches its upper limit of 15% for the dentin slab with the maximum thickness (i.e., *RDT* = 2.5 mm).

In Figure 12, the flux for three different dentinal discs in terms of porosity values is presented, while the *RDT*, the diffusion coefficient and the initial substance concentration are kept constant (i.e., *RDT* = 1.5 mm, R_s = 4.4 nm and M_0 = 0.01 mg/mL). From Figure 12, it can be observed that the magnitude of the porosity value does not practically affect the calculated molar flux through the dentinal discs.

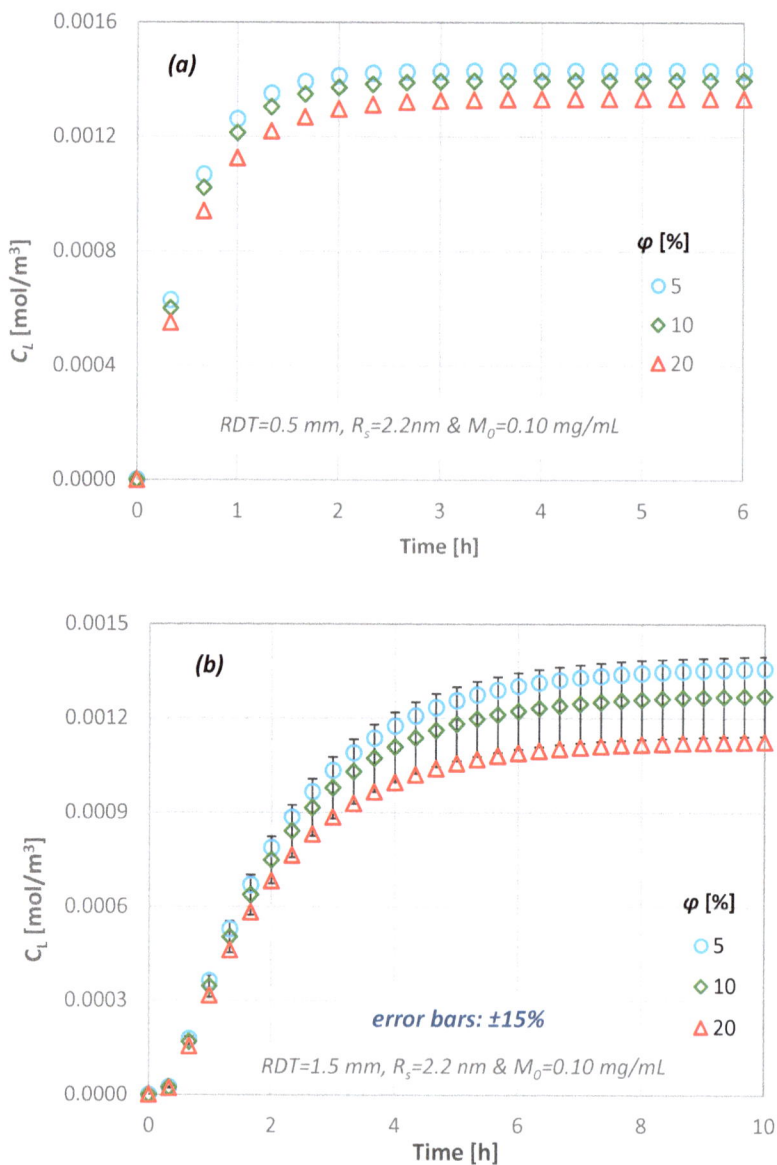

Figure 11. Effect of the dentinal disc porosity value on the diffusion characteristics for: (**a**) $RDT = 0.5$ mm, $R_S = 2.2$ nm and (**b**) $RDT = 1.5$ mm, $R_S = 2.2$ nm.

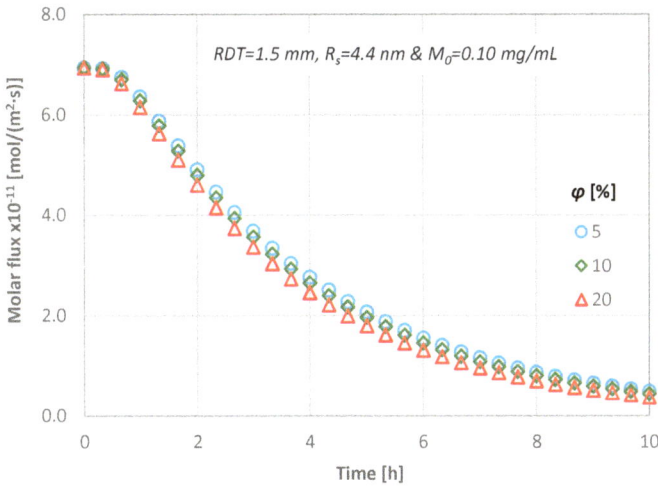

Figure 12. Temporal molar flux distribution for three different porous discs (RDT = 1.5 mm, R_s = 4.4 nm and M_0 = 0.10 mg/mL).

4.5. Effect of the Initial Diffusate Concentration

The transport of a substance similar to the therapeutic agent BMP-7, through a dentinal disc of RDT = 1.5 mm to the pulpal side under different values of initial concentration, has also been studied. The calculated concentrations at the bottom of the disc, C_L, are presented in Figure 13. One can observe that approximately 0.5 h is needed for the first molecules of the substance to reach the bottom end of a dentinal disc of 1.5 mm RDT. This is in agreement with the literature and especially with Pashley and Matthews [11] who have reported that diffusion is a low process and, depending on their size, it requires 30–120 min for the molecules to reach the pulp across a 1–2 mm dentin thickness.

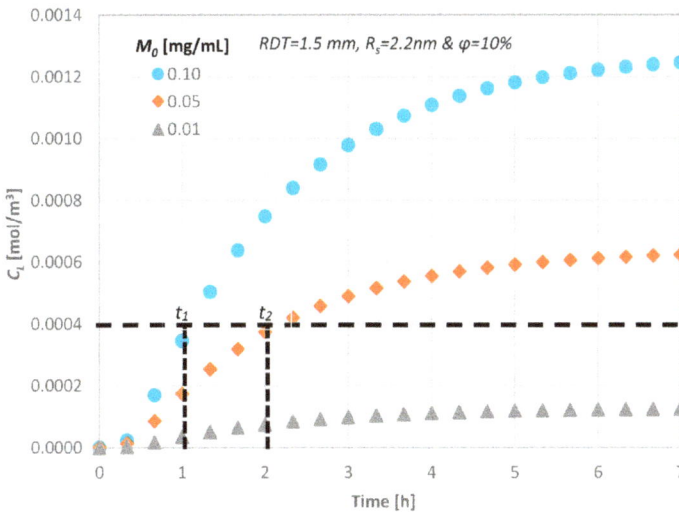

Figure 13. Effect of the initial substance concentration on the diffusion characteristics.

Obviously, the time required for the concentration of signaling molecules to attain a specific value at the bottom end is controlled by their initial concentration, i.e., by increasing the initial concentration of a potential therapeutic agent, the mass flux inside the disc increases and thus the signaling time of the molecules decreases (t_1 and t_2 in Figure 12). This is important for clinical practice of dentistry, since, by this procedure, the behavior of each therapeutic agent can be predicted prior to its application.

4.6. Prediction of the Drug Pulpal Concentration

In an effort to obtain a generalized correlation concerning the delivery of agents through porous domains to the dental pulp that could simulate human dentin tissues, we employed and modified a previous proposed correlation [17] concerning the diffusion through a typical dentinal tubule (Equation (6)):

$$\frac{C_L}{C_0} = 1.1 \exp\left(\frac{0.115}{t*} \cdot \ln(t*) + 10^{-3}L\right) \qquad (6)$$

Equation (6) is a simplified approach that predicts the temporal concentration distribution of the agent at the dentin–pulp junction in the ideal case where all dentinal tubules share the same geometrical characteristics and there is no agent consumption. Obviously, this cannot always adequately represent the actual case. For this reason, a new correlation for the prediction of the diffusion characteristics through dentinal discs is formulated to include all the key design parameters investigated in the present study. In the case where agents are consumed in the pulp, the shape of the molar concentration curvature differs from the one calculated through Equation (6), which assumes no flux (i.e., consumption of agents) at the bottom end of the dentinal tubules. The new equation (Equation (7)) is a 4th order polynomial that also takes into account the gradual decrease of the diffusate until its extinction:

$$C_L = C_0(10^{-10}R + 2.1)(125RDT + 1.3)0.11\varphi^{0.27}(1.1t*^4 - 5t*^3 + 4.9t*^2 + 0.06t* + 0.01) \qquad (7)$$

where R is the consumption rate in kg/(m^2·s) and RDT is expressed in m. The outcome of Equation (7) is in very good agreement with the available data from the computational simulations (overall uncertainty is less than ±15%). Figure 14 presents this comparison when an agent of R_s = 2.2 nm is transported through a 2.5 mm thick dentinal disc of φ = 10% for M_0 = 0.10 (Figure 14a) and 0.05 (Figure 14b) mg/mL, respectively.

If there is no consumption of the agent, Equation (7) is valid for $t* \leq 1$ (at the maximum C_L). Equation (7) permits the estimation of the necessary initial concentration of the therapeutic agent to be imposed for an efficient therapy to be achieved, i.e., the drug concentration at the pulp to reach a critical signaling value dictated by the dental clinical practice, when the maximum acceptable time of application and the consumption rate are given. From Figure 14, it can be observed that there is always a time margin during which the agent retains its maximum value at the pulp (i.e., ~2 h in the specific case of Figure 14; grey area). However, since the R values have been arbitrarily chosen, the scientific community that deals with phenomena related with the dental pulp irrigation and dentinal tissue regeneration must quantify the desirable consumption rate of biomolecules/proteins in the vicinity of the pulp.

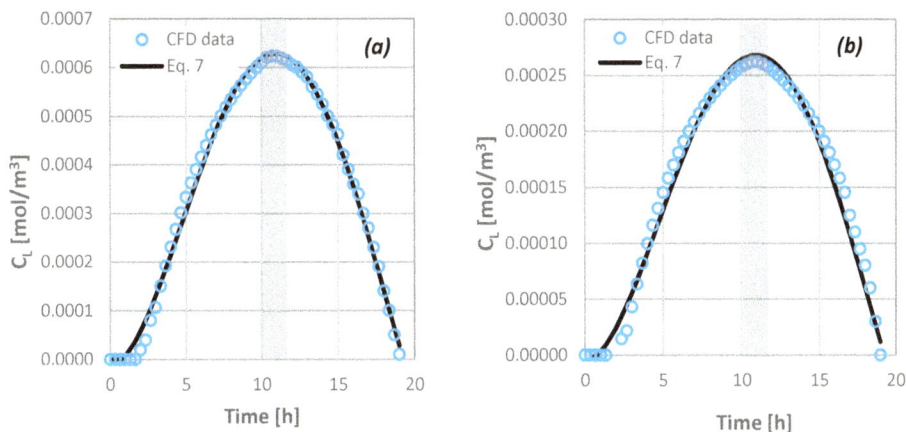

Figure 14. Comparison of computational fluid dynamics (CFD) data concerning the concentration of agents at the bottom end of the porous disc as a function of time with the outcome of Equation (7) for: $RDT = 2.5$ mm, $\varphi = 10\%$, $R_s = 2.2$ nm and (**a**) $M_0 = 0.10$ mg/mL and (**b**) $M_0 = 0.05$ mg/mL.

5. Conclusions

This study aims to investigate the key features that affect transdentinal drug delivery and consequently to determine the necessary quantitative and qualitative issues to be addressed before the application of effective treatment modalities in dental clinical practice. These issues are also related to the diffusion of potentially irritating molecules released from restorative materials. The study has been conducted using CFD simulations validated with relevant experimental data acquired by employing the novel, non-intrusive μ-LIF technique. More specifically, the effect of the molecular size and the initial concentration of various compounds on the transdentinal diffusion characteristics were investigated. The geometrical characteristics (i.e., the porosity of the tissue and its thickness) as well as the consumption rate of these agents at the pulpal side were also included in a parametric study based on the Design of Experiments (DOE) methodology.

The computational results reveal that:

- the transdentinal diffusion of drugs is mainly affected by the molecular size and the RDT, as it was expected,
- a porosity change of 5% to 20% results in less than $\pm15\%$ C_L difference,
- a variation of the agent consumption rate at the pulpal side between 0 and 10^{-10} kg/(m^2·s), leads to a 100% C_L decrease, while the consumption time is 18–25 h.

Finally, in the framework of thorough investigation of the physico-chemical and clinical parameters that influence the transdentinal delivery of potential irritating substances or therapeutic biomolecules in non-exposed dentinal cavities, we propose a new model. Given the geometrical characteristics of a dentinal cavity (i.e., porosity and RDT) in addition to the type of applied molecules, their critical pulpal concentration and the rate of their consumption at the pulp, the proposed model is able to estimate the initial concentration to be imposed if the desirable critical time of application is known and vice versa.

Acknowledgments: The authors would like to thank the Lab technician Asterios Lekkas and Christos Gogos for the construction and installation of the experimental setup. This research has been co-financed by the EU (ESF) and Hellenic National Funds Program "Education and Lifelong Learning" (NSRF)-Research Funding Program: "ARISTEIA"/"EXCELLENCE" (Grant No. 1904).

Author Contributions: Spiros V. Paras and Dimitris Tziafas had the initial conception of this work; Spiros V. Paras and Agathoklis D. Passos designed the experiments; Agathoklis D. Passos performed the experiments acquired and analyzed the data and drafted the work; Spiros V. Paras and Aikaterini A. Mouza interpreted the results and revised the manuscript.

Conflicts of Interest: The authors declare no conflict of interest.

Nomenclature

C	Molar concentration, mol/m^3
C_L	Molar concentration at the bottom end, mol/m^3
C_o	Initial molar concentration, mol/m^3
C_∞	Final molar concentration (for $t = L^2/D$), mol/m^3
D	Coefficient of diffusion, m^2/s
j	Concentration flux, $mol/m^2 \cdot s$
L	Length, m
M_0	Mass concentration, g/m^3
M_w	Molecular weight, g/mol
R	Agent consumption rate, $kg/(m^2 \cdot s)$
RDT	Remaining dentinal thickness, m
R_s	Stokes radius, m
r	Radius of conduit, m
T	Temperature, $^\circ C$
t^*	Dt/L^2, dimensionless
t	Time, s
U	Velocity, m/s
x	Distance, m
μ	Viscosity, $g/cm \cdot s$
φ	Porosity, %

References

1. De Peralta, T.L.; Nör, J.E. Regeneration of the living pulp. In *The Dental Pulp: Biology, Pathology, and Regenerative Therapies*; Springer: Berlin/Heidelberg, Germany, 2014; pp. 237–250.
2. Tziafas, D.; Smith, A.J.; Lesot, H. Designing new treatment strategies in vital pulp therapy. *J. Dent.* **2000**, *28*, 77–92. [CrossRef]
3. Rutherford, B.; Spangberg, L. Transdentinal stimulation of reparative dentine formation by osteogenic protein-1 in monkeys. *Arch. Oral Biol.* **1995**, *40*, 681–683. [CrossRef]
4. Kalyva, M.; Papadimitriou, S.; Tziafas, D. Transdentinal stimulation of tertiary dentine formation and intratubular mineralization by growth factors. *Int. Endod. J.* **2010**, *43*, 382–392. [CrossRef] [PubMed]
5. Smith, A.J.; Tobias, R.S.; Murray, P.E. Transdentinal stimulation of reactionary dentinogenesis in ferrets by dentine matrix components. *J. Dent.* **2001**, *29*, 341–346. [CrossRef]
6. Pashley, D.H. *Dentin Permeability: Theory and Practice*; CRC Press Inc.: Boca Raton, FL, USA, 1990.
7. Pashley, D.H. Dentin permeability and dentin sensitivity. *Proc. Finn. Dent. Soc.* **1992**, *88* (Suppl. S1), 31–37. [PubMed]
8. Murray, P.E.; Smith, A.J. Remaining dentine thickness and human pulp responses. *Int. Endod. J.* **2003**, *36*, 33–43. [CrossRef] [PubMed]
9. Tak, O.; Usumez, A. Diffusion of HEMA from resin cements through different dentin thicknesses in vitro. *Am. J. Dent.* **2015**, *28*, 285–291. [PubMed]
10. Smith, A.J. Vitality of the dentin–pulp complex in health and disease: Growth factors as key mediators. *J. Dent. Educ.* **2003**, *67*, 678–689. [PubMed]
11. Pashley, D.H.; Matthews, W.G. The effects of outward forced convective flow on inward diffusion in human dentine in vitro. *Arch. Oral Biol.* **1993**, *38*, 577–582. [CrossRef]
12. Hanks, C.T.; Wataha, J.C. Permeability of biological and synthetic molecules through dentine. *J. Oral Rehabil.* **1994**, *21*, 475–487. [CrossRef] [PubMed]
13. Boutsioukis, C.; Lambrianidis, T.; Kastrinakis, E. Irrigant flow within a prepared root canal using various flow rates: A computational fluid dynamics study. *Int. Endod. J.* **2009**, *42*, 144–155. [CrossRef] [PubMed]

14. Boutsioukis, C.; Verhaagen, B.; Versluis, M.; Kastrinakis, E.; van der Sluis, L.W. Irrigant flow in the root canal: Experimental validation of an unsteady computational fluid dynamics model using high-speed imaging. *Int. Endod. J.* **2010**, *43*, 393–403. [CrossRef] [PubMed]

15. Gao, Y.; Haapasalo, M. Development and validation of a three-dimensional computational fluid dynamics model of root canal irrigation. *J. Endod.* **2009**, *35*, 1282–1287. [CrossRef] [PubMed]

16. Su, K.C.; Chuang, S.F. An investigation of dentinal fluid flow in dental pulp during food mastication: Simulation of fluid-structure interaction. *Biomech. Model. Mechanobiol.* **2014**, *13*, 527–535. [CrossRef] [PubMed]

17. Passos, A.D.; Tziafas, D.; Mouza, A.A.; Paras, S.V. Study of the transdentinal diffusion of bioactive molecules. *Med. Eng. Phys.* **2016**, *38*, 1408–1415. [CrossRef] [PubMed]

18. Ingle, J.I.; Bakland, L.K.; Baumgartner, J.C. *Ingle's Endodontics 6*; BC Decker: Hamilton, ON, Canada, 2008.

19. Garberoglio, R.; Brannstrom, M. Scanning electron microscopic investigation of human dentinal tubules. *Arch. Oral Biol.* **1976**, *21*, 355–362. [CrossRef]

20. Andrew, D.; Matthews, B. Displacement of the contents of dentinal tubules and sensory transduction in intradental nerves of the cat. *J. Physiol.* **2000**, *529 Pt 3*, 791–802. [CrossRef] [PubMed]

21. Fearnhead, R.W. Histological evidence for the innervation of human dentine. *J. Anat.* **1957**, *91*, 267–277. [PubMed]

22. Hildebrand, C.; Fried, K. Teeth and tooth nerves. *Prog. Neurobiol.* **1995**, *45*, 165–222. [CrossRef]

23. Gendron, P.O.; Avaltroni, F.; Wilkinson, K.J. Diffusion coefficients of several rhodamine derivatives as determined by pulsed field gradient-nuclear magnetic resonance and fluorescence correlation spectroscopy. *J. Fluoresc.* **2008**, *18*, 1093–1101. [CrossRef] [PubMed]

24. Bindhu, C.V.; Harilal, S.S. Effect of the excitation source on the quantum-yield measurements of rhodamine B laser dye studied using thermal-lens technique. *Anal. Sci.* **2001**, *17*, 141–144. [CrossRef] [PubMed]

25. Sakakibara, J.; Adrian, R.J. Whole field measurement of temperature in water using two-color laser induced fluorescence. *Exp. Fluids* **1999**, *26*, 7–15. [CrossRef]

26. Murphy, D.B. *Fundamentals of Light Microscopy and Electronic Imaging*; Wiley: Hoboken, NJ, USA, 2002.

27. Mortensen, N.A.; Okkels, F.; Bruus, H. Reexamination of Hagen-Poiseuille flow: Shape dependence of the hydraulic resistance in microchannels. *Phys. Rev. E* **2005**, *71*, 057301. [CrossRef] [PubMed]

28. Paterson, M.S. The equivalent channel model for permeability and resistivity in fluid-saturated rock—A re-appraisal. *Mech. Mater.* **1983**, *2*, 345–352. [CrossRef]

29. Panda, M.N.; Lake, L.W. Estimation of single-phase permeability from parameters of particle-size distribution. *AAPG Bull. Am. Assoc. Pet. Geol.* **1994**, *78*, 1028–1039.

30. Erickson, H.P. Size and shape of protein molecules at the nanometer level determined by sedimentation, gel filtration, and electron microscopy. *Biol. Proced. Online* **2009**, *11*, 32–51. [CrossRef] [PubMed]

31. Kishen, A.; Vedantam, S. Hydromechanics in dentine: Role of dentinal tubules and hydrostatic pressure on mechanical stress-strain distribution. *Dent. Mater.* **2007**, *23*, 1296–1306. [CrossRef] [PubMed]

32. Stogiannis, I.A.; Mouza, A.A.; Paras, S.V. Study of a micro-structured PHE for the thermal management of a fuel cell. *Appl. Therm. Eng.* **2013**, *59*, 717–724. [CrossRef]

33. Outhwaite, W.C.; Livingston, M.J.; Pashley, D.H. Effects of changes in surface area, thickness, temperature and post-extraction time on human dentine permeability. *Arch. Oral Biol.* **1976**, *21*, 599–603. [CrossRef]

fluids

MDPI

Article

Thermal Fluid Analysis of Cold Plasma Methane Reformer

Sarvenaz Sobhansarbandi [1], Lizon Maharjan [2], Babak Fahimi [2] and Fatemeh Hassanipour [3,*]

[1] Department of Civil and Mechanical Engineering, The University of Missouri-Kansas City, Kansas City, MO 64110, USA; sarvenaz@umkc.edu

[2] Department of Electrical Engineering, The University of Texas at Dallas, Richardson, TX 75080, USA; Lizon.Maharjan@utdallas.edu (L.M.); fahimi@utdallas.edu (B.F.)

[3] Department of Mechanical Engineering, The University of Texas at Dallas, Richardson, TX 75080, USA

* Correspondence: fatemeh@utdallas.edu; Tel.: +1-972-883-2914

Received: 3 February 2018; Accepted: 23 April 2018; Published: 1 May 2018

Abstract: One of the most important methods of methane utilization is the conversion to synthesis gas (syngas). However, conventional ways of reforming methane usually require very high temperature, therefore non-thermal (non-equilibrium) plasma methane reforming is an attractive alternative. In this study, a novel plasma based reformer named 3D Gliding Arc Vortex Reformer (3D-GAVR) was investigated for partial oxidation of methane to produce syngas. The tangential input creates a vortex in the plasma zone and an expanded plasma presides within the entire area between the two electrodes. Using this method, the experimental results show that hydrogen can be produced for as low as \$4.45 per kg with flow rates of around 1 L per minute. The maximum methane conversion percentage which is achieved by this technology is up to 62.38%. In addition, a computational fluid dynamics (CFD) modeling is conducted for a cold plasma reformer chamber named reverse vortex flow gliding arc reactor (RVF-GA) to investigate the effects of geometry and configuration on the reformer performance. In this modified reformer, an axial air input port is added to the top of the reaction vessel while the premixed reactants can enter the cylindrical reaction zone through tangential jets. The CFD results show that a reverse vortex flow (RVF) scheme can be created which has an outer swirling rotation along with a low pressure area at its center with some component of axial flow. The reversed vortex flow utilizes the uniform temperature and heat flux distribution inside the cylinder, and enhances the gas mixtures leading to expedition of the chemical reaction and the rate of hydrogen production.

Keywords: partial oxidation of methane; synthesis gas; cold plasma; gliding arc discharge; computational fluid dynamics modeling

---

1. Introduction

Employment of fuel cells [1–3] for power generation has been recognized as an important area where hydrogen could be used to achieve higher energy efficiencies and greatly reduce emissions. However, since hydrogen is not readily available, production of H_2 from reforming of different types of fuel such as methanol, ethanol, glycerol, methane (Natural gas/Biogas), E85, gasoline, diesel/biodiesel, etc. has been recently studied. There are several conventional reforming technologies, including steam reforming, partial oxidation, auto-thermal reforming, methanol reforming and catalytic cracking, to address these needs.

Hydrogen is predominantly produced using a centralized large scale catalyst reforming plant. Normally, the process requires high pressure (up to 25 bar) and high temperature (above 800 °C) within a large facility. However, transportation of hydrogen and distribution infrastructure over a large area would be prohibitively expensive. This lead to a more favorable distributed approach where hydrogen is produced on-site and on-demand, especially through the reforming of methane, which is readily available to most consumers. The conventional ways of reforming methane usually require very high temperature, which leads to developing combined heat and power (CHP) technologies. For CHP systems, solid oxide fuel cells (SOFCs) with very high operating temperature (typically 700–1000 °C) are used [1]. The CHP systems based on SOFCs usually have long set-up time and high operational temperature, making them unsuitable for some the residential applications [2]. However, such technology (utility based SOFC) is one of the most efficient and environmental-friendly technologies in large-scale and have reached pilot-scale demonstration stages in many regions such as the US, Europe and Japan [4]. The SOFCs high operating temperature generates high quality heat byproduct which can be used for co-generation, or in combined cycle applications. This type of system is flexible with choice of fuel and has very low emissions by removing the danger of carbon monoxide in exhaust gases, since any produced CO is converted to CO_2 at the high operating temperature [4]. In comparison, proton exchange membrane fuel cell (PEMFC) operates at lower operating temperature and it is less sensitive to operating conditions [5–7], but they require hydrogen input at purity level of 99% or higher. A series of PEMFC-based CHP have been proposed in the literature [3,6,8,9]. In some works on PEMFC-based CHP systems, a cold plasma reforming of methane is employed, as reported by Barelli and Ottaviano [10]. In their work, such CHP system will be further extended with heat processor and multiport power electronics interface (MPEI) to integrate all the possible power sources and electrical loads found in residential applications. The PEMFCs are very sensitive to CO poisoning, and often require regular hydrogen purging for maintenance. SOFCs on the other hand are not vulnerable to CO poisoning and can take syngas directly as input without any purification. This eliminates capital and operational costs associated with hydrogen separation and purification, and provides attractive application for plasma assisted reformation systems. Consequently, the objective of this study is producing the hydrogen as the product of synthesis gas (syngas) with the lowest price for the application in SOFC systems.

One of the most significant methods of producing syngas is through conversion of methane, which has an important application in the petrochemical industry. Syngas is a combination of hydrogen and carbon monoxide, and can be applied in a variety of petrochemical processes such as methanol production via the so-called Fischer–Tropsch synthesis [11]. Synthesis gas production can be obtained by three primary methods for methane reforming. The first method is the methane steam reforming, a direct reaction between steam and methane which yield to a high fraction of hydrogen. Generally, this reaction happens over catalysts with high temperatures of around 425–550 °C. The reaction is highly endothermic and therefore consumes a high rate of energy [12]. The second method is the methane reforming with carbon dioxide. The thermodynamic and equilibrium characteristics of this reaction is similar to the steam reforming process, and it is strongly endothermic. However, this method produces syngas with a lower H_2/CO ratio. The principal trouble is from the deactivation of the catalyst utilized since the coke is deposited under the normal reaction conditions [13]. In the third method, which is the interest of this study, the partial oxidation of methane or oxygen-enhanced reforming produce syngas with a H_2/CO ratio close to 2. This reaction utilizes a small amount of oxygen to produce the high exothermic reaction without the application of a catalyst. Since this reaction generates heat, the temperature may rise to 1300–1400 °C [14].

Minimizing the use of steam is attracting great interest in recent years due to the significant drawbacks such as endothermic reactions, a 3/1 H_2/CO ratio as product, steam corrosion issues, and costs in handling the excess of H_2O. The process has thus moved from steam reforming to "wet" oxidation, and much research has been devoted to direct "dry" oxidation of CH_4/O_2 mixtures recently. A reactor utilizes this reaction would be much more energy efficient than the energy intensive steam reforming process, since the direct oxidation reaction is slightly exothermic. This indicates that a single stage process for syngas generation would be a viable alternative to steam reforming [15].

In non-thermal (non-equilibrium) plasma, the plasma is characterized by high temperature of the electron ($\sim 10^5$ K) while keeping a low bulk temperature of the gas phase ($< 10^3$ K) [16].

This high rate of energy and abundant electrons of the plasma activate reactants and initiate radical reactions. The plasma (discharge) initiation may be very quick (microsecond scale) with utilizing high voltage excitation. The advantages of this reaction is low operating temperature, fast start-up, transient performance, and eliminating the need to use expensive catalytic materials [17]. Different types of plasmas have been utilized in reforming of hydrocarbons and oxygenated hydrocarbons for hydrogen rich gas production [10,17–19]. Among these types, a high-frequency pulse plasma has demonstrated high-energy efficiency which can operate at relatively lower temperatures than arc discharge plasmas [20]. In auto-thermal reforming plasmas [21], high net efficiency and good volumetric throughput is achieved, however, oxygen necessitates either air separation or results in dilution with nitrogen for this type of plasma. Auto-thermal plasmas have significantly lower electrical energy consumption, which has a great potential for the application of this type in some circumstances (e.g., mobile fuel cell systems) while operating at higher temperatures.

Several studies investigate the methane reforming utilizing different types of plasmas, specifically with non-equilibrium plasma. Gliding arc discharge is a new method for non-equilibrium plasma generation with high efficiency and environmental friendliness, which is the target of this study. In this study, the partial oxidation of methane to produce syngas using a novel plasma based reformer was investigated. The effects of several operational parameters on the methane conversion and the selectivity of products were considered. In addition, the computational fluid dynamics modeling of another innovative cold plasma reformer chamber named a reverse vortex flow gliding arc reactor (RVF-GA), where the geometry and inlets of the 3D-GAVR are modified, was conducted to investigate the effect of geometry size and configuration for efficient performance.

2. Experimental Modeling

2.1. Partial Oxidation

The chemical process considered for reformation in this study is partial oxidation. In industrial practices, partial oxidation is often followed by Water Gas Shift (WGS) reaction, which increases the production yield of H_2 and eliminates toxic CO by converting it to CO_2. However, WGS does not affect the findings of this study and is not included. The reaction for partial oxidation is given in the following equation where the reaction is an exothermic reaction with enthalpy of -36.1 kJ/mol [22].

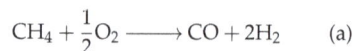

$$CH_4 + \frac{1}{2}O_2 \longrightarrow CO + 2H_2 \qquad \text{(a)}$$

It is to be noted that atmospheric air consists of 78.09% of N_2 which is not reactive and serves as the carrier gas, and 20.95% of O_2. Therefore, air is used instead of compressed oxygen for partial oxidation. The flow rate of air needs to be 4.77 times higher than the required flow rate of O_2.

2.2. 3D Gliding Arc Vortex Reformer

Presented in this study is a novel plasma based reformer hereafter referred as 3D Gliding Arc Vortex Reformer (3D-GAVR). The fundamentals of this reformer are based on gliding arc discharge, as described by Czernichowski [22]. A few variations of gliding arc discharge have been proposed [23]. Unlike the previously proposed reformer topologies, 3D-GAVR expands the plasma zone and introduces vortex type air flow which further intensifies the plasma. The geometry of the reformer and cross-sectional area are shown in Figure 1a,b, respectively. Figure 1c,d shows the top and side views of the 3D-GAVR, respectively.

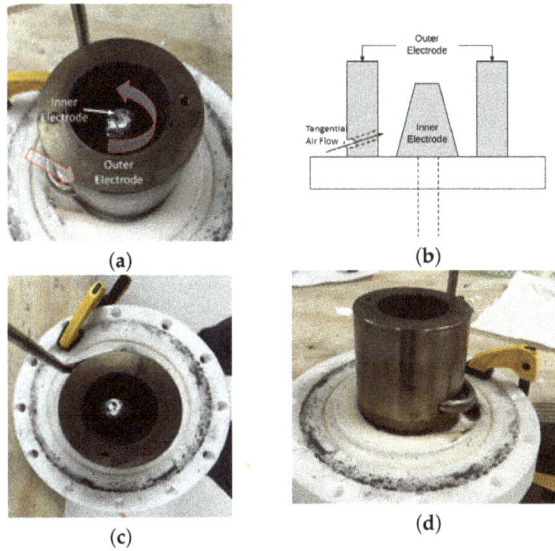

Figure 1. (a) Geometry of 3D-GAVR prototype; (b) CSA of 3D-GAVR; (c) top view 3D-GAVR; and (d) side view 3D-GAVR.

Following the structure of gliding arc, the distance between two electrodes is shortest at the bottom and longest at the top. This helps in arc propagation and keeps the arc in non-thermal region. When the voltage is applied, the arc initiates at the bottom due to smallest resistance and starts moving upwards. However, due to cylindrical electrodes, the movement occurs in rotational and upward direction simultaneously. The direction of rotation is dictated by the direction of input mixture flow. The tangential input creates a vortex in the plasma zone, as shown by red arrows in Figure 1a. This turbulent movement of mixture inside the reformer creates an expanded plasma that incorporates entire area between the two electrodes, hence allowing for more efficient reformation than traditional gliding arc structures. Furthermore, the increase in airflow increases the intensity of plasma rather than decreasing it, suggesting an operational flow rate of that is higher than other topologies such as pin and plate. Therefore, the suggested reformer can provide a solution that has higher efficiencies at higher flow rates. Operation of 3D-GAVR is depicted in Figure 2.

Figure 2. 3D-GAVR operation.

In addition, the thick electrodes, higher flow rates, and absence of sharp/pointed edges on the electrodes substantially reduces the electrode deterioration, and consequently maintenance time and cost. Although carbon should not be a product of an ideal partial oxidation reaction, the results do not always adhere to theoretical assumptions and some carbon deposition is often noticed. The given reformer topology is more tolerant towards carbon deposition than the counterparts. Moreover, excessive carbon deposition can be solved by slight increase in the input air flow.

2.3. Experimental Setup

The objective of the experiment was to pass a mixture of air and methane through 3D-GAVR, perform reformation, and detect the composition of the produced syngas. The setup prepared for the experiment is presented in Figure 3a, and allows for online detection of syngas composition without stopping the experiment. This step is important for maintaining consistency, especially during the comparative analysis that involves changes in specific parameters.

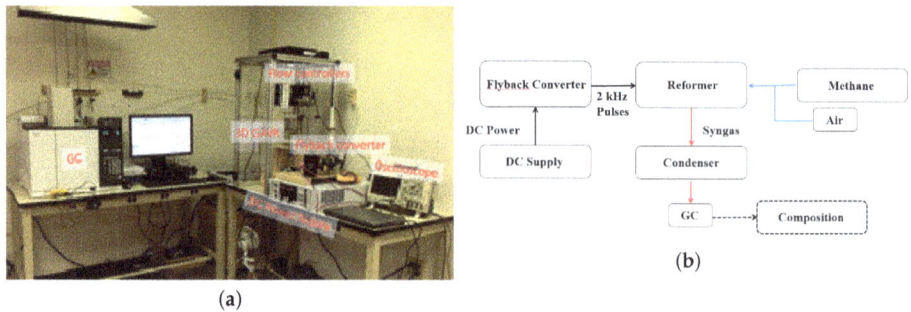

(a) (b)

Figure 3. (**a**) Test setup; and (**b**) system lock diagram of setup.

The setup description is presented in the form of block diagram in Figure 3b. The electric power required for reformation is provided through a flyback converter which was designed to provide up to 6 kV pulses at 2 kHz. However, the experiment was performed with pulses of peak voltage around 3 kV. The input power to the flyback converter is provided by Magna power TSA600 DC power supply. The compressed air and methane cylinders have been used to supply mixture into 3D-GAVR. The flow is controlled using individual flow controller, as shown in Figure 3a. Post reformation syngas is then transferred to Gas Chromatogram (GC) for detection of gas composition. The results of GC are presented in the following section. A condenser is placed between GC and the reformer, as shown in Figure 3b, to eliminate moisture as required for the safety of GC. The reformer is allowed to operate for a few minutes to reach steady state operation and GC valve is opened to collect sample syngas. The excess gas is safely released using the vent. Each test involved collection of three samples with intervals of 10 min to ensure accuracy of the readings. For sensitivity analysis, the parameters were changed and the procedures were repeated.

2.4. Calculation Method

The parameters collected from the experimental analysis are listed in Table 1, which summarizes the output presented in Figure 4 along with the input to the system. Table 1 is mentioned as a mere example to clarify the origin of data in Table 2. It is to be noted that this is not an optimal operation point. The input methane and air flow are measured using respective flow meters, and the input voltage and current are the values read by DC power supply, therefore the efficiency of the flyback converter is included in the calculation. The output of the cold plasma system is syngas since water gas shift has not been performed. The composition of syngas was detected using Gas Chromatogram (GC) Agilent 7890B (Agilent Technologies, Santa Clara, CA, USA). The output of the GC is the molar percent contents of different constituents of the output syngas. A typical result from GC is shown in Figure 4, which consists of molar percentages of CH_4, CO, CO_2, H_2, N_2 and O_2, as shown in Reaction (b). This dataset is used as input to the database that performs calculations for cost in $ per kg and flow rate of hydrogen in liters per minute (L/min). N_2 and O_2 do not affect these calculations, hence are not included in Tables 1 and 2. These calculations are made by first calculating the cost rate of the methane in $/s, following Equations (4) and (5). The cost rate of electricity is calculated using Equation (6). The cost of produced H_2 and the rate of production of H_2 are important parameters that serve as the basis of comparisons; these parameters are calculated based on Equations (9) and (10). The cost calculation of H_2 includes cost of input methane and input electricity to the system. The input CH_4 flow rate was converted to liters per second and the cost rate of $0.366/ccf at 20 psia was used to calculate the price per second of CH_4. Similarly, input power in watts was converted to kWh per second and electric cost rate of $0.09/kWh was utilized to calculate the price per second of electricity in Equation (9). Hence, the cost of H_2 per kg is also a good representation of the efficiency of the system.

The Alicat flow meter utilized has limitations related to number of components in the composition of gas, hence, the output flow rate could not be measured accurately. This was compensated by estimating the molar flow rate of hydrogen using mass balance and molar flow rate of carbon. Mass balance suggests that the total number of moles of carbon in the input and output side of the reaction should be equal, consequently, the molar flow rate of carbon should also be equal. The compositions of experimental input and output gas mixture are presented in Reaction (b). Air was used instead of oxygen, which introduces inert nitrogen to both input and output side of the reaction. More importantly, ideal reactions such as Reaction (a) is not possible in plasma assisted systems, therefore, output gas consists of CO_2, and unreacted CH_4 in addition to expected CO and H_2. Considering Reaction (b), the number of moles of carbon in reactant side CH_4 should be equal to total number of moles of carbon in product shown in Equation (1). Using the percent values of CO, CO_2 and unreacted CH_4 obtained from the GC readings as shown in Figure 4, molar flow rate of output carbon and hydrogen can be calculated using Equations (2) and (3), respectively.

$$CH_4 + Air \longrightarrow CO + \text{unreacted } CH_4 + N_2 + O_2 + H_2 \qquad \text{(b)}$$

$$\text{Percent composition of moles C (\%C)} = \%CO_2 + \%CO + \% \text{ unreacted } CH_4 \qquad (1)$$

$$\text{Molar flow rate of C } (\frac{\text{mol}}{\text{min}}) = \text{Molar flow rate of } CH_4 \ (\frac{\text{mol}}{\text{min}}) \qquad (2)$$

$$\text{Molar flow rate of } H_2 \ (\frac{\text{mol}}{\text{min}}) = \text{Molar flow rate of } CH_4 \ (\frac{\text{mol}}{\text{min}}) \times (\frac{\%H_2}{\%C}) \qquad (3)$$

$$\text{Volumetric flow of } CH_4 \ (\frac{\text{ccf}}{\text{s}}) = \frac{(\text{Flow in } \frac{L}{\text{min}})}{60} \times (\frac{\text{ccf}}{L}) \qquad (4)$$

$$\text{Cost rate of } CH_4 \ (\frac{\$}{s}) = (\frac{\text{ccf}}{s}) \times (\text{price per ccf}) \times (\frac{\text{Volume at operating pressure}}{\text{Volume at atm pressure}}) \qquad (5)$$

$$\text{Electricity cost rate } (\frac{\$}{s}) = \frac{\text{Input power (W)}}{1000 \times 3600} \times (\text{price per kWh}) \qquad (6)$$

$$\text{Molar flow rate of } CH_4 \ (\frac{\text{mole}}{\text{min}}) = \text{Vol. flow rate of } CH_4 \ (\frac{L}{\text{min}}) \times (\frac{g}{L}) \times (\frac{\text{mol}}{g}) \qquad (7)$$

$$\text{Molar flow rate of } H_2 \ (\frac{\text{mole}}{\text{min}}) = \text{Molar flow rate of } CH_4(\frac{\text{mol}}{\text{min}}) \times (\frac{\%H_2}{\%C}) \qquad (8)$$

$$\text{Cost of produced } H_2 \ (\frac{\$}{\text{kg}}) = \frac{\text{Cost rate of } CH_4 \ (\frac{\$}{s}) + \text{Electricity cost rate } (\frac{\$}{s})}{\text{Molar flow rate } H_2 \ (\frac{\text{mol}}{\text{min}}) \times (\frac{g}{\text{mol}}) \times (\frac{\text{kg}}{1000 \times 60 \text{ s}})} \qquad (9)$$

$$\text{Rate of } H_{2 \ production} \ (\frac{L}{\text{min}}) = \text{Molar flow rate of } H_2 \ (\frac{\text{mol}}{\text{min}}) \times (\frac{g}{\text{mol}}) \times (\frac{L}{g}) \qquad (10)$$

Table 1. Experimentally obtained parameters.

CH$_4$ Flow (L/min)	Air Flow (L/min)	V	I	%CH$_4$	%CO	%CO$_2$	%H$_2$
4.167	10	120.1	0.9	23.88	1.87	0.35	4.99

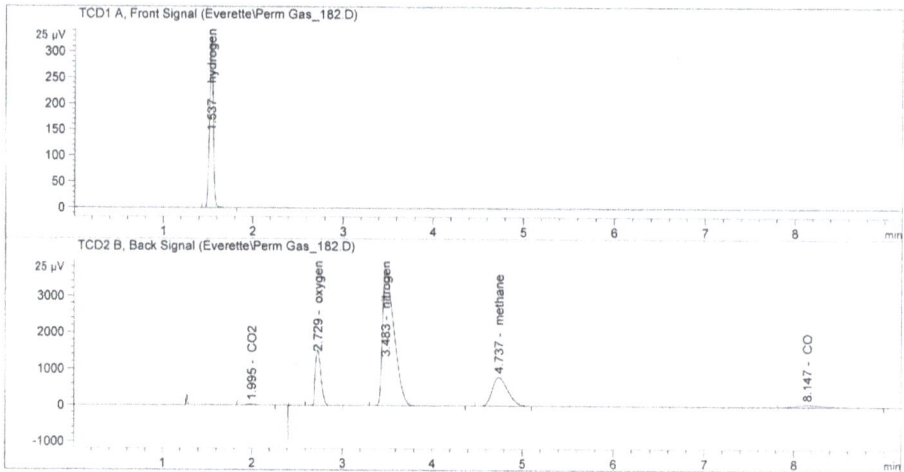

Figure 4. GC compositions percentage.

2.5. Results

The gas compositions are collected from the experiments and the calculations for price and production rate of hydrogen are made as described in previous section. The experimental analysis has been divided into two distinct portions. In the first analysis, the variation of power and flow rate for ideal Air and CH$_4$ with flow rate ratio of 2.4:1, respectively, are investigated. Since molar volumes of all gases is 22.4 L, the flow rate ratio can be derived by considering coefficient of O$_2$ per mole of CH$_4$ in Reaction (a), and the percent content of oxygen in air (20.95%) which will be $1/2 \times 1/0.2095 = 2.4$. The cost and flow rate of hydrogen for each test were calculated and the results are provided in Table 2. It should be noted that each data point is obtained by repeating each test three times to ensure consistency.

Figure 5 presents the summary of Table 2 which presents dependency of the cost of the produced H$_2$ in input power and total input flow. As indicated, most of the points with lowest cost lie around the region where the total flow is 7.65–8.925 L/min, which corresponds to CH$_4$ flow rate of 2.25–2.625 L/min and Air flow of 5.4–6.3 L/min, and the power ranged 135–164 W. The most efficient point is encircled in Table 2 and was chosen for the second portion of the experiment. The second portion involves keeping the power constant and varying the flow rate ratio of Air:CH$_4$.

Table 2. Experimental results—constant flow rate ratio of 2.4:1, where price per ccf of methane psia = $0.366 and Price per kWh of electricity = $0.09.

CH$_4$ (L/min)	Air (L/min)	Total Flow	Power	%CH$_4$	%CO	%CO$_2$	%H$_2$	$\frac{\$}{kg}$	H2($\frac{L}{min}$)	$\frac{L}{kWh}$	$\frac{H_2}{CH_4}$ Ratio	CH$_4$ %conv.
1.5	3.6	5.1	67.45	24.96	2.16	0.27	6.27	8.54	0.32	282.45	0.25&8.86	
1.5	3.6	5.1	83.029	20.19	3.78	0.53	9.88	5.31	0.56	404.17	0.49	17.62
1.5	3.6	5.1	98.9	21.65	3.10	0.41	8.30	7.07	0.46	277.64	0.38	13.96
1.5	3.6	5.1	155.93	19.20	3.75	0.63	9.25	7.68	0.54	209.64	0.48	18.55
1.875	4.5	6.375	67.45	22.91	2.56	0.41	6.67	6.95	0.45	397.62	0.29	11.46
1.875	4.5	6.375	164	18.51	4.11	0.74	10.27	6.19	0.76	278.89	0.55	20.75
1.875	4.5	6.375	138.8	16.33	5.87	1.50	7.07	8.30	0.52	223.68	0.43	31.09
2.25	5.4	7.65	108.09	21.26	3.08	0.51	8.08	6.18	0.68	375.48	0.38	14.45
2.25	5.4	7.65	127.4	20.45	3.35	0.59	8.66	6.10	0.74	347.84	0.42	16.17
2.25	5.4	7.65	148.51	20.14	3.59	0.65	9.17	6.20	0.78	316.52	0.46	17.38
2.25	5.4	7.65	153.6	19.33	3.94	0.73	10.18	5.60	0.88	344.84	0.53	19.46
2.25	5.4	7.65	156.6	19.14	3.96	0.74	10.15	5.64	0.89	339.31	0.53	19.72
2.25	5.4	7.65	164	18.90	4.08	0.77	10.50	5.56	0.92	336.68	0.56	20.43
2.25	5.4	7.65	134.55	15.48	4.07	1.10	8.39	5.47	0.85	377.14	0.54	25.04
2.625	6.3	8.925	164	18.86	4.23	0.84	10.63	5.11	1.08	394.84	0.56	21.18
2.625	6.3	8.925	137.16	15.49	4.39	1.20	8.96	4.91	1.03	451.49	0.58	26.52
3	7.2	10.2	138.8	13.57	4.36	1.35	5.20	7.33	0.75	323.69	0.38	29.61
3.75	9	12.75	67.45	25.49	1.42	0.29	3.81	10.48	0.49	431.76	0.15	6.29
4.167	10	14.167	108.09	22.72	2.22	0.49	5.32	7.70	0.81	447.17	0.23	10.64

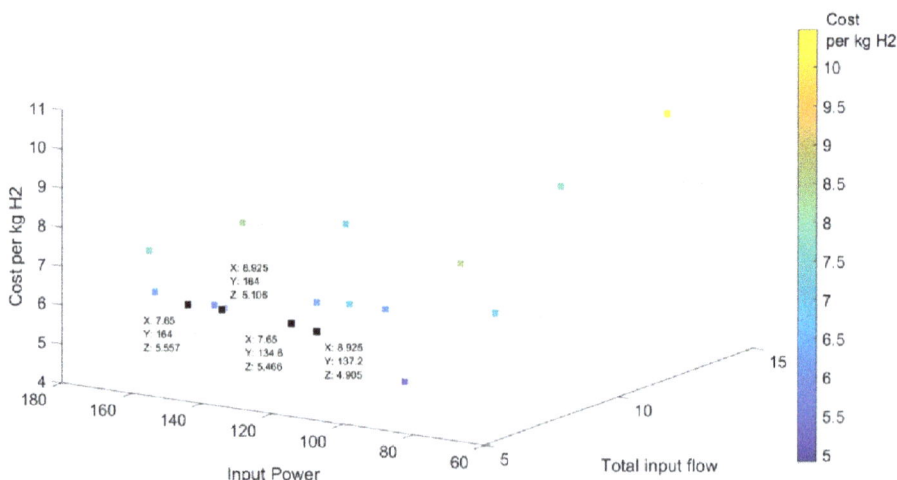

Figure 5. Change is cost of hydrogen with total flow and input power.

The results of second portion of the experiment are listed in Table 3. This section of testing involved starting at the most efficient point of the first part of the test and varying the Air:CH$_4$ flow rate ratio. The ratio was varied from 2.4:1 to 3.09:1. The lower limit was decided by theoretical limits as described earlier. Going below this value resulted in carbon deposition. The higher limit was governed by thermal limitation of chamber's Teflon end caps. The results of the test are explained in Figure 6a,b. The cost of the produced H$_2$ decreases, and the flow rate of H$_2$ increases, with approximate linearity, as the Air:CH$_4$ ratio increases. Thus, increasing the ratio has positive effects on both cost and rate of production within this range. However, it is important to realize that increasing the ratio further will eventually lead to combustion and formation of water vapor causing drastic decrease in H$_2$ production. The %H$_2$ in the output mixture decreases as the ratio increases, which contradicts the trend of H$_2$ flow rate. This can be explained by considering that the percent value is not an absolute quantity: this quantity depends on the flow rate of other output gases. As the ratio is increased, the flow rate of input air is increased while the flow of CH$_4$ is kept the same. Although higher amount of H$_2$ is produced, the rate of increase of H$_2$ is smaller than the rate of increase in components introduced due to increase in input air flow. Therefore, the percent H$_2$ decreases even though the flow rate of the produced H$_2$ increases.

Hence, H$_2$ can be produced for as low as $4.45 per kg with flow rates of around 1 L per minute. The maximum methane conversion percentage which is achieved by this technology is up to 62.38%. The non-optimized dimensions of 3D-GAVR is around 7 cm in radius and 15 cm in height with significant room for reduction. Therefore, further optimization of performance and size, and cascaded performance of multiple reformers can provide an attractive solution for on-site hydrogen generation in H$_2$ dispensing stations. This will help eliminate distribution cost and reduce storage costs which make up a significant portion of current hydrogen fuel cost [24].

Moreover, a comparison was made between the results from this study and the other studies in the literature. The results are presented in Figure 7a–c, where the histograms are presented by Petitpas et al. [17] and the black dot on the y-axis represents the performance of the proposed 3DGAVR.

Table 3. Results of experimental modeling, where price per ccf of methane psia = $0.366 and price per kWh of electricity = $0.09.

CH$_4$ (L/min)	Air (L/min)	V	I	%CH$_4$	%CO	%CO$_2$	%H$_2$	$\frac{\$}{kg}$	H$_2(\frac{L}{min})$	$\frac{H_2}{CH_4}$	CH$_4$ %conv.
2.625	8.1	67.9	2.02	6.06	5.78	3.57	6.95	4.62	1.10	1.15	6.68
2.625	8.1	67.9	2.02	5.71	6.00	3.64	7.20	4.45	1.14	1.26	62.83
2.625	7.3	67.9	2.02	11.71	4.68	1.77	8.03	4.71	1.07	0.69	35.52
2.625	7.3	67.9	2.02	11.51	4.78	1.85	8.05	4.70	1.08	0.70	36.56
2.625	7.3	67.9	2.02	11.39	4.88	1.89	8.18	4.63	1.09	0.72	37.27
2.625	6.3	67.9	2.02	15.27	4.45	1.25	8.98	4.87	1.04	0.59	27.19
2.625	6.3	67.9	2.02	15.49	4.39	1.20	8.96	4.91	1.03	0.58	26.52
2.625	6.3	67.9	2.02	15.58	4.34	1.18	8.88	4.96	1.02	0.57	26.17

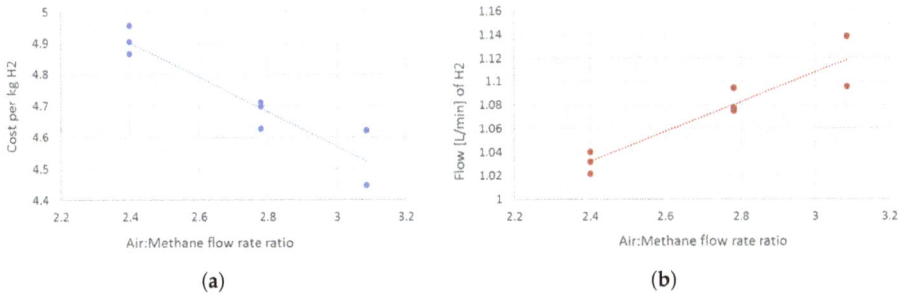

(a)

(b)

Figure 6. (a) Variance of cost per kg of generated H$_2$; and (b) variance of Flow rate of generated H$_2$ with Air:Methane ratio.

The values were obtained using the method presented by Petitpas et al. [17] for fair comparison. Figure 7a compares the methane conversion rate of 3DGAVR with previous publications. The conversion rate percent, obtained by Equation (11), is 62.8%. As shown in Figure 7a, this conversion rate is higher than most of the systems that use Methane as fuel (represented by green histograms). The efficiency of 3DGAVR is calculated to be 10.76% using Equation (12), which is significantly lower than most of the systems, as presented in Figure 7b. However, considering only the systems that use Methane as fuel, this efficiency is still higher than 42% of the listed results. Lastly, the specific energy requirement for 3DGAVR is 88.9 kJ/mol using Equation (13). As seen in Figure 7c, this value is competitive, even with consideration of entire fuel spectrum.

$$\chi = \frac{[CO + CO_2 + CH_4 + \cdots]_{produced}}{n \times [C_nH_m]_{injected}} \tag{11}$$

$$\eta = \frac{(H_2 + CO)_{produced} \times LHV(H_2)}{\text{Input plasma energy} + \text{fuel injected} \times LHV(\text{fuel})} \tag{12}$$

$$\nu = \frac{\text{Input plasma power}}{[H_2 + CO]_{produced}} \tag{13}$$

Consequently, the computational fluid dynamics modeling of another innovative cold plasma reformer chamber called a reverse vortex flow gliding arc reactor (RVF-GA), where the geometry and inlets of the 3D-GAVR are modified, was conducted to investigate the effect of geometry size and configuration for efficient performance.

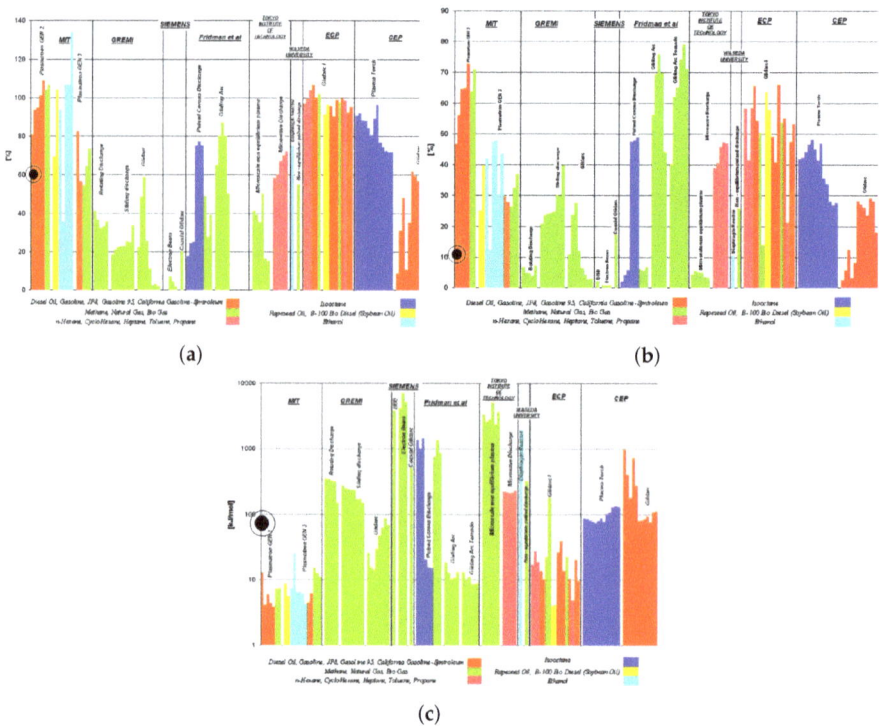

Figure 7. Performance parameters and comparison of 3DGAVR: (a) conversion rate; (b) efficiency; and (c) specific energy requirement. Reproduced with permission from [17].

3. Computational Fluid Dynamics Modeling

The obtained results from the experimental analysis enabled us to do further investigation to find the effect of geometry size and configuration of the reformer chamber in increasing the efficiency of the system. A comprehensive multi-physics computational fluid dynamics (CFD) modeling of the cold plasma chamber, where the geometry and inlets of the 3D-GAVR are modified, was conducted on the chamber named as a reverse vortex flow gliding arc reactor (RVF-GA) model [18]. A schematic of the RVF-GA reactor is shown in Figure 8, where it is designed in a way that the preheated air and vaporized fuel can be mixed in a small region of the reactor. The geometry is designed in a way to provide an axial air input at the top of the vessel, which yields to introducing more amount of air to the reaction zone and consequently enable us to investigate an extensive range of O:C ratio. The high voltage electrode and the grounded exhaust nozzle are separated by utilizing a dielectric ring made from boron nitride. The reactants are premixed and entered the cylindrical reaction zone through tangential jets that are assembled within the dielectric ring. The pointed tip of the high voltage electrode, which is located in front of the tangential swirl jets, is region where the gliding arc discharge is initiated. This configuration enables the maximum interaction between the reactant stream and the plasma discharge and also to be the driving force which causes discharge rotation and elongation. The internal volume of the reaction vessel is approximately 0.4 L (with inner diameter of 4.3 cm) and there is a small post-plasma treatment volume, which is essentially a 0.5 inch diameter stainless steel exhaust tube that has the length of approximately 30 inch [18,25,26].

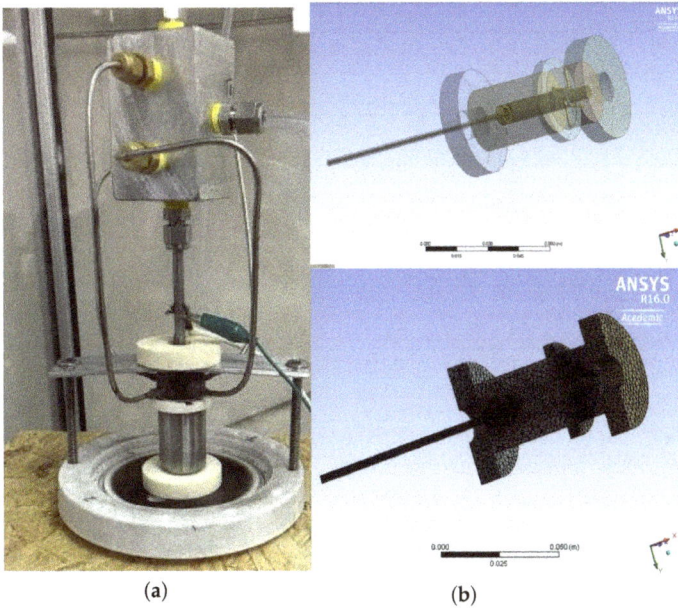

Figure 8. (**a**) The modified cold plasma chamber setup; and (**b**) geometry and meshed model in ANSYS FLUENT.

To perform current CFD analysis, a regular personal computer (PC) with Intel(R) Core(TM) i7-4770 CPU, 64-bit operating system was used. Commercial software (ANSYS FLUENT 16) was used to simulate the chamber and study the heat generation and heat transfer inside the chamber. The analysis began by highlighting the exact dimensions and configurations of the chamber shown in Figure 8. Bottom and top cylinder, input and output cap, the ceramic layer and the plunger are considered in the model. Double precision method with four processor parallel solver are applied. The unstructured meshes with physics preference of CFD and solver preference of Fluent was applied on the body, and the accuracy of meshing was checked by a slow transition and high smoothing mode with the total number of 7×10^5 elements and 125,000 nodes. The mesh refinement (gradient adaption tool) was applied to improve the mesh quality, by searching for the locations where a mesh refinement would be useful. The goal was to achieve a mesh independent solution which results in the optimal mesh for the solution (Figure 8).

The governing equations of continuity, momentum conservation and energy equation were solved for the model to describe the flow of the fluid inside of the chamber. The continuity equation can be written as follows:

$$\nabla \cdot (\rho \, \overrightarrow{v}) = -\frac{\partial \rho}{\partial t} \tag{14}$$

where ρ is the density of the air (1.225 kg/m^3), v is the velocity and t is time. The momentum equation (Navier–Stokes equation) for this model takes the following form:

$$\frac{\partial}{\partial t}(\rho v) + \nabla \cdot (\rho \, \overrightarrow{v} \, \overrightarrow{v}) = -\nabla p + \nabla \cdot (\overline{\overline{\tau}}) + \rho \, \overrightarrow{g} + \overrightarrow{F} \tag{15}$$

where p is the static pressure, $\overset{=}{\tau}$ is the stress tensor and $\rho \overrightarrow{g}$ and $\overrightarrow{F}$ are the gravitational body force and external body forces, respectively (where $\overrightarrow{F}$ contains other model-dependent source terms). The stress tensor $\overset{=}{\tau}$ is given by:

$$\overset{=}{\tau} = \mu[(\nabla \overrightarrow{v} + \nabla \overrightarrow{v}^{T}) - \frac{2}{3}\nabla . \overrightarrow{v} I] \tag{16}$$

where μ is the molecular viscosity of air, and I is the unit tensor where the viscosity of the air is 1.79×10^{-5} kg/m·s.

The energy model with the viscous model is used, which can be represented as the general form of:

$$\frac{\partial}{\partial t}(\rho E) + \nabla . (\overrightarrow{v}(\rho E + p)) = \nabla . \left(k_{eff} \nabla T - \sum_{j} h_j \overrightarrow{J_j} + (\overset{=}{\tau}_{eff} . \overrightarrow{v}) \right) + S_h \tag{17}$$

where k_{eff} is the thermal conductivity of the air as 0.02 W/mK. The second and third terms on the big parenthesis on right side are species diffusion and viscous dissipation, respectively. h_j and $\overrightarrow{J}_j$ are sensible enthalpy and diffusion flux, respectively. S_h in Equation (14) is the heat of chemical reaction, however, as the preliminary modeling, the chemical reaction is not considered, therefore that value is equal to zero for the purpose of this modeling.

Simulation is performed using the precise boundary conditions. The defined materials are air as the fluid, and the alumina-based-ceramic and Kovar as solids, which are set as the cell zone conditions. The boundary conditions are specified as follows:

- The flow rate of the axial top inlet port of the air is set around 12.65 SLPM which is applied as an input air.
- The mixture flow rates are assigned around 12.35 SLPM in total (5 SLPM air, and 7.35 SLPM methane for future modeling), which is divided into the four tangential swirl jets equally. For this modeling, this value is assigned just as the air entering the chamber through the swirl jets.
- The input power is 550 W and it is set as the input heat flux of the arc which is around 894 kW/m^2. This heat flux is applied at the two surfaces where the arc is created during the process.

The SIMPLE algorithm is applied and represented by pressure–velocity coupling method under relaxation factors of pressure as 0.3, density as 1, body forces as 1, momentum as 0.7 and energy as 1. The solution of the flow field is presumed to be converged when the continuity and momentum equations residuals reduced to less than 10^{-6}. Comprehensive convergence tests were done to investigate the time step size.

Figures 9 and 10 show the obtained results for the velocity vectors from simulation in 2D and 3D, respectively, and Figure 11 shows the pressure distribution along the reformer chamber where the created arc has the dimension of 5 mm thickness and it moves by time change. The results show that a reverse vortex flow (RVF) scheme is created which has an outer swirling rotation along with a low pressure area at its center with some component of axial flow where this flow regime is similar to a natural tornado. Due do the smaller size of the exhaust diaphragm compared with the internal diameter of the vessel cylinder, it acts as a barrier which generates vortex flow to reverse direction and flow upwards initially into the cylindrical vessel. This rotational velocity profile can enhance the gas mixtures, and as a result increasing the rate of chemical reaction of methane and oxygen. The Reynolds number is calculated by the average axial velocity near the outlet, where the dimension is similar to a pipe, as follows:

$$Re = \frac{u L}{\nu} = \frac{2.5 \times 0.03}{0.0000156} \simeq 4800 \tag{18}$$

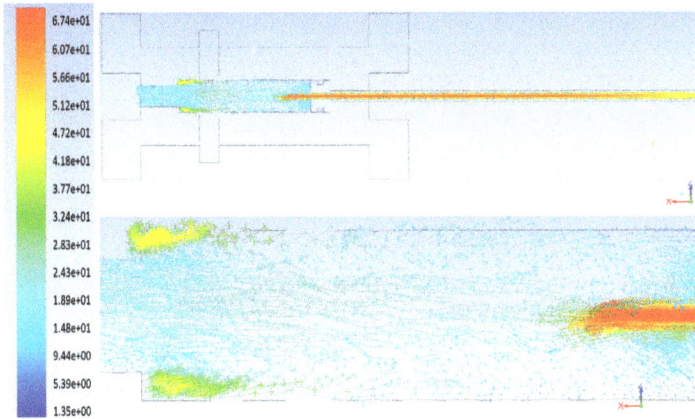

Figure 9. Velocity vectors of the gas mixture (m/s).

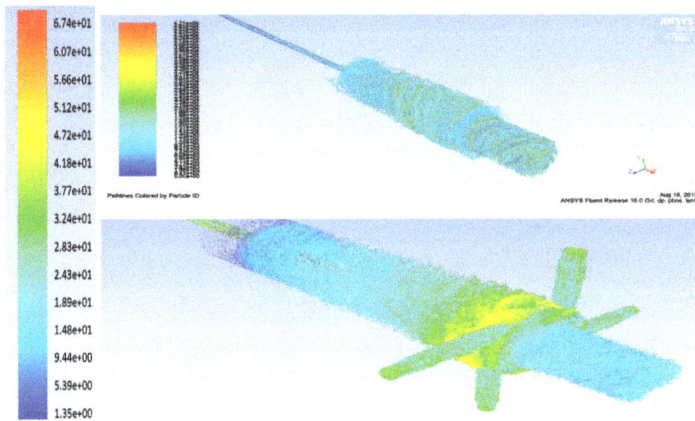

Figure 10. Velocity vectors of the gas mixture (m/s).

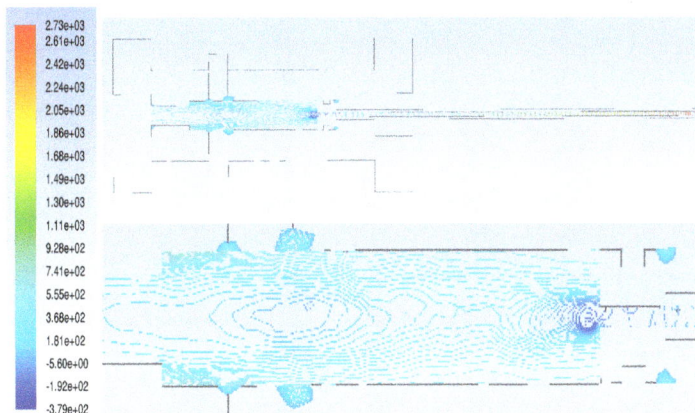

Figure 11. Pressure distribution (Pascal).

Figure 12 shows the temperature profile of the gas mixture at different time steps. The profile shows an interesting temperature range around the arc just in front of the tangential swirl jets where the applied heat flux can provide the required temperature for the chemical reaction and as such a catalyst is no longer required for this reaction.

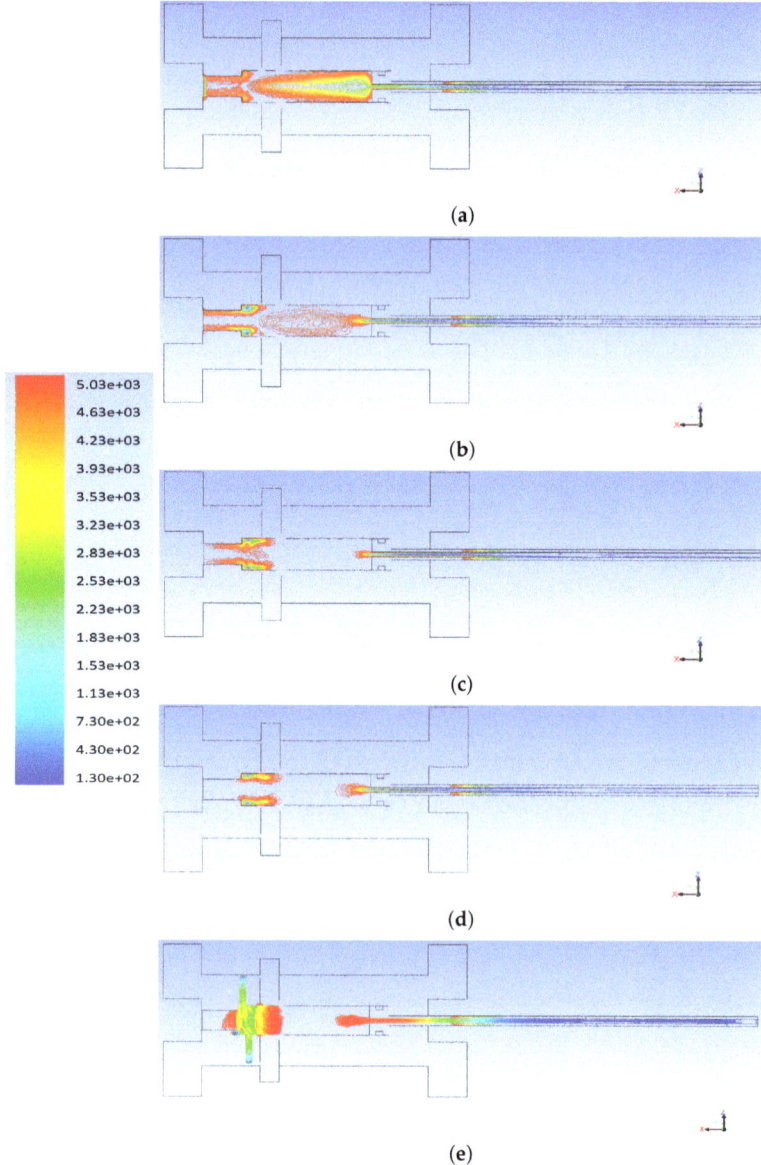

(a)

(b)

(c)

(d)

(e)

Figure 12. Temperature profile of the gas mixture: (**a**) contours of static temperature (K) at iteration #1; (**b**) contours of static temperature (K) at iteration #25; (**c**) contours of static temperature (K) at iteration #50; (**d**) contours of static temperature (K) at iteration #100–1000; and (**e**) contours of total temperature (K) at iteration #1000.

The obtained results of velocity distribution and the temperature profile with respect to position are shown in Figure 13, where the high velocity value is due to the RVF inside of the chamber which creates a turbulent flow behavior of the mixture, and the high temperature value belongs to the applied heat flux. The reversed vortex flow utilizes the uniform temperature and heat flux distribution inside the cylinder, which helps to expedite the chemical reaction and the rate of hydrogen production. Modifying the inlet's positions will increase the rate of reversed vortex flow, and consequently the uniform power distribution. It is found that the RVF compared with forward vortex flow (FVF) which is described in the previous section can perform more efficiently by minimizing heat losses through the cylindrical vessel walls.

By further improving the simulation model, such as applying the chemical reaction, by species transport and reactions and choosing the mixture material, one can reach the precise amount of hydrogen production for the proposed system as well as improving the heat transfer/storage and consequently an optimized reformer chamber model, which are not in the scope of this study.

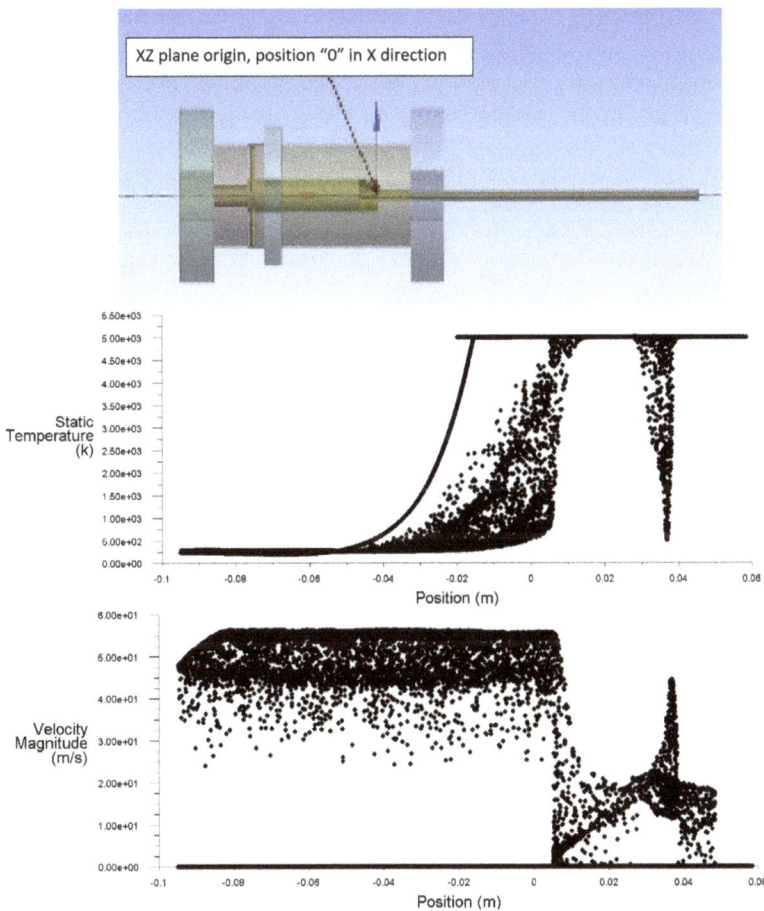

Figure 13. Velocity–temperature profile vs. position.

4. Conclusions

In this study, the partial oxidation of methane to produce syngas using a novel plasma based reformer named 3D Gliding Arc Vortex Reformer (3D-GAVR) was investigated. The fundamentals of this reformer are based on gliding arc discharge. The tangential input creates a vortex in the plasma zone and an expanded plasma that incorporates entire area between the two electrodes, thus allowing for more efficient reformation than traditional gliding arc structures. The effects of several operational parameters on the methane conversion and the selectivity of products were considered. The gas compositions are collected from the experiments and the calculations for price and production rate of hydrogen are made. The results show that hydrogen can be produced for as low as $4.45 per kg with flow rates of around 1 L per minute. The maximum methane conversion percentage that is achieved by this technology is up to 62.38%. Therefore, the suggested reformer can provide a solution that has higher efficiencies at higher flow rates. The non-optimized dimensions of 3D-GAVR is around 7 cm in radius and 15 cm in height with a lot of room for reduction. Hence, further optimization of performance and size, and cascaded performance of multiple reformers can provide an attractive solution for on-site hydrogen generation in H_2 dispensing stations. Consequently, a computational fluid dynamics (CFD) modeling of another innovative cold plasma reformer chamber named a reverse vortex flow gliding arc reactor (RVF-GA), where the geometry and inlets of the 3D-GAVR are modified, was conducted to investigate the effect of geometry size and configuration for efficient performance. In this reformer, an axial air input port is added to the top of the reaction vessel to allow for more air to be introduced into the reaction zone and the premixed reactants can enter the cylindrical reaction zone through tangential jets that are machined within the dielectric ring. The results show that a reverse vortex flow (RVF) scheme is created which has an outer swirling rotation along with a low pressure area at its center with some component of axial flow. In general, this flow scheme is very similar to a natural tornado. The exhaust diaphragm is much smaller than the internal diameter of the vessel cylinder and acts as a barrier that causes vortex flow to reverse direction and initially flow upwards into the cylindrical vessel. The reversed vortex flow utilizes the uniform temperature and heat flux distribution inside the cylinder, and enhances the gas mixtures which helps to expedite the chemical reaction and the rate of hydrogen production compared with forward vortex flow.

Author Contributions: S.S. and L.M. conceived, designed, performed and analyzed the experiments. S.S. also contributed to the computational modeling and analysis. S.S. and L.M. wrote the paper. B.F. contributed to the reagents and materials, and F.H. contributed to providing the computational tools and leading the team.

Acknowledgments: The authors acknowledge financial support from Everette Energy LLC. They would also like to thank Matthew McDonough from Sandia National lab for his helpful advice and useful discussion.

Conflicts of Interest: The authors declare no conflict of interest.

References

1. Aleknaviciute, I. Plasma Assisted Decomposition of Methane and Propane and Cracking of Liquid Hexadecane. Ph.D. Thesis, Brunel University, Uxbridge, UK, 2014.
2. Tao, X.; Bai, M.; Li, X.; Long, H.; Shang, S.; Yin, Y.; Dai, X. CH_4-CO_2 reforming by plasma—Challenges and opportunities. *Prog. Energy Combust. Sci.* **2011**, *37*, 113–124. [CrossRef]
3. Zhou, Z.; Zhang, J.; Ye, T.; Zhao, P.; Xia, W. Hydrogen production by reforming methane in a corona inducing dielectric barrier discharge and catalyst hybrid reactor. *Chin. Sci. Bull.* **2011**, *56*, 2162–2166. [CrossRef]
4. Stambouli, A.B.; Traversa, E. Solid oxide fuel cells (SOFCs): A review of an environmentally clean and efficient source of energy. *Renew. Sustain. Energy Rev.* **2002**, *6*, 433–455. [CrossRef]
5. Jiang, W.; Fahimi, B. Multiport power electronic interface-concept, modeling, and design. *IEEE Trans. Power Electron.* **2011**, *26*, 1890–1900. [CrossRef]
6. Barelli, L.; Bidini, G.; Gallorini, F.; Ottaviano, A. An energetic–exergetic comparison between PEMFC and SOFC-based micro-CHP systems. *Int. J. Hydrog. Energy* **2011**, *36*, 3206–3214. [CrossRef]
7. Gandiglio, M.; Lanzini, A.; Santarelli, M.; Leone, P. Design and optimization of a proton exchange membrane fuel cell CHP system for residential use. *Energy Build.* **2014**, *69*, 381–393. [CrossRef]
8. Oh, S.D.; Kim, K.Y.; Oh, S.B.; Kwak, H.Y. Optimal operation of a 1-kW PEMFC-based CHP system for residential applications. *Appl. Energy* **2012**, *95*, 93–101. [CrossRef]

9. Arsalis, A.; Nielsen, M.P.; Kær, S.K. Modeling and parametric study of a 1 kW e HT-PEMFC-based residential micro-CHP system. *Int. J. Hydrog. Energy* **2011**, *36*, 5010–5020. [CrossRef]

10. Barelli, L.; Ottaviano, A. Solid oxide fuel cell technology coupled with methane dry reforming: A viable option for high efficiency plant with reduced CO_2 emissions. *Energy* **2014**, *71*, 118–129. [CrossRef]

11. Sreethawong, T.; Thakonpatthanakun, P.; Chavadej, S. Partial oxidation of methane with air for synthesis gas production in a multistage gliding arc discharge system. *Int. J. Hydrog. Energy* **2007**, *32*, 1067–1079. [CrossRef]

12. O'Connor, A.M.; Ross, J.R. The effect of O_2 addition on the carbon dioxide reforming of methane over Pt/ZrO_2 catalysts. *Catal. Today* **1998**, *46*, 203–210. [CrossRef]

13. Wang, S.; Lu, G.; Millar, G.J. Carbon dioxide reforming of methane to produce synthesis gas over metal-supported catalysts: State of the art. *Energy Fuels* **1996**, *10*, 896–904. [CrossRef]

14. Jing, Q.; Lou, H.; Fei, J.; Hou, Z.; Zheng, X. Syngas production from reforming of methane with CO_2 and O_2 over $Ni/SrO–SiO_2$ catalysts in a fluidized bed reactor. *Int. J. Hydrog. Energy* **2004**, *29*, 1245–1251. [CrossRef]

15. Bharadwaj, S.; Schmidt, L. Catalytic partial oxidation of natural gas to syngas. *Fuel Process. Technol.* **1995**, *42*, 109–127. [CrossRef]

16. Odeyemi, O.O. Generation of Hydrogen-Rich Gas Using Non Equilibrium Plasma Discharges. PhD Thesis, Drexel University, Philadelphia, PA, USA, 2013.

17. Petitpas, G.; Rollier, J.D.; Darmon, A.; Gonzalez-Aguilar, J.; Metkemeijer, R.; Fulcheri, L. A comparative study of non-thermal plasma assisted reforming technologies. *Int. J. Hydrog. Energy* **2007**, *32*, 2848–2867. [CrossRef]

18. Gallagher, M.J.; Geiger, R.; Polevich, A.; Rabinovich, A.; Gutsol, A.; Fridman, A. On-board plasma-assisted conversion of heavy hydrocarbons into synthesis gas. *Fuel* **2010**, *89*, 1187–1192. [CrossRef]

19. Deminsky, M.; Jivotov, V.; Potapkin, B.; Rusanov, V. Plasma-assisted production of hydrogen from hydrocarbons. *Pure Appl. Chem.* **2002**, *74*, 413–418. [CrossRef]

20. Aleknaviciute, I.; Karayiannis, T.; Collins, M.; Xanthos, C. Towards clean and sustainable distributed energy system: The potential of integrated PEMFC-CHP. *Int. J. Low-Carbon Technol.* **2016**, *11*, 296–304. [CrossRef]

21. Giddey, S.; Badwal, S.; Kulkarni, A.; Munnings, C. A comprehensive review of direct carbon fuel cell technology. *Prog. Energy Combust. Sci.* **2012**, *38*, 360–399. [CrossRef]

22. Czernichowski, A. GlidArc assisted preparation of the synthesis gas from natural and waste hydrocarbons gases. *Oil Gas Sci. Technol.* **2001**, *56*, 181–198. [CrossRef]

23. Diatczyk, J.; Komarzyniec, G.; Stryczewska, H. Power consumption of gliding arc discharge plasma reactor. *Int. J. Plasma Environ. Sci. Technol.* **2011**, *5*, 12–16.

24. Bonner, B. Current Hydrogen Cost. Available online: https://www.hydrogen.energy.gov/pdfs/htac_oct13_10_bonner.pdf (accessed on 26 April 2018).

25. Gutsol, A.; Bakken, J. A new vortex method of plasma insulation and explanation of the Ranque effect. *J. Phys. D Appl. Phys.* **1998**, *31*, 704. [CrossRef]

26. Gutsol, A.; Larjo, J.; Hernberg, R. Comparative calorimetric study of ICP generator with forward-vortex and reverse-vortex stabilization. *Plasma Chem. Plasma Process.* **2002**, *22*, 351–369. [CrossRef]

fluids

MDPI

Communication

Thermal Jacket Design Using Cellulose Aerogels for Heat Insulation Application of Water Bottles

Hai M. Duong *, Ziyang Colin Xie, Koh Hong Wei, Ng Gek Nian, Kenneth Tan, Hong Jie Lim, An Hua Li, Ka-Shing Chung and Wen Zhen Lim

Department of Mechanical Engineering, National University of Singapore, Singapore 9 Engineering Drive 1, EA-07-08, Singapore 117575, Singapore; colinxie@u.nus.edu (Z.C.X.); a0124203@u.nus.edu (K.H.W.); a0125030@u.nus.edu (N.G.N.); a0124334@u.nus.edu (K.T.); a0124479@u.nus.edu (H.J.L.); anhua.li@u.nus.edu (A.H.L.); chungkashing@u.nus.edu (K.-S.C.); limwenzhen@yahoo.com.sg (W.Z.L.)
* Correspondence: mpedhm@nus.edu.sg; Tel.: +65-9769-9600

Received: 2 November 2017; Accepted: 17 November 2017; Published: 23 November 2017

Abstract: Thermal jacket design using eco-friendly cellulose fibers from recycled paper waste is developed in this report. Neoprene as an outmost layer, cellulose aerogels in the middle and Nylon as an innermost layer can form the best sandwiched laminate using the zigzag stitching method for thermal jacket development. The temperature of the ice slurry inside the water bottle covered with the designed thermal jackets remains at 0.1 °C even after 4 h, which is the average duration of an outfield exercise. Interestingly, the insulation performance of the designed thermal jackets is much better than the commercial insulated water bottles like FLOE bottles and is very competition to that of vacuum flasks for a same period of 4 h and ambient conditions.

Keywords: cellulose aerogel; thermal conductivity; water bottle; thermal jacket design; heat insulation

1. Introduction

Soldiers in the Armed Forces (AF) regularly can embark on vigorous physical activities in a hot and humid training environment. The military canteen is an essential piece of equipment to contain fluid, which provides sufficient rehydration for each soldier during training and prevents the onset of heat-related injuries [1]. Heat injury is often a consequence of prolonged activity under hot and humid conditions, which is exacerbated by dehydration and failure to replenish lost bodily fluids. Its severity ranges from mild heat cramps to heat exhaustion and serious heat stroke, which is a medical emergency and can potentially be fatal if not treated immediately [1].

Despite an extensive list of heat management guidelines, heat injuries still occur from time to time in the AF. Certainly, it can be done more to prevent the very occurrence of heat injury during training. Several sports science studies [2,3] have stated that lower core body temperature and longer work tolerance time can be achieved by ingesting ice slurry before and during physical exercise. Ingestion of ice slurry can regulate body temperature more effectively than consumption of chilled water. This is because of the higher cooling capacity of ice particles, which can absorb heat at a near constant phase-change temperature [2,3]. Therefore, a lightweight and highly insulated water bottle to store the ice slurry is desired.

The current AF military canteen is made of high-density polyethylene (HDPE) material, which is lightweight, durable, and low cost [4]. However, HDPE has rather high thermal conductivity and maintains poorly chilled beverages [4]. Commercial FLOE bottles (Teknicool Ltd, Auckland, New Zealand) are designed for sports activities such as cycling. The FLOE bottles have an air gap between the inner and outer materials of the bottles, which can reduce heat transfer through conduction. But the FLOE bottle cannot keep ice slurry for a prolonged period in outfield [5]. By creating a layer of vacuum between the inner and outer surface, commercial vacuum flasks can eliminate heat transfer via conduction and convection and maintain ice retention for up to a few days [6]. However, there is little or no available research on the applicability of the vacuum flasks under the walking and running conditions of military purposes. The vacuum flasks also having high cost and heavy weight may not be the best alternative replacing the current military canteens [5,6].

As the commercial thermal insulated bottles are not optimized for military applications in terms of thermal insulating properties, mechanical properties and cost consideration, another engineering solution is to develop a lightweight and cost-effective thermal jacket wrapped on the existing military canteens. Beer Koozie utilizes a foam sleeve wrapped on bottles to provide thermal insulation and prevent condensation on the surface of the bottle, which can slow down heat transfer [7]. The effect of condensation shows that the temperature changes of water inside the bottle in an environment with 40% and 85% relative humidity is 39 °C and 44 °C, respectively [1,7]. So Koozie sleeves cannot keep the low temperature of the military canteen water.

Among common insulation materials such as polystyrene foam [7,8], using cellulose aerogels from paper waste [9–12] can be a better insulator choice. The cellulose aerogels are environmentally friendly and are made of the cellulose fibers from recycled waste paper through freeze-drying processes [9–12]. The cellulose aerogels have the low thermal conductivity (0.03–0.04 W/m·K) and less toxic emissions and require less harmful chemicals during the fabrication [9–12]. The cellulose aerogels has the lowest density of 0.04 g/cm^3, which is lighter than other common insulation materials, and perform similar insulation results [10,12].

So this communication focuses on using cellulose aerogel to design an insulated jacket for the military canteens to prolong ice slurry to active military personnel in training or operation. However, the cellulose aerogels have low tensile strength, and their structures can be damaged easily upon rough handling [9–12]. It is thus necessary to sandwich the cellulose aerogels between two protective layers so that the insulated jacket can be more durable. The commercial synthetic fabrics and 10-cm thick cellulose aerogels are chosen. It is clear that increasing the thickness can increase the thermal resistance of the insulated jacket [10,12]. However, the thick insulated jacket has negative implications on its weight and user-friendliness. The final jacket design is aimed to meet following points: (i) The design of the insulated jacket can conform well to the contours of the military canteens, (ii) The cellulose aerogel needs to be properly fitted and secured within the two protective layers to prevent its structure damage, (iii) The bottle must be inserted and removed from the jacket easily, and (iv) Heat transfer should be minimized and cannot counter the insulating effects of the cellulose aerogel. The heat insulation performance of the developed insulated jacket is also compared with commercial FLOE sport bottles and vacuum flask bottles in this work.

2. Results

From Figure 1 below, the vacuum flask yields the best thermal insulating results with the least temperature increase after 4 h, followed by the military canteen that is wrapped by the insulated jacket. The temperature of the content inside both bottles is approx. 0.1–0.2 °C after 4 h, which proves that they can maintain the ice slurry inside chilled. The insulation performance of the FLOE bottle comes next whereby the significant temperature rise is observed after 1 h. The military canteen without the insulated jacket shows a significant increase in temperature after 30 min, the worst heat insulation performance. The FLOE bottle and the military canteen without the insulated jacket cannot keep the ice slurry inside after 3 h and 1 h, respectively. Despite having the best result, the vacuum flask posed some concerns in terms of heavier weight load and higher costs when compared with the other 3 in Appendix A Table A1 Hence it is not ideal to replace the current military canteens that are used by the soldiers with the vacuum flask bottles. From the Figure 1 and the Appendix A, it is concluded that the thermal jacket using the cellulose aerogels is designed successfully and that the simple but effective engineering solution for the heat insulation of the military canteens is validated.

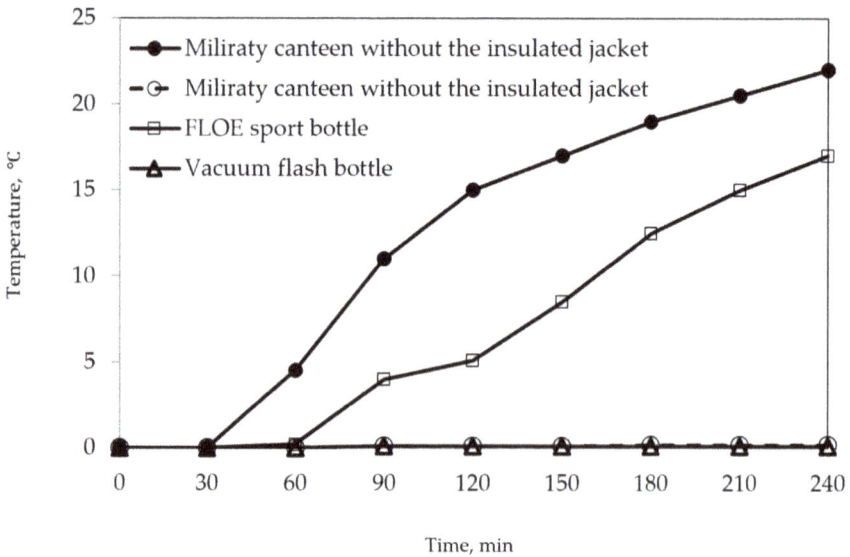

Figure 1. Heat insulation performance of various insulated bottles during 4 h testing.

3. Discussion

3.1. Advantages of the Developed Thermal Jacket

This developed insulated jacket can address the issues of previous thermal jacket designs. Assembling the design from multiple pieces ensures that the insulated jacket can fit very well the complicated contours of the bottle. The aerogel is tightly secured in place between two fabric layers to resolve the collapsing. It is much easier to insert or remove the military canteen, as it merely needs to be pulled out or pushed in before closing the zipper of the insulated jacket. The zipper length is optimized to reduce the heat transfer from the zipper. The insulated jacket also uses environmentally friendly cellulose fibers from recycled paper waste. When considering the high amounts of paper waste that is generated annually, recycling helps to reduce material wastage and lower greenhouse emissions. This green initiative can also reflect positively on social responsibility.

The mass production cost is estimated in the Appendix A Table A2. Figure 2 shows the cost estimation and the ice slurry temperature after 4 h of four different bottles. It is evident that the insulated military canteen has the preferred combination of the low price and the excellent heat insulation performance. Due to the ultra-lightweight cellulose aerogel that is used as the central layer, the total weight of the insulated jacket is 2.5 times and 1.4 time lighter than that of the vacuum flask the FLOE bottles, respectively. The total unladed weight of the insulated jacket and the military canteen is 360 g, of which 200 g is the weight of the designed jacket. In comparison, the vacuum flask with the same capacity and no ice slurry weighs 900g, while two FLOE bottles with 500 mL capacity for each weighs 500 g in total.

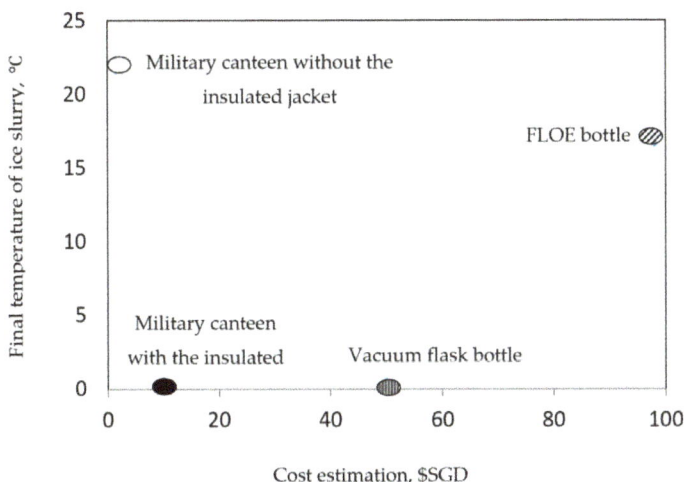

Figure 2. Comparison of the cost and heat insulation performance after 4 h of various bottles having the same 1-L capacity.

3.2. Limitations of Thermal Jacket Design

The insulated military canteen may be considered bulky by few end-users because the insulated jacket adds approx. 3 cm to the thickness of the military canteen. This is largely due to the 1.5 cm thick aerogel layer and 0.3 cm neoprene outer layer in the jacket design. Further design improvement may reduce more the thickness of the jacket and improve the insulated military canteen ergonomics. Currently we cannot determine the accurate technical service life of the insulated military canteen. So cyclic fatigue testing, pierce resistance, abrasion resistance, tear resistance, and ultraviolet (UV) resistance tests are desired in the future work.

Finally, there is no capacity for mass laboratory production of the cellulose aerogels used in this insulated jacket. In this project, the cellulose aerogels were laboriously produced in a small lab scale. The electrical overheads contribute to 60% of the total cost in the Appendix A Table A2, which is an incredibly high proportion for a mass manufactured product. Hence a better aerogel production method must be found for the mass production to be feasible.

4. Materials and Methods

4.1. Fabrication of Cellulose Aerogels

12 g of cellulose fibers is added into 1.2 L of the deionized (DI) water, stirred thoroughly and placed in an ultrasonic processor for 15 min at maximum power. After 15 min, the obtained suspension is removed from the ultrasonic processor to cool down to room temperature using a water bath. Kymene cross-linker is added to the mixture and it is subjected to additional 5×3 min cycles of sonication. After the completed sonication process, the mixture is poured into the mould and is placed in the freezer overnight. The next step is to place the sample in the freeze dryer at a temperature of $-91\,°C$ to obtain the cellulose aerogel. After the freeze-drying process, the aerogel sample is placed in the oven for the cross-linking process. After 3 h at 120 °C, the aerogel sample can achieve its finalized thickness and shape as seen in Figure 3b. Mode details of the morphology and the properties of the cellulose aerogel can be found on our previous works [9–12].

4.2. Design of Sandwich Structures

For the internal layer, Nylon in Figure 3a is chosen over the others due to the lowest thermal conductivity value high durability, low cost, high availability and smooth texture [13]. Heavy-duty fabric of Neoprene, as seen in Figure 3c, is chosen as the external layer of the thermal jacket due to the low thermal conductivity, excellent durability, mostly weather proof, high abrasion and tear resistance [14]. More details of the layer material properties can be found in the Appendix A Table A3. Figure 3d shows the proposed sandwich layer using different materials forming the insulated jacket.

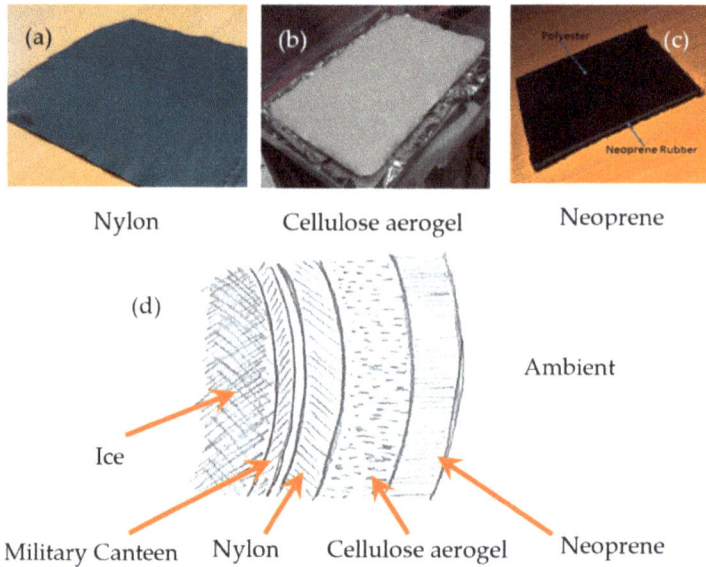

Figure 3. Materials of (**a**) nylon, (**b**) cellulose aerogel, (**c**) neoprene are used for the insulated jacket design, and (**d**) the sandwich structure consists of Neoprene as the outmost layer, the cellulose aerogel within and Nylon as the innermost layer.

4.3. Prototype Fabrication Process of the Thermal Jacket

The prototype fabrication has the following steps: (i) Cutting the inner and outer fabric into specifically-shaped panels in Figure 4a, (ii) Sewing the inner fabric pieces together to form an inner jacket, (iii) Wrapping the aerogel securely around the inner jacket, (iv) Sewing the outer fabric pieces and zipper together over the aerogel layer and with the inner jacket, and (v) Sewing the collar onto the neck of the assembly to complete the product. Straight stitch should not be applied as the straight stitch can tear the fabric easily because Neoprene and Nylon are stretchable. Hence, the zig-zag stitch method is preferred. The zipper is applied to allow the military canteen to be inserted and removed easily from the insulated jacket. The collar is used to reinforce the neck area. The components come together to form a final thermal jacket are shown in Figure 4b.

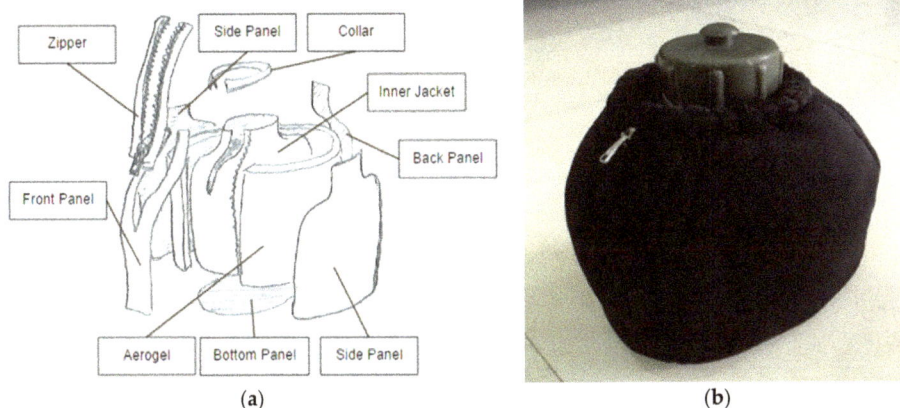

Figure 4. (a) Sketch of the insulated jacket and (b) final insulated jacket on the military canteen.

4.4. Characterization

The heat insulation test under static condition was conducted in a laboratory with an ambient temperature of 25 °C for 4 h. The content of the bottle is ice slurry, which comprises 1/3 water and 2/3 crushed ice and the temperature of the content of the bottle is recorded every 30 min. Four bottles including the military canteen without the insulated jacket, the military canteen wrapped with the insulated jacket, the commercial FLOE bottle and the vacuum flask bottle were used in the experiments for comparison.

Water contact angle measurements are carried out on a VCA Optima goniometer (AST Products Inc., Billerica, MA, USA) to evaluate the water repellence of the methyltrimethoxysilane (MTMS)-coated aerogels. During the test, water drops of 0.5 μL were controlled by the syringe system of the tester and were dispensed drop by drop onto the surface of the samples. The contact angle values were calculated using the contact angle meter software, which analyses the droplet shape in the image. For each sample, measurements were repeated at several different positions, and the contact angle was ultimately determined by averaging these contact angle values from the various measurements. In order to verify the moisture absorbency for the shortlisted fabrics, the VCA Optima was used to measure the contact angle of each fabric to determine which material will be most suitable for the shell and lining of the product.

The C-Therm TCi Thermal Conductivity Analyser (C-Therm Technologies Ltd., Fredericton, NB, Canada) was used for this work. It employs the Modified Transient Plane Source (MTPS) technique, which uses a one-sided interfacial heat reflectance sensor to apply a momentary constant heat source to the specimen and measure the K value from the rate of increase in the sensor voltage. The results achieved by the MTPS method are similar in accuracy to the traditional guarded hot plate apparatus.

5. Conclusions

After fabricating prototypes and analyzing the flaws of our design, the cost-effective insulated jacket with the effective heat insulation performance is developed successfully for mass production. Interestingly, the military canteen wrapped with the cellulose aerogel can match the insulation performance of the vacuum flask bottles after 4 h, which is the average duration of an outfield exercise. The materials chosen in this work can withstand the harsh conditions of a jungle environment. Currently, soldiers have been equipped three-liter polyethylene water pouches. The insulated jacket developed in this work can be used for the water pouch so that soldiers may carry up to four liters of ice slurry during outfield operations instead of one liter of the insulated military canteen. Finally, the design concept of the insulated military canteen may be extended to other commercial bottles as well.

Acknowledgments: The authors would like to thank C-265-000-049-001 FB Funding and DSO National Laboratory for their research support.

Author Contributions: X.Z.C., T.K. and W.Z.L. conceived and designed the experiments; K.H.W. and N.G.N. performed the experiments; H.M.D. and W.Z.L. analyzed the data; H.J.L., K.S.C. and A.H.L. contributed reagents/materials/analysis tools; H.M.D wrote the paper.

Conflicts of Interest: The authors declare no conflict of interest.

Appendix A

Table A1. Property comparison of various bottles.

Parameters	Military Canteen [4]	Floe Bottle [6]	Vacuum Flask [5]
Thermal conductivity, W/m·K	0.33	0.14	-
Cost, S$	$1.52	$98.00	$50.00
Capacity, L	1.0	1.0	1.0
Weight, g	160	500	900

Table A2. Estimated cost of manufacturing the insulated jacket.

S/N	Items	Unit Cost, S$	Quantity	Cost, S$
1	Military Canteen Bottle	S$1.52/bottle	1	1.52
2	Polyester Fabric	S$1.40/m^2	0.100	0.14
3	Recycled Cellulose	S$4.36/kg	0.0287	0.13
4	Kymene	S$1.05/kg	0.0057	0.01
5	MTMS	S$2.24/kg	0.143	0.32
6	Zipper	S$0.46/m	0.100	0.05
7	Neoprene (3-mm thickness)	S$3.02/m^2	0.100	0.30
8	Manpower	S$2.93/h	0.5	1.47
9	Electricity	S$0.13/kWh	47.8	6.21
10	Water	S$0.73/m^3	0.0029	0.05
	Total cost estimation			**S$10.14**

Table A3. Insulating Material Selection.

Material Selection Criteria	Cellulose Aerogel [9–12]	Neoprene Fabric [14]	Nylon Fabric [13]
Thermal conductivity (W/m·K)	0.03	0.047	0.046
Density (g/cm^3) [6]	0.04	-	-
Durability	Fair	Excellent	Good
Stretchability	Fair	Excellent	Excellent
Low Cost and availability	Yes but no availability in the market	Yes and availability in the market	Yes and availability in the market

References

1. Heat Stroke: Symptoms and Treatment. Available online: https://www.webmd.com/a-to-z-guides/heat-stroke-symptoms-and-treatment#1 (accessed on 1 November 2017).

2. Lee, J.K.W.; Kennefick, R.W.; Cheuvront, S.N. Novel cooling strategies for military training and operations. *J. Strength Cond. Res.* **2015**, *29*, S77–S81. [CrossRef] [PubMed]
3. Dugas, J. Ice slurry ingestion increases running time in the heat. *Clin. J. Sport Med.* **2011**, *21*, 541–542. [CrossRef] [PubMed]
4. Seah, K.H. *Fundamental Principles of Manufacturing*, 1st ed.; Cobee Publishing House: Singapore, 2015; ISBN 9789810959050.
5. Best Insulated Water Bottles—Our Top Picks, Hydration Anywhere. Available online: https://hydrationanywhere. com/best-insulated-water-bottle-our-top-picks/ (accessed on 1 November 2017).
6. Floe Bottle. Available online: https://www.floebottle.com/ (accessed on 1 November 2017).
7. Beer Koozies Do Help Keep Your Drink Cold—But Not The Way You Think They Do. Available online: http:// www.ibtimes.com/beer-koozies-do-help-keep-your-drink-cold-not-way-you-think-they-do-1220253 (accessed on 1 November 2017).
8. Five Most Common Thermal Insulation Materials. Available online: https://www.thermaxxjackets.com/5-most-common-thermal-insulation-materials/ (accessed on 1 November 2017).
9. Feng, J.; Nguyen, S.T.; Fan, Z.; Duong, H.M. Advanced fabrication and oil absorption properties of super-hydrophobic recycled cellulose aerogels. *Chem. Eng. J.* **2015**, *270*, 168–175. [CrossRef]
10. Duong, H.M.; Nguyen, S.T. *Nano and Biotech Based Materials for Energy Building Efficiency*; Springer International Publishing: Cham, Switzerland, 2016; pp. 411–427, ISBN 978-3-319-27505-5.
11. Nguyen, S.T.; Feng, J.; Le, N.; Le, A.T.; Hoang, N.; Duong, H.M. Cellulose aerogel from paper waste for crude oil spill cleaning. *Ind. Eng. Chem. Res.* **2013**, *52*, 18386–18391. [CrossRef]
12. Nguyen, S.T.; Duong, H.M.; Tan, V.B.C.; Ng, S.K.; Wong, J.P.W.; Feng, J. Advanced thermal insulation and absorption properties of recycled cellulose aerogels. *Colloids Surf. A Physicochem. Eng. Asp.* **2014**, *445*, 128–134. [CrossRef]
13. Nylon vs. Polyester. Available online: http://www.diffen.com/difference/Nylon_vs_Polyester (accessed on 1 November 2017).
14. Neoprene Rubber Polymer. Available online: http://www.warco.com/polymer/neoprene/ (accessed on 1 November 2017).

![fluids logo] *fluids*

MDPI

Article

Fluid Dynamics and Mass Transfer in Spacer-Filled Membrane Channels: Effect of Uniform Channel-Gap Reduction Due to Fouling

Chrysafenia P. Koutsou [1], Anastasios J. Karabelas [1,*] and Margaritis Kostoglou [1,2]

[1] Chemical Process and Energy Resources Institute, Centre for Research and Technology–Hellas, P.O. Box 60361, 6th km Charilaou-Thermi Road, Thermi, Thessaloniki GR570-01, Greece; ckoutsou@alexandros.cperi.certh.gr (C.P.K.); kostoglu@chem.auth.gr (M.K.)
[2] Chemistry Department, Aristotle University of Thessaloniki, Thessaloniki 54124, Greece
*   Correspondence: karabaj@cperi.certh.gr; Tel.: +30-23-1049-8181; Fax: +30-23-1049-8189

Received: 13 December 2017; Accepted: 26 January 2018; Published: 2 February 2018

Abstract: The time-varying flow field in spacer-filled channels of spiral-wound membrane (SWM) modules is mainly due to the development of fouling layers on the membranes that modify the channel geometry. The present study is part of an approach to tackling this extremely difficult dynamic problem at a small spatial scale, by uncoupling the fluid dynamics and mass transfer from the fouling-layer growth process. Therefore, fluid dynamics and mass transfer are studied for a spacer-filled channel whose geometry is altered by a uniform deposit thickness h. For this purpose, 3D direct numerical simulations are performed employing the "unit cell" approach with periodic boundary conditions. Specific thickness values are considered in the range 2.5–10% of the spacer-filament diameter D as well as other conditions of practical significance. The qualitative characteristics of the altered flow field are found to be very similar to those of the reference geometry with no gap reduction. For a given flow rate, the pressure drop, time-average wall-shear stresses and mass-transfer coefficients significantly increase with increasing thickness h due to reduced channel-gap, as expected. Correlations are obtained, applicable at the "unit cell" scale, of the friction factor f and Sherwood number Sh, which exhibit similar functional dependence of f and Sh on the Reynolds and Schmidt numbers as in the reference no-fouling case. In these correlations the effect of channel-gap reduction is incorporated, permitting predictions in the studied range of fouling-layer thickness $(h/D) = 0$–0.10. The usefulness of the new results and correlations is discussed in the context of ongoing research toward improved modeling and dynamic simulation of SWM-module operation.

Keywords: spacer-filled membrane channels; channel-gap reduction; membrane fouling; direct numerical simulations; flow characteristics; mass transfer

---

1. Introduction

The spiral-wound membrane (SWM) module is the dominant element of the reverse osmosis (RO) and nanofiltration (NF) water-treatment plants, which are comprised of pressure vessels, each containing several SWM modules in series. Therefore, significant research and development (R&D) efforts have been invested aiming to optimize SWM-module design (for various applications) and improve its overall performance (e.g., [1,2]). The operation at steady state of a desalination SWM module, and of the entire RO plant, is characterized by an inherent spatial variability of all process parameters [3,4] due to the passage/filtration of purified water from the feed-side channels to the permeate side. However, in RO and NF plants there is also temporal variability due to fouling [5,6], usually caused by rejected organic and/or inorganic impurities depositing on the membrane surfaces, which leads to deterioration in plant performance; this undesirable variability (manifested in feed-pressure increase, for constant recovery, and in reduced salt rejection) is extremely difficult to predict, although it is necessary for developing optimal plant design and operation strategies. A direct attack of the full 3D dynamic problem, enabling simulation and prediction of the SWM and pressure-vessel performance from first principles, is impossible at present due to its complexity and considering the lack of understanding of some key physico-chemical interactions affecting fouling (e.g., [7–9]). Nevertheless, some efforts in that direction have been reported (e.g., [5,10]), in which significant simplifying assumptions have been made.

To address this complicated problem in a systematic manner, researchers have numerically analyzed, first, the flow field and mass transfer in restricted spatial domains of the spacer-filled membrane channels, as summarized in [11]. Significant work along these lines has been reported (focusing on small or intermediate spatial scales) aiming to understand the effect of spacer characteristics on flow and mass transfer (e.g., [12–14]) in order to optimize the feed-spacer geometry (e.g., [15–17]) and to study fouling, including bio-fouling (e.g., [18,19]) phenomena. These detailed studies of the flow field in the spacer-filled channels, focusing on a "unit cell" formed by the net-type spacers (also considering periodicity), have proven to be quite fruitful; in addition to improved basic understanding, they have yielded correlations of pressure drop and mass transfer applicable at "unit cell" scale [20,21]. Comprehensive modeling of flow and mass transfer throughout a membrane sheet (or a SWM module) can rely on such correlations in order to develop a reliable simulator of SWM performance at steady state [3,4,22,23] and to further a dynamic model and respective simulator accounting for a developing fouling layer [6,24]; the dynamic simulator, as a general-purpose tool, would enable reliable projections of the performance of a pressure vessel and of an entire desalination plant.

In practically all the aforementioned studies, a fairly idealized feed-side channel geometry is considered, with the membrane-channel gap characterized by the nominal thickness of the net-type spacer which is assumed to be in line- or point-contact with the bounding flat membranes. However, there are two main effects in practice that cause the real membrane channel gap to be smaller than the "reference gap", with significant negative impact on the process; i.e., the pressure applied on the spacer-filled membrane envelopes during winding in SWM-module fabrication [1] and the foulants depositing on the membranes [8,25]. The former leads to the deformation of the membrane and/or the spacer, thereby effectively reducing the channel gap [1]; the latter effect manifests itself as deposit/fouling layers gradually developing on the active membrane surfaces. In general, the fouling layers considered here (in addition to the added resistance to permeation) tend to modify the detailed geometrical characteristics of the spacer-filled flow-channels, impacting on the transport phenomena therein. Therefore, this effect of channel-gap reduction merits particular attention for the improved understanding and comprehensive process modeling outlined above.

The scope of this work is to study in detail the effect of thickness h of a fouling layer (assumed to develop uniformly on the membranes) on hydrodynamics and mass transfer in the spacer-filled channels, under conditions representative of those prevailing in RO desalination plants. For this purpose, 3D-flow simulations were performed in "unit cells", with geometry representative of typical commercial spacers, by employing advanced numerical techniques described in previous publications [20,21]. In the analysis and the reported results and correlations (for pressure drop and mass transfer), an additional parameter (h/D) is introduced, involving the fouling-layer thickness h made dimensionless with the diameter D of spacer filaments, the latter considered to be cylindrical.

2. Mathematical Formulation and Numerical Simulations

The base case of the problem at hand is the flow field within flat narrow channels (bounded by membranes) whose gap/height is determined by the nominal feed-spacer thickness. This geometry is commonly considered a satisfactory approximation of spacer-filled channels in spiral-wound elements, when no fouling or other channel deformation exists [20]. In the basic (or initial) geometry, shown in Figure 1a, a flat narrow channel is formed by the membrane-bounding surfaces, in contact with two layers of straight cylindrical filaments (comprising the spacer). The filaments of each layer are parallel and equidistant, while the two superposed layers have different orientation and "intersect" at a certain angle β. All cylindrical filaments have the same diameter (*D*), so that the channel gap or height (*H*) is twice the filament diameter (i.e., *H* = 2*D*). Another parameter of interest is the ratio (*L*/*D*) of distance L between the axes of symmetry of parallel filaments over the filament diameter *D*. It will be noted that recent studies (e.g., [26]) take into consideration the irregularities of the filament cylindrical shape, developing during filament extrusion and spacer fabrication. These irregularities, and their impact on the flow field, have to be taken into account in future studies, particularly in the context of fouling-layer development. The particular problem studied here involves fouling layers developing on both bounding membrane surfaces, but not on the spacer surfaces (Figure 1b). The thickness h of these fouling layers is considered to develop uniformly on both membrane surfaces, resulting in a flow channel with reduced gap, and reduced spacer-filament surfaces exposed to the flow, compared to the initial configuration, as indicated in Figure 1b.

As in previous similar studies [20,21] it is considered that the flow exhibits periodicity in each unit cell AXBY formed by four neighboring filaments (shown in Figure 1a); therefore, the detailed flow-field study is restricted to a computational domain indicated as ABCD in Figure 1a. The rationale and usefulness of adopting periodic boundary conditions to describe the flow field in the present case, as well as other types of transport phenomena in geometries with a periodic character, is discussed elsewhere [20,21].

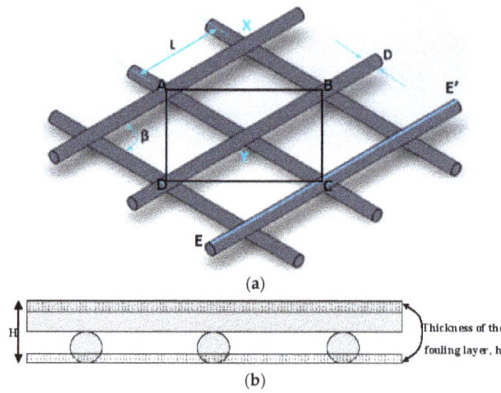

(a)

(b)

Figure 1. (**a**) The basic geometry of the feed-spacer considered, indicating the 3D unit cells, and the computational domain ABCD; (**b**) a cross-sectional view of the spacer-filled flow channel, along the EE′ plane, with filament crossing angle β = 90°, and fouling layers of thickness *h*, reducing the channel gap *H*.

Flow development and mass transport are governed by the Navier–Stokes, continuity and mass-transfer equations, considering that the fluid is Newtonian and incompressible:

$$\frac{\partial u}{\partial t} + u \cdot \nabla u = -\nabla P + \frac{1}{Re} \nabla^2 u \tag{1a}$$

$$\nabla u = 0 \tag{1b}$$

$$\frac{\partial C}{\partial t} + u \cdot \nabla C = \frac{1}{Pe} \nabla^2 C \tag{1c}$$

The channel walls are assumed to be impermeable, which may be justified considering that the RO permeation velocities are much smaller than the feed-flow velocities. Consequently, no-slip and no penetration boundary conditions are considered on the channel walls and on the cylindrical filament surfaces. Additionally, for the mass-transfer simulations, uniform membrane wall/surface concentration, C_w, is assumed.

It is emphasized that the results of this study are obtained through a direct numerical simulation (DNS) involving Equation 1a–c, without the introduction of any approximation regarding flow development or turbulence modeling. A commercial CFD code (FLUENT, v.6.2.16, ANSYS, Canonsburg, PA, USA) is used, which employs the finite-volume method. In each simulation the governing equations are integrated in time by imposing a constant mean-pressure gradient until the flow and the concentration reach a statistically steady state. More details regarding the fluid flow and mass transfer simulations are provided elsewhere [20,21].

The physical parameters of the fluid flow and mass transport in the retentate spacer-filled channels are the dimensionless Reynolds, Schmidt and Sherwood numbers defined here as follows:

$$Re = \frac{D'U}{\nu} \tag{2a}$$

$$Sc = \frac{\nu}{D_c} \tag{2b}$$

$$Sh = \frac{kD'}{D_c} \tag{2c}$$

where

D': $D-h$;

D: Spacer filament diameter (m);

h: Fouling layer thickness (m);

U: Superficial axial velocity in the retentate channel of gap $(H-2h)$ (m/s);

ν: Water kinematic viscosity (m²/s);

D_c: Species/salt diffusivity (m²/s);

k: Mass-transfer coefficient (m/s).

It should be noted that in the geometry studied (i.e., channel of width w and gap $2D'$), the superficial velocity and *Re* number are given as $U = Q/[w \cdot 2D']$ and $Re = [D' \cdot U]/\nu = [Q/2w\nu]$. The latter expression shows that, for constant flow rate Q, the Reynolds number remains constant for different gaps, because the reduction of channel gap $(2D')$ is compensated by an increase of velocity U. Consequently, one can conveniently present and compare results on pressure drop and other quantities (for various gaps $2D'$) as a function of either *Re* number or flow rate Q.

Under typical conditions prevailing in spiral-wound membrane elements, the velocity in the retentate channels created by two adjacent membrane leaves (separated by the spacer) does not exceed ~0.35 m/s whereas the allowable pressure drop, recommended by manufacturers, should not exceed 0.6 bar/m [27]. Therefore, the Reynolds number, defined on the basis of the superficial velocity and the spacer filament diameter D', is commonly less than 200. Typical Schmidt numbers for brackish and sea water are of the order 103. In the series of mass transfer simulations reported herein, the Schmidt number values varied from 1 to 100; calculations at higher *Sc* were not performed due to limitations associated with available computer power. Table 1 summarizes the conditions of performed simulations.

To select a realistic range of fouling-layer thickness h for these computations, the results of fairly extensive RO membrane fouling studies, performed in this laboratory (e.g., [8,28–30]) have been used for guidance. In particular, the results of fouling studies with alginates [30–32], which are representative of a major class of common RO foulants [8,30,33], suggest that the developing fouling layers are quite coherent gels that tend to deposit fairly evenly on the RO membranes [32]. Moreover, these layers exhibit viscoelastic properties and a significant yield stress that (at least in the cases studied [32]) appears to be greater than typical mean shear stresses prevailing in the spacer-filled channels [20]; this implies that, once formed, such layers would not be detached by the flow. Measured alginate fouling-layer thickness h on a RO membrane, of magnitude order ~10% of the channel gap ([32], Table 1) was associated with a substantial trans-membrane pressure (TMP) increase (e.g., >20%), which is considered excessive in practice, requiring membrane cleaning [27]. Therefore, the range of h values selected for this work was 0–10% of filament diameter; specifically, numerical simulations were performed for four values of fouling-layer thickness (h/D = 0.025, 0.05, 0.075, 0.1) for the typical spacer geometry with characteristics: $L/D = 8$ and $\beta = 90°$ (Table 1, Figure 1). The main flow was directed along the diagonal of the spacer unit cell; i.e., along the line AB (Figure 1).

Table 1. Conditions of performed simulations in this study. Spacer geometry: $L/D = 8$, $\beta = 90°$. *Sc* number range: 1–100.

	Case 1	Case 2	Case 3	Case 4
Fouling-layer thickness, h/D	0.025	0.05	0.075	0.1
Re number range	108–200	104–193	101–187	97–180

3. Results and Discussion

3.1. Qualitative Flow Features

For all the cases considered, simulations reveal some generic qualitative flow features which are very similar with those of the (fouling-free) base case [20]. Typical instantaneous images obtained from the numerical simulations are included in Figures 2 and 3; the chromatic scale of local velocities is such that the highest and smallest velocities are designated by red and purple color, respectively. In Figure 2, fluid-particle path lines are shown, for the Case 2 of this study where the fouling-layer thickness h is 5% of the spacer-filament diameter, D. As in the fouling-free case, the main flow feature is a free vortex along the diagonal of the unit cell, i.e., approximately aligned with the main flow direction. This vortex, which seems to be enhanced as the channel-gap is reduced and the Re number increases (Figure 3), results from interaction/merging of two different flow streams entering the unit cell from different directions, thus creating a surface of high shear in the vicinity of the horizontal mid-plane of the unit cell. A second characteristic is the flow separation in the downstream side of each cylindrical filament, which is due to the three-dimensional nature of the flow and the component of pressure gradient parallel to the cylinders; this leads to spiral paths along the respective cylinder directions, as shown in Figure 2. Another feature, also evident in Figure 2, is the presence of closed recirculation regions attached to each cylinder, which result from the interaction of the vortices along the cylinder with the central free vortex.

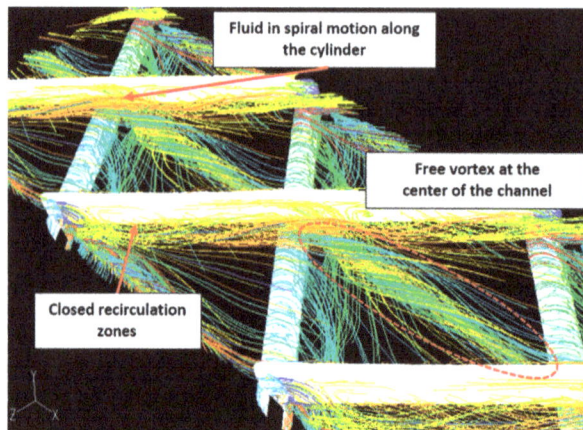

Figure 2. An instantaneous view of fluid-particle pathlines showing the major qualitative characteristics of the flow; $h/D = 0.05$, $Re = 193$.

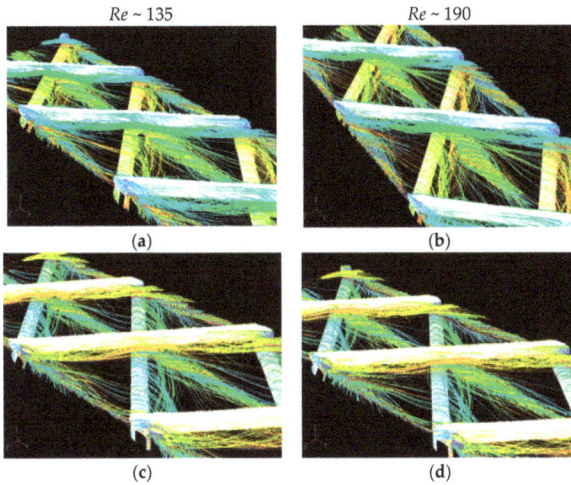

Figure 3. An instantaneous view of fluid-particle pathlines for different fouling-layer thicknesses (**a**,**b**) $h/D = 0.025$; (**c**,**d**) $h/D = 0.10$; and Reynolds number (**a**,**c**) *Re* ~135; (**b**), (**d**) *Re* ~ 190.

It is of considerable interest to examine the effect of channel-gap reduction on the transition to unsteady flow, which obviously impacts on wall shear-stress fluctuations, wall mass transfer and concentration polarization. For the reference (no fouling) case, a thorough data analysis regarding the evolving flow field, including spectra of fluctuating wall velocities and shear stresses, is presented by Koutsou et al. [20]. Among other results, it is shown there that the transition to unsteady flow occurs at rather low *Re* numbers; i.e., in the narrow range 35 to 45, for the studied fairly broad range of spacer geometric parameters. The present results confirm the general trends observed previously [20]; moreover, as one might have expected, the reduction of gap tends to stabilize the flow. Indeed, as Figure 4 indicates (as well as other data not presented here), the transition to unsteady flow occurs at increasing Re number with decreasing channel gap $2D'$ (or increasing h). For a fouling-layer thickness $h/D = 0.1$ (Figure 4), the time series of transverse velocity component v (plotted for increasing Re numbers), clearly suggests that flow unsteadiness appears above $Re = 60$; with increasing *Re* number, the transitional unsteady flow [20] tends to become chaotic and turbulent (at high *Re*).

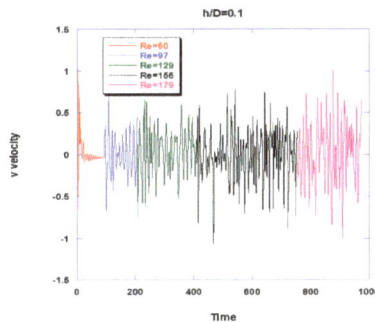

Figure 4. Time series of the instantaneous normalized transverse (v) velocity component for progressively increased *Re* numbers; $h/D = 0.1$. The v-velocity component is normalized with U and the time with D'/U.

3.2. Effect of Channel-Gap Reduction on Pressure Drop

A key objective of the present simulations is to determine the effect of a uniform fouling layer on the pressure drop in the retentate channels. The pressure drop correlation for the "clean channel" ($h = 0$) and fixed geometric spacer parameters (β, L/D) has already been reported [20] and it has the form: $f = 0.85Re^{-0.19}$ The present study also aims to incorporate into the aforementioned correlation the effect of a uniform fouling-layer thickness (h/D) by an appropriate modification.

Figure 5 depicts the calculated pressure drop versus flow rate, for fouling-layer thickness from 2.5–10% of the filament diameter D, in comparison with the corresponding data of the clean channel case ($h/D = 0$). The flow rate Q corresponds to the computational domain (Figure 1a), for the range of superficial (or cross-flow) velocities U of practical interest, i.e., ~0.10 to ~0.20 m/s. It is clear from Figure 5 that (as expected) the pressure drop tends to increase with increasing thickness h and the corresponding channel-gap reduction. The constriction of the free flow, due to reduction of the effective gap ($H-2h$) of the retentate channel, results in increased mean velocity, thus leading to increased pressure loss due to increased contributions of shear stresses and form drag. It should be noted that experimental data, presented elsewhere [20], on pressure drop corresponding to the clean channel case ($h/D = 0$), for spacer with $L/D = 8$ and $\beta = 90°$, show very good agreement with the respective results of numerical simulations, thus providing support to the validity of the latter.

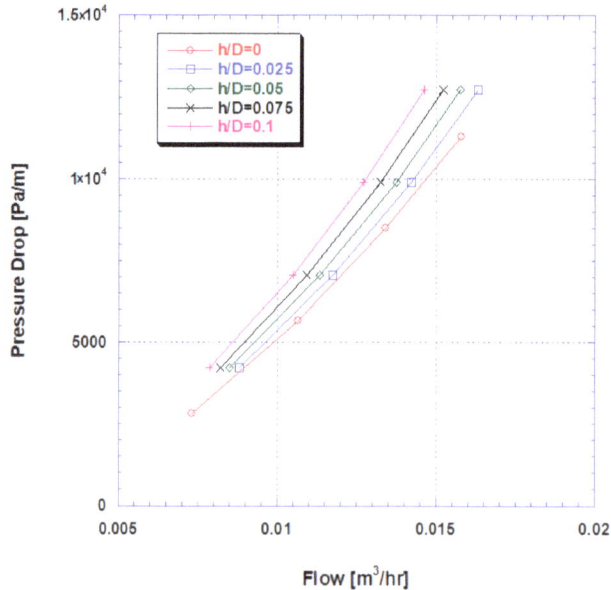

Figure 5. Computational results on the effect of the fouling-layer thickness *h* on pressure drop in the retentate channel. The flow rate corresponds to the computational domain (Figure 1), for the range of superficial velocities *U* of interest.

The data plotted in Figure 5 are made dimensionless by defining as follows the friction factor f, where U is the superficial cross-flow velocity corresponding to the effective gap of the retentate channel ($H - 2h$):

$$f = \frac{\left[\frac{dP}{dL}\right]}{\frac{\rho U^2}{D'}} \tag{3}$$

The computational results in dimensionless form, included in Figure 6, suggest that the dependence of friction factor f on Re number, for all the values of fouling-layer thickness h examined, is quite close to that holding for the no-fouling geometry. However, the friction factor computed on the basis of D' ($D' = D - h$) tends to decrease systematically with increasing fouling-layer thickness, which should be accounted for in an appropriate correlation.

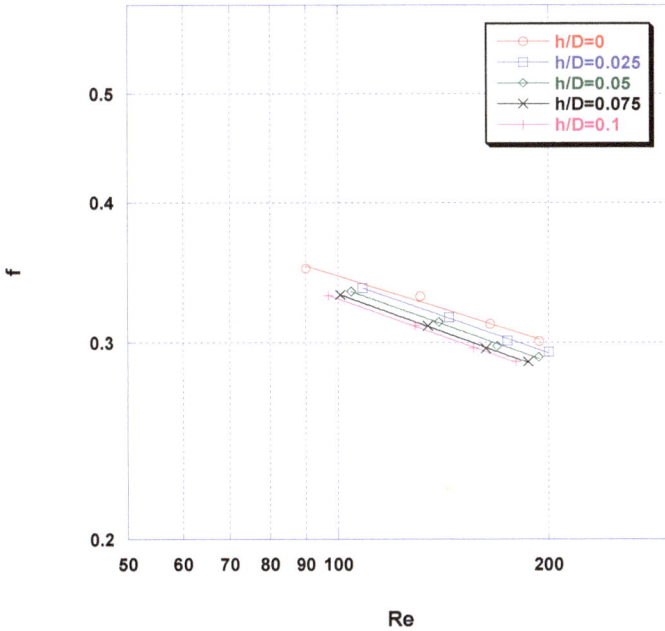

Figure 6. Friction factor versus Reynolds number for various thicknesses of fouling layer.

As shown in Appendix A, the data of friction factor f, plotted versus the ratio h/D (Figure A1), show practically the same dependence of f on the ratio h/D (with an exponent approx. -0.03) for all the cases examined in this work. Therefore, the new data may be correlated by a function of the form (Figure A2):

$$f = 0.85 \left(\frac{h}{D}\right)^{-0.03} Re^{-0.22} \tag{4}$$

This expression is similar to the correlation for the "no-fouling" case (i.e., $f = 0.85 Re^{-0.19}$), obtained in a previous campaign [20], which only slightly differs in the Re dependence. As outlined in the Appendix, this close similarity allows the following correlation (of a somewhat different form), also covering the base case ($h = 0$), to be obtained:

$$f = 0.85 Re^{-0.19} \left(1 - 0.783 \frac{h}{D}\right) \tag{5}$$

where Re is defined as in Equation (2a). It should be pointed out that this expression (applicable in the range $(h/D) = 0$ to ~0.15), can be conveniently incorporated in comprehensive dynamic process simulators (e.g., [24]) which are under development. Finally, it is interesting to note that the results of this study (data in Figure 5 and the respective correlation, Equation (5)) show that for a 10% reduction of the spacer-filled channel gap (i.e., $(h/D) = 0.1$) the pressure drop increase amounts to ~26%, compared to the base case.

3.3. Effect of Fouling-Layer Thickness on Shear Stress at the Membrane Surface

From the detailed numerical flow simulations, dimensionless time-averaged wall shear stresses have also been obtained, which are directly related to the mass-transfer coefficients in the spacer-filled membrane channels. In Figure 7, typical results are depicted regarding the distribution of time-averaged shear stress on the membrane surface. The images of the spatial shear stress distributions included in Figure 7 (for approximately the same Reynolds number) correspond to fouling-layer thickness in the range 0–10% of the spacer-filament diameter. Purple and red colors represent minimum and maximum shear-stress values, respectively. One observes in Figure 7 that, for an increasing uniform fouling-layer thickness h (in the range of parameter values studied), the distribution of shear stresses is qualitatively very similar, for all the cases examined; however, as expected, the high-shear regions tend to increase with increasing h. Shear stresses are at a minimum (zero) at the contact of linear strips of spacer filaments with the surfaces of the retentate channel. High shear-stress values are observed at the narrow regions between the spacer filaments and the channel walls, where the velocity is increased. The maximum shear-stress values are attributed to the recirculation zones (marked in Figure 2), the presence of which results in further constriction of the fluid (as shown in Figures 20 and 21 of [20]), thus intensifying the stress at that location of the retentate channel.

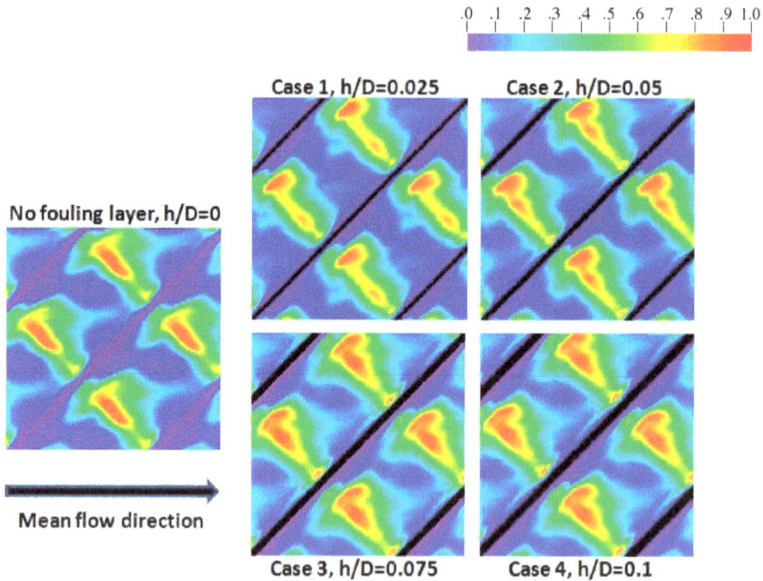

Figure 7. A view of spatial distribution of the local time-averaged shear stress on the membrane surface for the no-fouling geometry ($h/D = 0$) compared with distributions for progressively increased fouling-layer thickness; Re ~135. In the normalized chromatic scale, the values of 0 and 1 correspond to the minimum and maximum of the plotted variable, respectively.

Quantitative comparison of results for the "no-fouling" case with those corresponding to cases of various fouling-layer thicknesses h, shown in Figure 8, suggests that as thickness h increases, the distribution of the local shear stresses tends to become more uniform (with elimination of a secondary peak) and somewhat broader, extending toward higher values. This is attributed to the increased mean velocity, due to the reduction of the effective gap ($H-2h$) of the channel, which results in increased velocities close to the retentate channel walls, with concomitant increase of the wall shear stress. This is also evident in the data of Figure 9, where more uniform and broader shear-stress distributions prevail, toward higher values with increasing Re number. Additionally, the increased wall shear stresses with increasing h contribute to the computed increase of pressure drop in the channel.

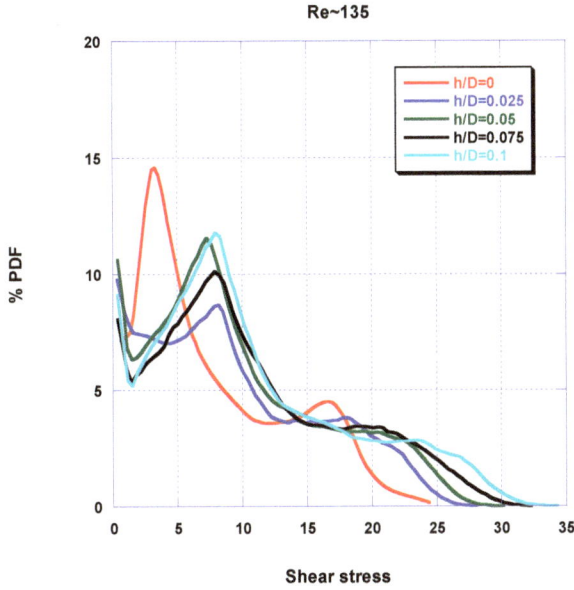

Figure 8. Percent probability density function of the dimensionless local time-averaged shear stresses on the membrane surface for the cases examined in this work. *Re* ~135. Shear stresses are normalized with the quantity $\mu U/D$.

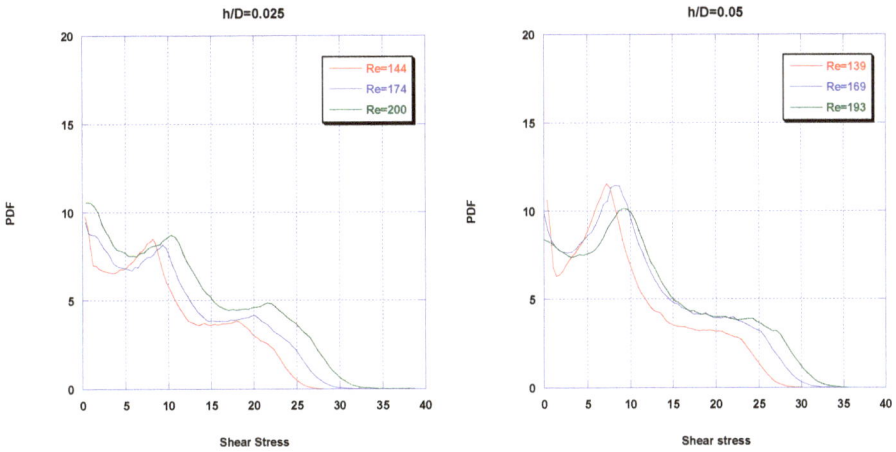

Figure 9. *Cont.*

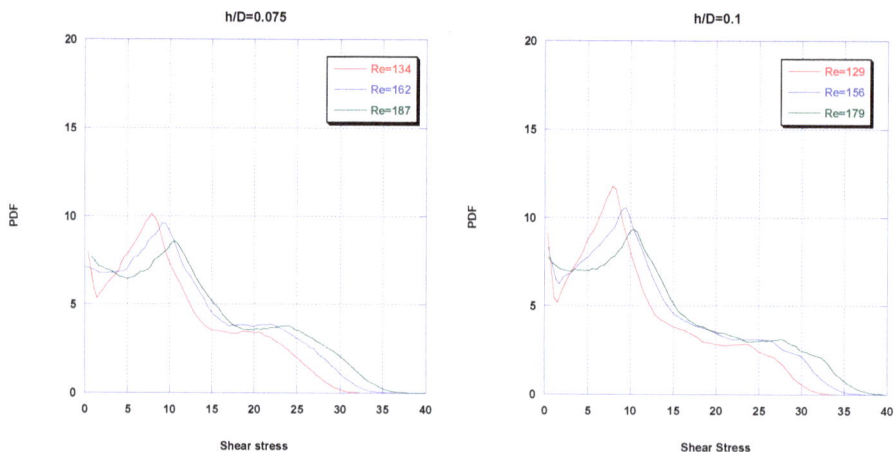

Figure 9. Probability density function of the dimensionless local, time-averaged shear stresses on the membrane surface for the four cases of fouling-layer thicknesses examined in this work. Shear stresses are normalized with the quantity $\mu U/D$.

3.4. Effect of Fouling-Layer Thickness on Mass Transfer

Mass transfer simulations of the spacer-filled channel were carried out in the Schmidt number range 1–100, which allowed a modest extrapolation to greater *Sc* numbers in correlating the numerical data. It is noted that for the systems of practical interest (i.e., saline waters), the Schmidt number is of order 10^3 leading to very thin concentration boundary layers at channel walls, which require a very fine grid at the wall, and entail increased computational load, exceeding the capabilities of available computational facility. Simulations of mass transfer were performed for the geometries listed in Table 1, i.e., for fouling-layer thickness h between 2.5–10% of the spacer-filament diameter.

Figure 10, depicts predicted distributions of local time-averaged Sherwood number *Sh* for fouling-layer thickness $(h/D) = 0.025$ and 0.10 in the *Sc* number range 1–100. *Sh* number distributions are generally skewed to the right; there is a significant percentage of near-zero values, due to the contact lines/areas of spacer filaments with the membrane surfaces comprising the retentate channel, and the nearly stagnant flow regions near and along these lines. At greater *Sc* number (leading to reduced boundary-layer thickness), the *Sh* number distributions tend to become wider. It is also interesting to note that, in the data plotted in Figure 10 (as well as in those not shown here for economy of space), with increasing thickness h/D the distributions tend to become normal due to the reduced proportion of near zero values; this reflects the change of flow-field geometry and the apparent reduction of almost stagnant flow zones at increased h/D. However, the reduction (with increasing h/D) of *Sh* number values at the high-value end of the distribution may be partly due to the non-dimensionalization; i.e., as shown in the *Sh* number definition (Equation (2c) involving the product (kD'), a likely increase of local mass-transfer coefficient k (at reduced D') is apparently smaller than the reduction of D' itself.

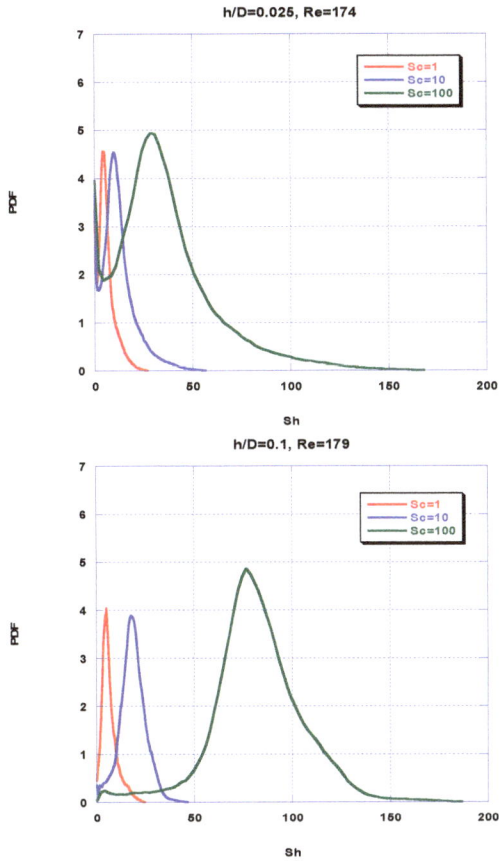

Figure 10. Probability density function of the time-averaged *Sh* number at the membrane surface for fouling-layer thickness (h/D) = 0.025 and 0.10 in the *Sc* number range 1–100.

As in the case of pressure drop, a key objective of the mass-transfer simulations is to develop a sound correlation of *Sh* number with fouling-layer thickness, including the *Re* and *Sc* number dependencies. In Figure A4 (Appendix B), the measured space and time-average Sherwood number is plotted as a function of *Sc* number, for the range of Reynolds numbers and fouling-layer thickness examined in this work. These data sets, show that for all fouling-layer thicknesses tested, the *Sc* dependence of mean Sherwood number is practically constant, and equal to ~0.47.

This similar dependence of *Sh* on *Sc* allows re-plotting the data in terms of $Sh/Sc^{0.47}$ versus the ratio D'/D, to determine the D'/D dependence. The results presented in Figure 11a show that for various *Re* numbers between 100 and 200 (where the commercial spiral-wound elements usually operate), there is similar dependence on the ratio D'/D with an exponent close to -1.29 in all cases; moreover, as one would expect, the space average *Sh* number tends to increase with gap reduction. Therefore, the dimensionless quantity $[Sh/Sc^{0.47}(D'/D)^{-1.29}]$, plotted versus *Re* number in Figure 11b, allows a fair correlation to be obtained:

$$Sh = 0.13\left(\frac{D'}{D}\right)^{-1.29} Re^{0.66} Sc^{0.47} \tag{6}$$

However, this expression can be recast (employing a Taylor expansion) in the following form to facilitate applications (including modeling) that involve explicitly the thickness h:

$$Sh = 0.13Re^{0.66}Sc^{0.47}\left(1 + 1.29\frac{h}{D}\right) \tag{7}$$

Finally, it should be noted that the results of this study (and the respective correlation in Equation (7)) show that for a 10% reduction of the spacer-filled channel gap (i.e., $(h/D) = 0.1$) the space-average mass-transfer coefficient increases by approximately 30%, compared to the base case.

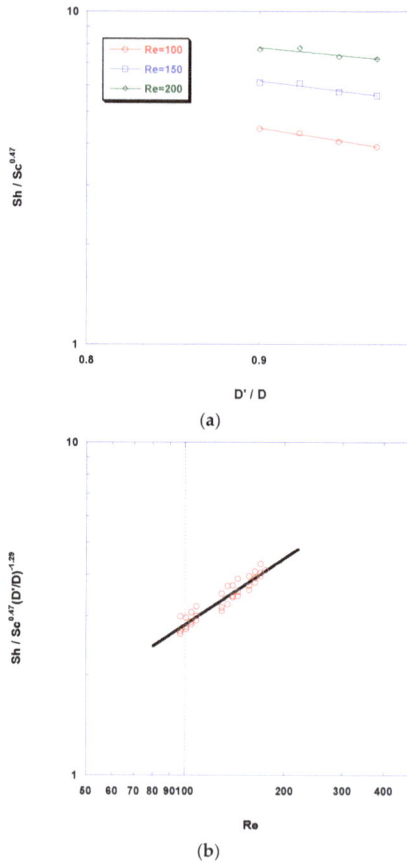

(a)

(b)

Figure 11. (**a**) Time and space-average *Sh* number, on the membrane surface, as a function of ratio D'/D; (**b**) correlation of dimensionless *Sh* number parameter (involving *Sc* and D'/D) as a function of Re number.

It should be pointed out that experimental data (at high *Sc*) and similar numerical results (for spacer type $L/D = 8$, $\beta = 90°$), obtained by the present authors [21] with the reference unobstructed channel geometry ($h = 0$), are well correlated by the following expression:

$$Sh = 0.16Re^{0.605}\,Sc^{0.42} \tag{8}$$

which is quite close to the above correlation (Equation (7)) regarding the *Sh* dependence on *Sc* and *Re* numbers. This close similarity provides a validation of the numerical results presented herein.

4. Discussion

Prediction from first principles of the spatial-temporal flow-field evolution, under simultaneous development of fouling layers, throughout the desalination membrane sheets (in spacer-filled narrow channels), is a formidable task, considered impossible at present. Aiming to develop a sound approximate method to tackle this problem, by un-coupling fouling-layer evolution from the fluid mechanical problem, this study considers the effect of uniformly developing fouling layers on fluid dynamics and mass transfer in spacer-filled channel walls. The assumption is made of fouling-layer uniformity at the "unit cell" spatial scale, realizing that it may be realistic only for some types of foulants but not for all of them. Indeed, as outlined in the Introduction, common foulants such as polysaccharides (including extracellular polymeric substances (EPS) in biofilms) seem to develop fairly coherent gels on membranes [8,32], impossible to detach by fluid-shear forces prevailing during desalination [32], thus prone to be fairly uniform. However, for other foulant types such as humic acids alone, the fouling-layer thickness in spacer-filled channels appears to be non-uniform [21]. Similarly, it is reported that there is non-uniform layer thickness in biofouling experiments [34]. However, the simplifying assumption of thickness uniformity made here for rather thin layers is considered realistic in efforts to model and quantify channel-gap reduction effects on flow and convective mass transfer at a large spatial scale; this assumption is also made in other, similar studies [5]. Obviously, the topic of fouling-layer development at small as well as large spatial scales merits further attention.

The results obtained here, including the correlations for pressure drop and mass-transfer coefficient, are readily applicable to the case of gap reduction of spacer-filled narrow channels (by causes other than fouling) in order to obtain reliable estimates of the respective quantities. Such gap reduction, attributed to feed-channel deformation in the SWM-module fabrication, is reported by membrane manufacturers [1], but no means to accurately determine key parameters such as pressure drop are available for such cases to the authors' best knowledge.

The spacer geometric parameters employed in this study ($L/D = 8$, $\beta = 90°$), closely correspond to the commercial "28 mil" spacer, where $L/D \approx 8.5$. However, the similarity of the flow field and the dimensionless quantities employed, in the simulations and in the developed correlations, permit the use of the new results and correlations to the others frequently encountered in practice-spacer types, i.e., "26 mil" and "34 mil", where $L/D \approx 8.9$ and ~8.0, respectively.

The correlations obtained in this work can find practical use in parametric studies and in various benchmarking exercises. However, the usefulness of the new results and correlations on developing and/or improving comprehensive dynamic simulators of RO desalination plant performance, which is pursued by the present authors [4,6,24], is of particular interest and will be briefly assessed. The effect of even a modest reduction of channel gap on the pressure drop $\Delta P/\Delta x$ appears to be quite important, as discussed in the foregoing section; therefore, the $\Delta P/\Delta x$ correlation, including the channel-gap correction, can be properly adapted to SWM performance models and simulators (e.g., [22,24]). Regarding the mass-transfer correlation, Equations (6) or (7) show that the channel-gap reduction is also significantly affecting the mass-transfer coefficient k, even though there is some scatter (Figure 11b) in numerical-data correlation due to the fact that the exponent on D'/D is weakly dependent on *Re* number whereas a constant average value is chosen (−1.29). However, it is shown below that even a rough estimate of dissolved species mass-transfer coefficient due to fluid convection at the membrane surface (k) is sufficient in modeling species mass transfer in the presence of a fouling layer, because k during the desalination process tends to contribute insignificantly to the overall mass transfer K_t, compared to cake-induced mass-transfer reduction. To demonstrate the significance of the latter effect, one should consider the total mass-transfer coefficient, which is given as:

$$K_t = (h/D_e + 1/k)^{-1} \tag{9}$$

where D_e is the species effective diffusion coefficient in the fouling layer. Multiplying both sides of Equation (9) by D/D_c and substituting the correlation (Equation (7)) for k in the form $kD/D_c = Sh_0(1 + az)$ where $z = h/D$ and $a = 1.65$ one obtains:

$$\frac{K_t D}{D_c} = \left(\frac{z}{E} + \frac{1}{[Sh_0(1 + az)]} \right)^{-1} \sim \frac{Sh_0}{1 + \left(\frac{Sh_0}{E} - a \right)z} \tag{10}$$

Here $E = D_e/D_c$ is the ratio of the effective diffusivity in the fouling layer over the species diffusivity in the liquid phase. A comparison of the terms at the right-hand side of Equation (10) clearly shows that (as the fouling layer grows) the increase of mass transfer due to convection (k) contributes very little to the total mass-transfer variation. In particular, using the ratio $\alpha E/Sh_0$, one can show, for typical values of the parameters involved, that only 1–2% of the total mass-transfer variation is due to an increase of coefficient k, because of channel-gap reduction; therefore, the accuracy in the computation of k does not affect the estimates for total mass-transfer coefficient K_t, needed in comprehensive desalination-process modeling.

It is evident from this study that the reliable prediction of increasing fouling-layer thickness h and of transport phenomena related to fouling-layer properties (as a function of desalination-system parameters and foulants) is critical in the development of a sound dynamic process simulator [24]. Therefore, future work should be focused on measuring, understanding and modeling fouling-layer development. Significant efforts to understand, monitor and predict developing fouling-layer properties, including h, have already been reported (e.g., [8,9,35–38]) under idealized conditions. Furthermore, in the context of hydrodynamics and fouling evolution in spacer-filled channels, investigating the effect of the detailed geometry and filament shape of the spacer is of high priority. The present results with the common spacer type, comprised of unwoven filaments of cylindrical shape, for which there is extensive literature and experimental data, provide a kind of base case.

5. Conclusions

This study is mainly motivated by the need to simulate the time-varying flow field in spacer-filled desalination membrane channels, caused by the development of fouling layers on the membranes that modify the flow-field geometry and affect the transport phenomena at the membrane surface. As the complete dynamic problem of flow and mass transfer, under the simultaneous growth of fouling layers, cannot be tackled at present due its complexity, a simplified approach is pursued whereby at a small spatial scale the fluid dynamics and mass transfer are uncoupled from the fouling-layer growth process. Therefore, fluid dynamics and mass transfer are studied herein in detail for a spacer-filled narrow channel whose geometry is altered by a developed uniform fouling layer of thickness h on the membranes. For this purpose, 3D direct numerical simulations are performed (in a restricted spatial domain) employing the "unit cell" approach where periodic boundary conditions are applied. The specific thickness values considered are in the range 2.5–10% of the cylindrical spacer filament diameter D, which covers conditions of practical significance. The geometric parameters of the spacer employed in the simulations correspond to those of commonly encountered commercial spacers.

The new results reveal that the qualitative characteristics of the altered flow field in the spacer-filled channel (of reduced gap due to fouling) are very similar to those of the reference geometry, i.e., with no gap reduction. The main flow characteristics include a free vortex along the diagonal of the unit cell (roughly in the main flow direction) as well as spiral vortices and recirculation zones downstream of each spacer filament. It is interesting, however, that the gap reduction apparently tends to stabilize the flow and extend the transition to unsteady flow to somewhat greater Re numbers. As expected, for a given flow rate, pressure drop and time-averaged wall shear stress as well mass-transfer coefficient k tend to significantly increase with increasing fouling-layer thickness h due to the reduced effective channel-gap, leading to increased flow velocities.

Useful correlations are obtained from the numerical data on friction factor f and overall average *Sh* number, which exhibit similar functional dependence of *f* and *Sh* on *Re* and *Sc* numbers to the reference case of the unobstructed channel (*h* = 0). A noteworthy feature of these correlations is the incorporation of the effect of channel-gap reduction, which allows predictions in the range of fouling-layer thickness (*h/D*) = 0–0.10. The proposed correlations, in addition to other uses, can be employed to improve steady-state models and develop realistic dynamic models simulating the operation of SWM modules and RO/NF plants.

Author Contributions: Chrysafenia P. Koutsou performed the numerical simulations, collaborated in the development of correlations and drafted the manuscript. Anastasios J. Karabelas collaborated in planning the study, in developing the correlations and in editing/finalizing the manuscript. Margaritis Kostoglou collaborated in the planning and in modeling aspects of the study.

Conflicts of Interest: The authors declare no conflict of interest.

Appendix A. Pressure-Drop Correlation Applicable to Reduced Channel Gap

Data of friction factor f, as a function of the ratio h/D, plotted in Figure A1, show practically the same dependence of f on the ratio h/D (with an exponent approx. −0.03) for all the cases examined in this work. This particular dependence of dimensionless pressure drop on the ratio h/D allows re-plotting the data in terms of $f/(h/D)^{-0.03}$ versus Re, in order to determine the Re number dependence; the results are presented in Figure A2 and a correlation thus obtained is of the form:

$$\langle\frac{\mathrm{d}P}{\mathrm{d}L}\rangle = 0.85\left(\frac{h}{D}\right)^{-0.03} Re^{-0.22} \tag{11}$$

It will be noted, by comparing this correlation [Equation (A1)] with the corresponding correlation for the "no-fouling" case ($f = 0.85Re^{-0.19}$), obtained in a previous campaign [20], that they are very similar; i.e., with the same pre-exponential factor and only a slight difference in the Re dependence. As outlined in the following, this close similarity allows the development of a correlation of a somewhat different form, which will also cover the base case (*h* = 0), thus facilitating its application in practice as well as its inclusion in desalination-process simulators (e.g., [22,24]).

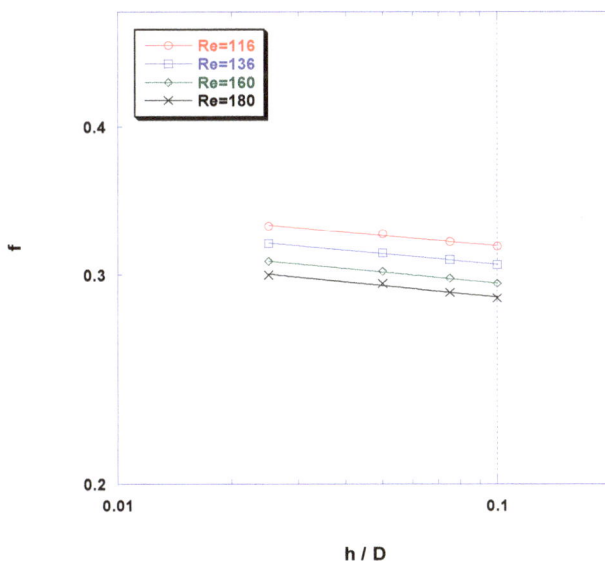

Figure A1. Dimensionless pressure drop (friction factor) as a function of the ratio h/D.

By plotting the dimensionless quantity $[(f/f_{h=0}) - 1]$ versus the fouling-layer thickness (h/D), one obtains a practically linear dependence on the ratio h/D (Figure A3), which leads to the following correlation:

$$f = f_{\frac{h}{D}=0}\left(1 - 0.783\frac{h}{D}\right) \rightarrow f = 0.85Re^{-0.19}\left(1 - 0.783\frac{h}{D}\right) \tag{12}$$

where Re is defined in Equation (3b).

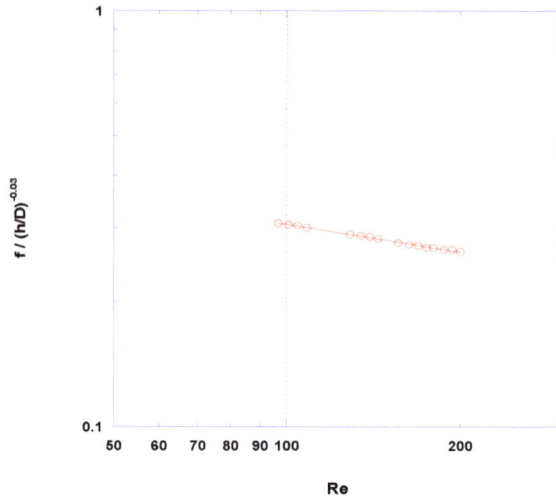

Figure A2. Dimensionless pressure drop (modified friction factor) as a function of the ratio h/D and Re number. Correlation: $f = 0.85(h/D)^{-0.03}Re^{-0.22}$.

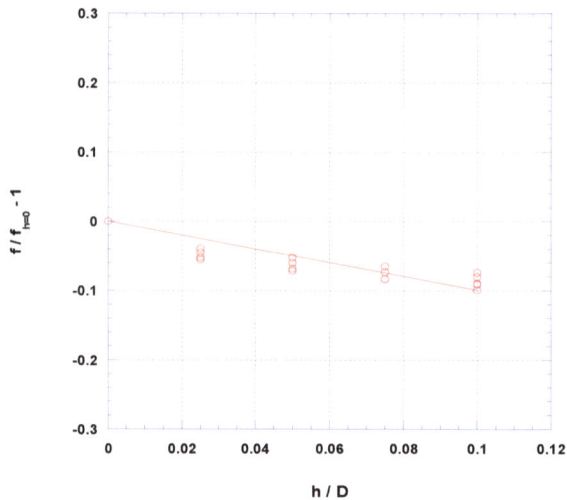

Figure A3. Variation of expression $[(f/f_o)-1]$ with dimensionless fouling-layer thickness.

Appendix B. Mass-Transfer Correlation Applicable to Reduced Channel Gap

In Figure A4 the measured space and time-average Sherwood number is plotted as a function of *Sc* number for the range of Reynolds numbers and fouling-layer thickness examined in this work. These data sets (as well as other similar data for $h/D = 0.025$ and 0.075 not included here), show that for all fouling-layer thicknesses tested, the *Sc* dependence of mean Sherwood number is practically constant, and equal to ~0.47. This similar dependence of *Sh* on *Sc* allows re-plotting the data in terms of $Sh/Sc^{0.47}$ versus the ratio D'/D, to determine the D'/D dependence, toward the development of a generalized correlation.

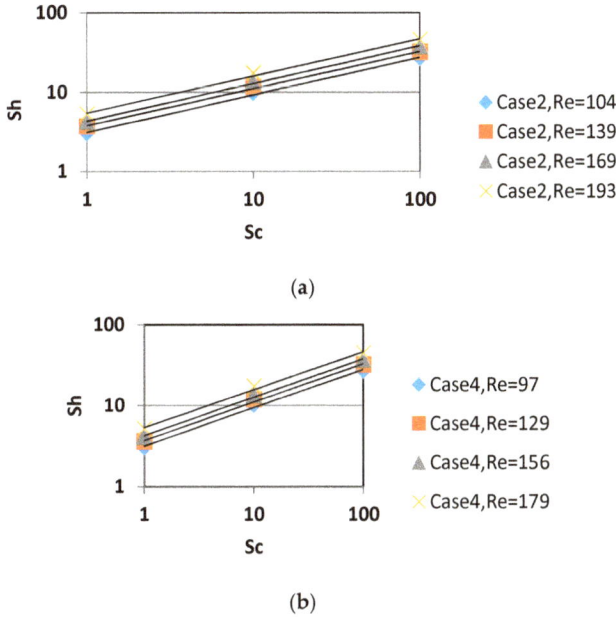

(a)

(b)

Figure A4. Time and space-average *Sh* number on the membrane surface as a function of *Sc* number for (**a**) $h/D = 0.05$; and (**b**) $h/D = 0.10$.

References

1. Johnson, J.; Busch, M. Engineering aspects of reverse osmosis module design. *Desalin. Water Treat.* **2010**, *15*, 236–248. [CrossRef]
2. Karabelas, A.J. Key issues for improving the design and operation of membrane modules for desalination plants. *Desalin. Water Treat.* **2014**, *52*, 1820–1832. [CrossRef]
3. Karabelas, A.J.; Koutsou, C.P.; Kostoglou, M. The effect of spiral wound membrane element design characteristics on its performance in steady state desalination—A parametric study. *Desalination* **2014**, *332*, 76–90. [CrossRef]
4. Koutsou, C.P.; Karabelas, A.J.; Kostoglou, M. Membrane desalination under constant water recovery—The effect of module design parameters on system performance. *Sep. Purif. Technol.* **2015**, *147*, 90–113. [CrossRef]
5. Hoek, E.M.V.; Allred, J.; Knoell, T.; Jeong, B.-H. Modeling the effects of fouling on full-scale reverse osmosis processes. *J. Membr. Sci.* **2008**, *314*, 33–49. [CrossRef]
6. Kostoglou, M.; Karabelas, A.J. A mathematical study of the evolution of fouling and operating parameters throughout membrane sheets comprising spiral wound modules. *Chem. Eng. J.* **2012**, *187*, 222–231. [CrossRef]
7. Tang, C.Y.; Chong, T.H.; Fane, A.G. Colloidal interactions and fouling of NF and RO membranes: A review. *Adv. Colloid Interface Sci.* **2011**, *164*, 126–143. [CrossRef] [PubMed]

8. Karabelas, A.J.; Sioutopoulos, D.C. New insights into organic gel fouling of reverse osmosis desalination membranes. *Desalination* **2015**, *368*, 114–126. [CrossRef]
9. Sim, S.T.V.; Taheri, A.H.; Chong, T.H.; Krantz, W.B.; Fane, A.G. Colloidal metastability and membrane fouling—Effects of cross flow velocity, flux, salinity and colloid concentration. *J. Membr. Sci.* **2014**, *469*, 174–187. [CrossRef]
10. Johannink, M.; Masilamani, K.; Mhamdi, A.; Roller, S.; Marquardt, W. Predictive pressure drop models for membrane channels with non-woven and woven spacers. *Desalination* **2015**, *376*, 41–54. [CrossRef]
11. Fimbres, G.A.; Wiley, D.E. Review of 3D CFD modelling of flow and mass transfer in narrow spacer-filled channels in membrane modules. *Chem. Eng. Process. Process Intensif.* **2010**, *49*, 759–781. [CrossRef]
12. Karode, S.K.; Kumar, A. Flow visualization through spacer filled channels by computational fluid mechanics, I. Pressure drop and shear stress calculations for flat sheet geometry. *J. Membr. Sci.* **2001**, *193*, 69–84. [CrossRef]
13. Schwinge, J.; Wiley, D.E.; Fletcher, D.F. A CFD study of unsteady flow in narrow spacer-filled channels for spiral-wound membrane modules. *Desalination* **2002**, *146*, 195–201. [CrossRef]
14. Geraldes, V.; Semio, V.; de Pinho, M.N. Hydrodynamics and concentration polarization in NF/RO spiralwound modules with ladder-type spacers. *Desalination* **2003**, *157*, 395–402. [CrossRef]
15. Schwinge, J.; Wiley, D.; Fane, A. Novel spacer design improves observed flux. *J. Membr. Sci.* **2004**, *229*, 53–61. [CrossRef]
16. Koutsou, C.P.; Karabelas, A.J. A novel retentate spacer geometry for improved spiral wound membrane (SWM) module performance. *J. Membr. Sci.* **2015**, *488*, 129–142. [CrossRef]
17. Li, F.; Meindersma, W.; de Haan, A.B.; Reith, T. Novel spacers for mass transfer enhancement in membrane separations. *J. Membr. Sci.* **2005**, *253*, 1–12. [CrossRef]
18. Picioreanu, C.; Vrouwenvelder, J.S.; Van Loosdrecht, M.C.M. Three-dimensional modeling of biofouling and fluid dynamics in feed spacer channels of membrane devices. *J. Membr. Sci.* **2009**, *345*, 340–354. [CrossRef]
19. Radu, A.I.; Vrouwenvelder, J.S.; van Loosdrecht, M.C.M.; Picioreanu, C. Modeling the effect of biofilm formation on reverse osmosis performance: Flux, feed channel pressure drop and solute passage. *J. Membr. Sci.* **2010**, *365*, 1–15. [CrossRef]
20. Koutsou, C.P.; Yiantsios, S.G.; Karabelas, A.J. Direct numerical simulation of flow in spacer-filled channels: Effect of spacer geometrical characteristics. *J. Membr. Sci.* **2007**, *291*, 53–69. [CrossRef]
21. Koutsou, C.P.; Yiantsios, S.G.; Karabelas, A.J. A numerical and experimental study of mass transfer in spacer-filled channels: Effects of spacer geometrical characteristics and Schmidt number. *J. Membr. Sci.* **2009**, *326*, 234–251. [CrossRef]
22. Kostoglou, M.; Karabelas, A.J. Comprehensive simulation of flat-sheet membrane element performance in steady state desalination. *Desalination* **2013**, *316*, 91–102. [CrossRef]
23. Kostoglou, M.; Karabelas, A.J. Mathematical analysis of the meso-scale flow field in spiral wound membrane modules. *Ind. Eng. Chem. Res.* **2011**, *50*, 4653–4666. [CrossRef]
24. Kostoglou, M.; Karabelas, A.J. Dynamic operation of flat sheet desalination-membrane elements: A comprehensive model accounting for organic fouling. *Comput. Chem. Eng.* **2016**, *93*, 1–12. [CrossRef]
25. Yiantsios, S.G.; Sioutopoulos, D.; Karabelas, A.J. Colloidal fouling of RO membranes: An overview of key issues and efforts to develop improved prediction techniques. *Desalination* **2005**, *183*, 257–272. [CrossRef]
26. Haaksman, V.A.; Siddiqui, A.; Schellenberg, C.; Kidwell, J.; Vrouwenvelder, J.S.; Picioreanu, C. Characterization of feed channel spacer performance using geometries obtained by X-ray computed tomography. *J. Membr. Sci.* **2017**, *522*, 124–139. [CrossRef]
27. *Hydranautics, RODESIGN, Hydranautics RO System Design Software*, version 6.4 (c); Pretreatment and design limits section. 1998.
28. Sioutopoulos, D.C.; Karabelas, A.J.; Yiantsios, S.G. Organic fouling of RO membranes: Investigating the correlation of RO and UF fouling resistances for predictive purposes. *Desalination* **2010**, *261*, 272–283. [CrossRef]
29. Sioutopoulos, D.C.; Yiantsios, S.G.; Karabelas, A.J. Relation between fouling characteristics of RO and UF membranes in experiments with colloidal organic and inorganic species. *J. Membr. Sci.* **2010**, *350*, 62–82. [CrossRef]
30. Sioutopoulos, D.C.; Karabelas, A.J. Correlation of organic fouling resistances in RO and UF membrane filtration under constant flux and constant pressure. *J. Membr. Sci.* **2012**, *407*, 34–46. [CrossRef]

31. Karabelas, A.J.; Sioutopoulos, D.C. Toward improvement of methods for predicting fouling of desalination membranes - The effect of permeate flux on specific fouling resistance. *Desalination* **2014**, *343*, 97–105. [CrossRef]

32. Sioutopoulos, D.C.; Goudoulas, T.B.; Kastrinakis, E.G.; Nychas, S.G.; Karabelas, A.J. Rheological and permeability characteristics of alginate fouling layers developing on reverse osmosis membranes during desalination. *J. Membr. Sci.* **2013**, *434*, 74–84. [CrossRef]

33. McCarthy, M.D.; Hedges, J.I.; Benner, R. The chemical composition of dissolved organic matter in seawater. *Chem. Geol.* **1993**, *107*, 503–507. [CrossRef]

34. Farhat, N.M.; Staal, M.; Siddiqui, A.; Borisov, S.M.; Bucs, S.S.; Vrouwenvelder, J.S. Early non-destructive biofouling detection and spatial distribution: Application of oxygen sensing optodes. *Water Res.* **2015**, *83*, 10–20. [CrossRef] [PubMed]

35. Taheri, A.H.; Sim, S.T.V.; Sim, L.N.; Chong, T.H.; Krantz, W.B.; Fane, A.G. Development of a new technique to predict reverse osmosis fouling. *J. Membr. Sci.* **2013**, *448*, 12–22. [CrossRef]

36. Gao, Y.; Haavisto, S.; Li, W.; Tang, C.Y.; Salmela, J.; Fane, A.G. Novel approach to characterizing the growth of a fouling layer during membrane filtration via optical coherence tomography. *Environ. Sci. Technol.* **2014**, *48*, 14273–14281. [CrossRef] [PubMed]

37. Mairal, A.P.; Greenberg, A.R.; Krantz, W.B. Investigation of membrane fouling and cleaning using ultrasonic time-domain reflectometry. *Desalination* **2000**, *130*, 45–60. [CrossRef]

38. Kujundizic, E.; Greenberg, A.R.; Fong, R.; Hernandez, M. Monitoring protein fouling on polymeric membranes using ultrasonic frequency-domain reflectometry. *Membranes* **2011**, *1*, 195–216. [CrossRef] [PubMed]

fluids

MDPI

Article

Computational Fluid Dynamics Simulations of Gas-Phase Radial Dispersion in Fixed Beds with Wall Effects

Anthony G. Dixon [1,*] and Nicholas J. Medeiros [1,2]

1 Department of Chemical Engineering, Worcester Polytechnic Institute, Worcester, MA 01609, USA
2 Currently at Coventor, 91140 Villebon-sur-Yvette, France; nicholas.medeiros01@gmail.com
* Correspondence: agdixon@wpi.edu; Tel.: +1-508-831-5350

Received: 8 September 2017; Accepted: 18 October 2017; Published: 21 October 2017

Abstract: The effective medium approach to radial fixed bed dispersion models, in which radial dispersion of mass is superimposed on axial plug flow, is based on a constant effective dispersion coefficient, D_T. For packed beds of a small tube-to-particle diameter ratio (N), the experimentally-observed decrease in this parameter near the tube wall is accounted for by a lumped resistance located at the tube wall, the wall mass transfer coefficient k_m. This work presents validated computational fluid dynamics (CFD) simulations to obtain detailed radial velocity and concentration profiles for eight different computer-generated packed tubes of spheres in the range $5.04 \leq N \leq 9.3$ and over a range of flow rates $87 \leq Re \leq 870$ where Re is based on superficial velocity and the particle diameter d_p. Initial runs with pure air gave axial velocity profiles $v_z(r)$ averaged over the length of the packing. Then, simulations with the tube wall coated with methane yielded radial concentration profiles. A model with only D_T could not describe the radial concentration profiles. The two-parameter model with D_T and k_m agreed better with the bed-center concentration profiles, but not with the sharp decreases in concentration close to the tube wall. A three-parameter model based on classical two-layer mixing length theory, with a wall-function for the decrease in transverse radial convective transport in the near-wall region, showed greatly improved ability to reproduce the near-wall concentration profiles.

Keywords: computational fluid dynamics; fixed bed; mass transfer; transverse dispersion

1. Introduction

Fixed bed reactors are tubes filled with catalytic packing material, on which a chemical reaction may occur. Fixed bed reactors are preferred for their simplified technology and operation, and the design of such beds has been based on effective medium models coupled with correlation-based transport parameters. Low tube-to-particle diameter ratio (N) beds are typically selected for highly exothermic or endothermic reactions, such as in partial oxidations or hydrocarbon reforming. The use of larger particles within a tube of relatively small diameter results in a reduced pressure drop across the bed and, thus, a more economically-favorable operation. However, this results in a higher void fraction near the wall of the bed, which induces significant flow and transport effects that are difficult to account for in conventional packed bed models. In this near wall area, boundary layers develop in which heat and mass transfer are dominated by molecular transport, and a high resistance to transport is present, causing strong changes near the wall.

Dispersion is a phenomenon in which a solute in a porous medium spreads as a result of the combined effects of molecular diffusion and convection in the interstices or the space available for fluid flow between particles. Typical applications of dispersion are, for example, in quantifying contaminant transport in soils or in modeling solute transport in fixed beds. The study of dispersion in packed bed reactors is fundamental to their design in that dispersive phenomena characterize reactant and product transport within the bed [1–4]. Dispersion may be modeled analogously to Fickian diffusion by replacing the diffusion coefficients with dispersion coefficients as seen in the following partial differential equation for species concentration:

$$D_L \frac{\partial^2 C}{\partial z^2} + \frac{1}{r} \frac{\partial}{\partial r} \left(D_T r \frac{\partial C}{\partial r} \right) - u \frac{\partial C}{\partial r} = \frac{\partial C}{\partial t} \tag{1}$$

In Equation (1), D_L is the axial or longitudinal dispersion coefficient and D_T is the radial or transverse dispersion coefficient, related to flow in the freestream and cross-stream directions, respectively. When dealing with dispersion, it is often useful to define dimensionless Péclet numbers, representing the ratio of advective to dispersive transport:

$$Pe_T = \frac{v_i \, d_p}{D_T} \tag{2}$$

and:

$$Pe_L = \frac{v_i \, d_p}{D_L} \tag{3}$$

where v_i is the interstitial velocity in the bed and d_p is the particle diameter.

The present study is focused on the transverse dispersion coefficient, rather than on the longitudinal coefficient. The latter is important for such applications as chromatography and the understanding of residence time distribution in fixed beds. Our main motivation for studying dispersion, however, is the analogy between convective heat and mass transfer, since radial heat transfer is one of the most important parameters in fixed bed modeling. Studies of mass dispersion can give insight into the modeling of fluid-phase convective heat transfer.

Strong radial temperature profiles are common in fixed beds of low N. In contrast, both random walk theory and early experimental results suggested that radial concentration profiles would be fairly flat, although radial dispersion near the tube wall could be reduced due to flow channeling there, leading to radial concentration profiles. Experimental studies of radial dispersion in fixed beds have been conducted for over 70 years (see, for example, [5–7]) and have for the most part been conducted by a centerline pulse injection of a tracer substance, with a small number of radial concentrations measured at the bed exit to avoid disturbing the bed packing [5–9]. These measurements are made on beds of relatively short length, before the solute has spread over the entire bed radius, to obtain non-uniform radial concentration profiles. The consequence is that for most studies, the wall effects are deliberately excluded, so as to obtain infinite-medium D_T values.

Opinions have been divided on whether there is a wall effect on fixed bed radial dispersion. An early study by Fahien and Smith [6] used a centerline point source, but allowed the dispersing solute to reach the tube walls. They found that D_T was not constant across the tube radius and that the decrease in D_T near the tube wall depended on the tube-to-particle diameter ratio, N. The authors correlated their results for radial average transverse dispersion coefficient by:

$$Pe_T = \frac{8}{1 + 19.4/N^2} \tag{4}$$

Wall effects on transverse dispersion were also found in the presence of thermal gradients by Schertz and Bischoff [10] who attributed the effect to changes in fluid viscosity with temperature. Gunn and Pryce [8], on the other hand, found no wall effect, although their tracer did not reach the tube walls. Han et al. [9] also concluded that the values of D_T did not depend on bed location, as they were not time dependent in their transient study. Foumeny et al. [11] from tracer experiments with internal bed measurements at different bed heights suggested a correlation with a term that explicitly accounted for wall effects and remarked upon the unreliability of data taken from the bed exit, as the flow patterns change rapidly there, and upon the high correlation between the longitudinal and transverse dispersion coefficients in standard models.

In order to investigate the wall effects on dispersion and obtain wall transport coefficients that could be used in heat transfer studies by analogy, several research groups performed experiments in which the tube wall was coated with a dissolving substance [12–19]. A wall mass transfer coefficient was introduced, either through a modified boundary condition on Equation (1) [12,15–17,19] or by the theory for limiting current measurements in an electrochemical system [13,14,18]. A detailed understanding of the transport processes near the tube wall could not be obtained by these methods, due to the small number of radial concentration measurements that could be made.

In an experimental and modeling study, Coelho and Guedes de Carvalho [20] studied transverse dispersion in granular beds. They used a two-region plug-flow model in which a near-wall region of thickness δ may be thought of as a laminar sub-layer in which molecular diffusion is the dominant mechanism. Outside of this layer, the transport in the bed is due solely to dispersion. They performed experiments with a soluble wall for different bed lengths and obtained values of the overall mass transfer coefficient k_{av} to which they were able to fit the two parameters D_T and δ. Using these parameters, they were able to predict overall mass transfer coefficients for other values of Re and L with reasonable agreement. An interesting result of their work was that for a long enough bed, the δ-region saturates with the soluble species and a one-parameter model suffices. The criterion for a one-parameter model to be valid is:

$$\frac{L}{d_p} > 0.625 \frac{Re \cdot Sc}{\varepsilon} \tag{5}$$

where ε is the bed voidage and Re is based on the interstitial velocity.

In following experimental studies, Guedes de Carvalho and Delgado [21,22] studied lateral dispersion in fixed beds of spheres by including in the bed a soluble cylinder of benzoic acid aligned with the flow of water entering the tube. They neglected any near-wall effects on dispersion, stating that the section of packing contacting the soluble slab actually indents onto this soluble section, removing the near-wall, high-voidage region that is typical in a fixed bed. Solute concentration in the outlet section of the bed was measured as a means to determine the rate of dissolution of the cylinder. Diffusion was treated as occurring in one dimension, given that the bed followed the criterion in Equation (5). The authors noted the strong dependence of D_T/D_m on Sc in the range $140 < Sc < 500$ and showed that dispersion behavior was generally independent of particle size. In later work by Delgado [4], both longitudinal and transverse dispersion in porous media were studied over a wide range of the Schmidt and Péclet numbers. The author supplied a set of equations each for the longitudinal and transverse dispersion coefficients, subdivided by different regimes of dispersion based on the molecular Péclet number: (1) diffusion regime ($Pe_m < 0.1$); (2) predominant diffusional regime ($0.1 < Pe_m < 4$); (3) predominantly mechanical dispersion ($4 < Pe_m$ and $Re < 10$); (4) pure mechanical dispersion ($10 < Re$, $Pe_m < 10^6$); and (5) dispersion beyond the validity of Darcy's law ($Pe_m > 10^6$). While this paper presents important correlative equations for dispersion, the experimental data were largely from beds above the low N range ($2 < N < 8$). The difficulties of obtaining detailed experimental measurements in complicated systems such as fixed beds has suggested the use of computational tools as a complementary method of obtaining insight into transport properties such as dispersion.

Computational fluid dynamics (CFD) is a numerical method-based computing approach for simulating fluid flow, mass and heat transport and reaction kinetics, among other physical phenomena. In general, the use of CFD requires converting the geometry of interest, here a fixed bed of spheres, into a number of small control volumes, collectively called the computational mesh, or grid. After supplying the appropriate boundary conditions related to flow and species transport and an initial, estimated solution for the system of interest, a complete numerical solution can then be obtained by iterative, convergence-guided, numerical techniques. With the recent introduction of high performance computing capabilities and the continued growth of such computing resources, CFD can now be used as an important tool for the design and simulation of fixed beds. The use of CFD allows the elucidation of certain flow and heat characteristics that cannot be obtained from experimental methods, such as the velocity distribution around and between particles or the temperature at any point within the bed.

An early use of CFD to simulate flow and dispersion in computer-generated random sphere packs was made by Schnitzlein [23]. The motivation for the study was his observation that radial dispersion should be the sum of molecular diffusion and fluid mechanical phenomena (eddy diffusion in local voids, which act as mixing cells, branching in the packing and channeling due to the placement of particles and the wall effect) and should not be driven by a concentration gradient. This point of view was similar to one suggested by Gunn [1] who described axial and radial dispersion in terms of probability theory, accounting for dispersion in the fast stream (convective-dominated) and the slow stream near the tube wall (diffusion-dominated). Gunn estimated the probability of a particle existing in the diffusion boundary layer or moving into the fast stream area of the bed. He stated that diffusion in the bed had little effect on convective radial dispersion and introduced a fluid-mechanical Péclet number based on convective-dispersion alone, which in the limit of a high Reynolds number tended to 11.

In Schnitzlein's work, a two-dimensional map of the void fraction was obtained from the computer-generated packing and used in the porous medium 2D Navier–Stokes equations to obtain a two-dimensional velocity field. The velocity field was then used in a 2D mass balance to generate a simulated dispersion experiment. This was then fit by a standard dispersion model to estimate D_T, which gave a limiting fluid-mechanical Péclet number of 28, which was probably due to low variation in local voidage, leading to underestimation of radial displacement of fluid. A second study by Schnitzlein [24] used the sphere pack to obtain a 3D network model with inertial terms, but no diffusion or wall friction, which gave a velocity distribution. Using this in the 2D tracer model for $Re > 100$ gave $Pe_T \cong 12$.

Magnico [25] simulated beds of $N = 5.96$ and 7.8 consisting of 326 and 620 spheres, respectively, with a focus on near-wall effects on radial dispersion. He used particle tracing without molecular diffusion and found that particles in a boundary layer at the tube wall of thickness $d_p/4$ moved longitudinally and tangentially, but not radially. He concluded that exchange with this layer was mainly diffusive. At higher flows, $Re = 200$, tracer particles were not easily transported through the boundary layer along the wall, signifying a molecular diffusion-dominated mechanism of mass transfer in this region.

Lagrangian particle tracking along with lattice Boltzmann methods (LBM) were used by Freund et al. [26] for a tube with $N = 9.3$ and $L/d_p = 27$. Asymptotic behavior was not reached as the tube was too short. Their computed values of D_T/D_m were in good agreement with literature experimental results [8,9] for $0.1 < Pe_m < 10^4$ with a tortuosity of $\tau = 1.45$. Soleymani et al. [27] simulated dispersion in laminar flow for a packing of 3000 spheres for various larger N values to get D_T in the post-Darcy region. Results were slightly lower than in [9]. Jafari et al. [28] used packings of >3000 particles with $N \cong 60$, with solute continuously injected at the inlet on the centerline, but they estimated D_L only.

Augier et al. [29] simulated liquid flow in a bed of 440 particles in the form of a square volume taken from the center of a larger bed, to eliminate wall effects. They simulated the radial dispersion experiment of Han et al. [9] in which a radial step profile is gradually flattened out, for $Re < 100$. Their results for Pe_T were larger than literature correlations [3,4,21,22]. This was attributed to the change in void fraction that occurred when the authors reduced their particle sizes by 2% to avoid meshing problems near the contact points. A correction factor was applied that improved the agreement.

In a recent study by Jourak et al. [30], radial dispersion coefficients were derived by supplying 3D concentration profiles to a 2D effective porous medium model and fitting the coefficients to the data. Their simulation experiments used a bed of regular and randomly packed particles, with a fixed concentration boundary condition at the tube wall, to mimic earlier tracer injection experiments. Results were presented for laminar flow with $0.1 < Re < 100$. The researchers recommended using beds of large width and length, noting that radial dispersion coefficients showed some length dependency. That is, the radial dispersion coefficient was found to decrease as the length of the bed increased. They also noted that a 2D effective medium approach typically predicts dispersion coefficients higher in magnitude than the 3D counterpart. This is due to higher intercellular fluid motion in the lateral direction and, therefore, a higher radial dispersion coefficient.

Dispersion in packed beds of low N is an area of research that demands attention. Previous studies of radial dispersion typically neglect the presence of the tube wall, and these studies are generally conducted in beds of large N. The development of the diffusion-dominated boundary layer induces significant flow effects in the bed, and research that includes this additional dispersion mechanism is important. In a review of previous studies, liquid phase flows were typically the fluid of interest in experimental situations. It is of interest to study dispersion in a gas-phase system, over a range of flow rates, covering different flow regimes from steady-state laminar at low Re through transition regimes, up to so-called turbulent flow at $Re > 600$, approximately. Experimental tracer injections are highly erroneous when deriving dispersion coefficients: concentration data points are selected from few radial positions, and beds of long length are studied, which yield concentration profiles with small gradients. It is therefore important to study dispersion in beds of developing flow also, in which a clear and total picture of the radial concentration profile is included.

The objective of the present work is therefore to evaluate the ability of standard dispersion models to describe developing concentration profiles in fixed beds of low N where wall effects are important. This is difficult to do experimentally because of the small number of measurements that can be taken, typically at the exit of the bed and not inside it. We shall do this by computational generation of a number of 3D fixed beds of spheres, with more than 1000 particles in each, over a range of $5.04 < N < 9.3$, followed by resolved-particle simulation of flow and diffusion of a solute from a coated wall in each bed, over a range of Re. These simulated "data" sets will then be used to provide a rigorous test of 2D effective medium models with a single dispersion parameter D_T, two parameters D_T and a wall mass transfer coefficient and, finally, a model derived from mixing length concepts that gives D_T as a function of radial position [31,32].

2. Computational Methodology

The finite volume method implemented in ANSYS Fluent v. 16.2 (ANSYS, Canonsburg, PA, USA) was used for the analysis in the present work. In finite volume methods, the computational domain is divided or discretized into a number of small control volumes. The collection of all control volumes, referred to as the mesh or computational grid, represents the geometry being simulated. The governing equations are integrated over the control volumes to produce algebraic equations for the dependent variables of interest (i.e., pressure, velocity, species mass fraction in this case). In conjunction with the necessary boundary conditions, a complete numerical solution can then be obtained by solving the resulting large set of nonlinear algebraic equations. This solution method is repeated iteratively until specified convergence criteria have been met and a converged solution is reached. Detailed descriptions of the governing equations and the numerical methods can be found in the ANSYS Fluent manual [33], so only a brief description of the main equations and methods is provided here.

2.1. Fluid Mechanics

The steady-state equation for continuity of mass is given by the following partial differential equation:

$$\nabla \cdot (\rho \vec{u}) = S_m \tag{6}$$

The source term S_m is mass added to the continuous phase through user-defined sources or through phase changes. For the simulations carried out in this work, this term was zero.

The steady-state momentum balance for the fluid in three dimensions is given by the Navier–Stokes equation:

$$\nabla \cdot (\rho \vec{u}\, \vec{u}) = -\nabla p + \nabla \cdot (\bar{\bar{\tau}}) + \rho \vec{g} + \vec{F} \tag{7}$$

In the above equation, p is the static pressure, and the term $\rho \vec{g}$ represents the gravitational body force. The last term $\vec{F}$ is used for external body forces acting on the fluid and in this study was zero. The stress tensor $\bar{\bar{\tau}}$ is defined as:

$$\bar{\bar{\tau}} = \mu \left[(\nabla \vec{u} + \nabla \vec{u}^T) - \frac{2}{3} \nabla \cdot \vec{u}\, \bar{\bar{I}} \right] \tag{8}$$

where μ is the molecular viscosity, $\bar{\bar{I}}$ is the identity tensor and the second term on the right gives the effect of volume dilation due to fluid motion.

2.2. Chemical Species Transport

The chemical species mass fractions Y_i are given by the steady-state convection-diffusion equations:

$$\nabla \cdot \left(\rho \vec{u} Y_i \right) = -\nabla \cdot \vec{J}_i + R_i + S_i \tag{9}$$

The last two terms on the right-hand side R_i and S_i denote the rate of generation of species i by chemical reaction and the rate of addition of species i through user-defined sources, respectively. Both were zero in this study.

The term $\vec{J}_i$ accounts for the species diffusion flux that occurs from a concentration gradient and is as follows:

$$\vec{J}_i = -\rho D_{i,m} \nabla Y_i \tag{10}$$

where $D_{i,m}$ is the diffusion of species i in the mixture m. To obtain the diffusion coefficient, Fickian diffusion was assumed using the dilute approximation method.

2.3. Geometry and Discretization

Eight beds of different N were generated, each packed with spherical particles of 0.0254 m diameter. The 3D random beds of spheres were produced using a modified soft-sphere collective rearrangement Monte Carlo algorithm [34]. The original algorithm was used along with a method to arrange the layer of spheres around the tube wall more realistically at the bottom of the tube [35]. Each bed consisted of three sections. The first was an empty section spanning 0.0762 m ($3d_p$) from the entrance of the bed, denoted as the "calming section." The following section was the packed section, whose length is specified in Table 1. The final section was another empty section, 0.254 m ($10d_p$) in length, used to mitigate backflow effects near the exit of the bed. The bulk void fractions in Table 1 were calculated using the center section of the bed, from $5d_p$ above the inlet to $3d_p$ below the outlet, to exclude end effects, except for $N = 5.04$, where a shorter length was used to avoid packing defects.

Table 1. Fixed bed configurations.

N	L/d_p	No. Spheres	Bed Voidage
5.04	50.13	1000	0.442
5.45	40.12	1000	0.430
5.96	36.36	1080	0.429
6.40	31.97	1113	0.427
7.04	29.18	1200	0.431
7.44	26.63	1250	0.418
7.99	23.15	1250	0.414
9.30	17.01	1250	0.408

The model geometry is shown for three of the beds in Figure 1. These beds represent the smallest, largest and an intermediate value of N. Since only fluid transport mechanisms were to be simulated, only the fluid domain needed to be meshed. Tetrahedral cells characterized the fluid domain, and the control volume linear size was 0.00127 m, which corresponded to $d_p/20$. Boundary or prism layers extended from the particle and tube wall surfaces into the fluid and were 2.54×10^{-5} m thick. This gave mesh counts of 30.2–40.3 M cells depending on the value of N. Mesh refinement studies were carried out on a smaller test column with $N = 3.96$, using tetrahedral cells of sizes $d_p/16.7$ (0.00152 m), $d_p/25$ (0.001016 m) and $d_p/33.3$ (0.000763 m), which gave mesh counts from 4.2–30.0 M cells. Excellent agreement was observed for void fraction and interstitial velocity radial profiles for all mesh sizes, with average differences on the order of 1%–2%. Further mesh refinement tests were made using three prism layers on the tube surfaces, again with excellent agreement between profiles.

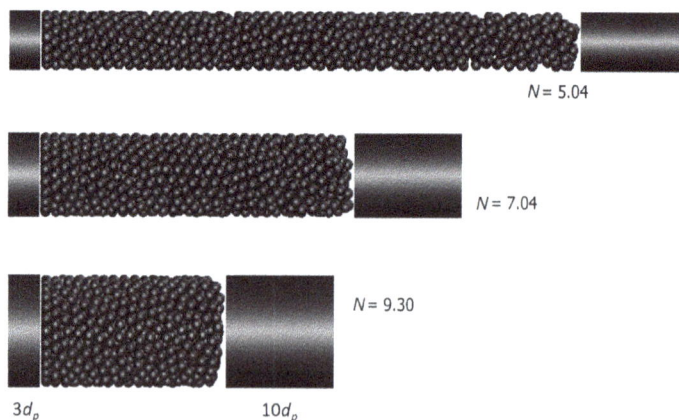

Figure 1. Cut-away views of three computer-generated beds: $N = 5.04$, $N = 7.04$ and $N = 9.3$ with the short unpacked inlet sections and the longer unpacked outlet sections.

The contact points between spheres and between the spheres and the tube wall give very narrow "fillets" of fluid, which cause problems in meshing. Various methods have been compared to deal with this difficulty [36], and a local modification using the "caps" contact point approach [37] was selected as being suitable for flow problems with no heat transfer. In our implementation of this method, a small cylinder of height 5.13×10^{-4} m was aligned with its axis along the vector connecting the particle centers and centered at the contact point between the particles, then subtracted from each particle and replaced by a small amount of fluid. At the particle-wall contacts, a similar procedure was followed; this cylinder was 6.15×10^{-4} m in height, slightly larger because of the opposite curvature of the tube wall. The bed void fraction change from this modification to the geometry was less than 0.5%. The modifications are shown schematically in Figure 2.

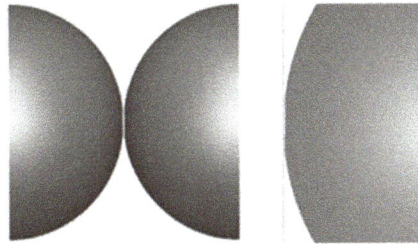

Figure 2. Treatment of contact points showing close-up views of caps approaches, for both particle-particle and particle-wall cases.

2.4. Boundary Conditions

To obtain velocity profiles in the bed, a uniform velocity boundary condition was set normal to the inlet of the tube, for each of the four particle Reynolds numbers used in this study, namely $Re = 87$, 348, 696 and 870. For these flow rates, direct numerical simulation was used. Air was specified as the fluid in the bed, with constant properties for isothermal conditions (i.e., $\mu = 1.7894 \times 10^{-5}$ kg/m/s, $\rho = 1.225$ kg/m^3). A constant pressure condition was specified at the exit of the tube as atmospheric pressure. No-slip boundary conditions were assigned to the tube walls and particle surfaces.

The focus of this work was to model dispersion and more specifically near-wall effects in low-N beds, by simulating methane diffusion from a coated tube wall into flowing air. To avoid unrepresentative concentration profiles caused by developing flow, the first $5d_p$ axial length of the bed was uncoated to allow flow to become established. The coated section was implemented by setting the methane species mass fraction as unity on the tube walls.

2.5. Computational Aspects for Computational Fluid Dynamics (CFD)

The semi-implicit pressure-linked equations (SIMPLE) method for pressure-velocity coupling was chosen [33]. The least squares cells approach was selected for the gradient method. For spatial discretization, the first-order upwind scheme was chosen for pressure, momentum and species for the first 200–300 solution iterations, to take advantage of the stability properties. After iterations stabilized, the discretization was switched to the more accurate second-order upwind for the remainder of the solution.

Solution convergence was attained by ensuring that column pressure drop remained constant over at least 1000 iterations. For species transport, a monitor was added to a 2-mm isosurface at the end of the packed section to track methane concentration on each iteration step. Once the concentration was no longer monotonically increasing, the solution was deemed converged. A complete solution for a single case required approximately 15,000–20,000 iterations, with solution times averaging 1000 iteration steps per 12 h. All simulations were completed on a Dell R620 PowerEdge Server (Dell, Round Rock, TX, USA) running a Windows server 2008 R2 (Microsoft, Redmond, WA, USA) operating system. The server contained 2 Intel Xeon E5-2680 CPUs (Intel, Santa Clara, CA, USA), each with 8 cores, and 128 GB of Random-access memory (RAM).

To obtain velocity and methane concentration profiles, cylindrical surfaces aligned axially with the bed were generated at various radial positions spanning the column diameter. The surfaces for axial velocity were clipped five particle diameters from the start of the bed packing and three particle diameters before the end of the packing, to prevent any end effects on the velocities. The resulting velocity profile was therefore averaged both axially and circumferentially. The surfaces for the methane concentrations were created similarly, with those surfaces clipped to 2 mm in height, at different axial positions, giving angularly-averaged methane concentration profiles across the diameter of the bed, at specified axial positions. Area-weighted averages were performed to collect both data sets for all runs.

2.6. Dispersion Models

To obtain dispersion coefficients based on the CFD 3D resolved-particle simulation results, 2D effective medium dispersion models based on Equation (1) were used and their mass transfer parameters fitted to the CFD-generated radial profiles of methane concentration. The finite element commercial code COMSOL Multiphysics® (COMSOL Inc., Stockholm, Sweden) was used. The two-dimensional domain was subdivided as shown in Figure 3 to reflect the corresponding sections in the 3D CFD model.

The radii of the beds simulated in COMSOL were the same as those in the CFD simulations. In Figure 3, the first two and the last sections of the bed represent those parts that were assigned a wall boundary condition of zero methane flux or zero methane concentration. These include the empty sections at the beginning and end of the tube. The "flux" section represents the portion of the 3D model that was assigned a non-zero wall mass fraction of methane. In the two-dimensional model, either a methane wall concentration or a non-zero wall methane flux condition is instead assigned, as detailed in the following subsections. The four horizontal internal boundaries were positioned at the z-values at which concentration radial profiles were extracted from the CFD simulations and were used as the dataset for fitting radial dispersion parameters. For each bed, the highest z-location was always three particle diameters below the top of the bed, to avoid exit effects, while sampling at the most developed profile location possible. The other three z-locations were originally approximately equidistant along the rest of the bed, but as results were analyzed, it was sometimes necessary to dispense with the lowest z-location and replace it with a location further down the bed, which corresponded to more developed concentration profiles.

The 2D domain was meshed with 32 layers of quadrilateral prisms on the tube wall surface, to capture the steep gradients there. The first layer thickness was 1/20 of the local element size, and a stretching factor of 1.3 was applied. The rest of the domain was meshed with triangular elements of sizes between 7.81×10^{-4} m and a maximum of 0.012 m. The coarse mesh sizes were used only in the bed center region, where the concentration profiles were relatively flat. Different mesh refinement settings were used within COMSOL® to check for mesh-independence, which led to the given mesh sizes.

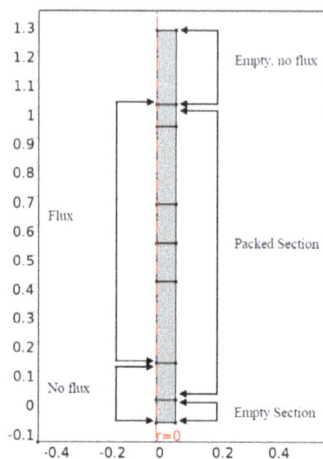

Figure 3. Two-dimensional representation of $N = 5.45$ packed bed for dispersion models.

Optimization studies were conducted for each of the 2D models, coupled to the transport models within the COMSOL Multiphysics® framework. The Levenberg-Marquardt least-squares optimization method was employed, which is commonly used for parameter fitting. The objective function was the sum of squares of deviations between the radial concentration profiles of methane, extracted from the CFD simulations at the four bed heights, and the corresponding methane radial concentration profiles from the 2D dispersion model under study. Each optimization was conducted for a given N and Re. The axial Péclet number remained constant at $Pe_a = 2$ in all runs. The study was complete when a minimization of the least-squares value of methane radial concentration deviations was achieved and an optimality tolerance (1×10^{-6}) was met.

2.6.1. Model 1: Constant D_T

In this model, we tried to reproduce the developing radial concentration profile with a constant transverse dispersion coefficient and a radially-varying interstitial velocity:

$$v_z(r)\frac{\partial c_A}{\partial z} = \frac{\bar{v}_z d_p}{Pe_L}\frac{\partial^2 c_A}{\partial z^2} + \frac{\bar{v}_z d_p}{Pe_T}\left(\frac{\partial^2 c_A}{\partial r^2} + \frac{1}{r}\frac{\partial c_A}{\partial r}\right) \tag{11}$$

and at the coated tube wall $r = R$:

$$c_A = c_W \tag{12}$$

Here, $Pe_L = \frac{\bar{v}_z d_p}{D_L} = 2$ and $Pe_T = \frac{\bar{v}_z d_p}{D_T}$. The only fitted parameter was Pe_T.

2.6.2. Model 2: Constant D_T and Wall k_m

In this model, we try to take account of the steep gradient near the tube wall by introducing a step change at the wall, governed by a wall mass transfer coefficient:

$$v_z(r)\frac{\partial c_A}{\partial z} = \frac{\bar{v}_z d_p}{Pe_L}\frac{\partial^2 c_A}{\partial z^2} + \frac{\bar{v}_z d_p}{Pe_T}\left(\frac{\partial^2 c_A}{\partial r^2} + \frac{1}{r}\frac{\partial c_A}{\partial r}\right) \tag{13}$$

and at the tube wall $r = R$:

$$-\frac{\partial c_A}{\partial r} = Bi_m \frac{1}{R}(c_A - c_W) \tag{14}$$

Here, $Pe_L = \frac{\bar{v}_z d_p}{D_L}$ (again equal to 2), $Pe_T = \frac{\bar{v}_z d_p}{D_T}$ and $Bi_m = \frac{k_m R}{D_T}$. The two fitted parameters were Pe_T and Bi_m.

2.6.3. Model 3: Radially-Varying D_T

In this model, we use a simplified model for the transverse dispersion coefficient that decreases sharply near the tube wall to take account of the strong decrease in D_T there, while being constant and neglecting any minor variations in the center of the packed bed:

$$v_z(r)\frac{\partial c_A}{\partial z} = \frac{\bar{v}_z d_p}{Pe_L}\frac{\partial^2 c_A}{\partial z^2} + \frac{1}{r}\frac{\partial}{\partial r}\left(r D_T(r)\frac{\partial c_A}{\partial r}\right) \tag{15}$$

and at the tube wall $r = R$:

$$c_A = c_W \tag{16}$$

Here, $Pe_L = \frac{\bar{v}_z d_p}{D_L} = 2$, and the simplified transverse dispersion coefficient is given by:

$$D_T(r) = D_m + \frac{\bar{v}_z d_p}{Pe_T}f(R - r) \tag{17}$$

where:

$$f(R-r) = \begin{cases} \left[\frac{(R-r)}{\hat{\delta} d_p} \right]^{n_m} & 0 < R - r \leq \hat{\delta} d_p \\ 1 & \hat{\delta} d_p < R - r \leq R \end{cases} \tag{18}$$

which is based on a mixing length model [31] and re-written following Winterberg et al. [32] with some differences. The three fitted parameters were Pe_T, δ and n_m.

The original version of the model [32] was developed to include thermal conductivity and was therefore based on superficial velocity in both the differential equation and the equations for radial variation in the parameters, which used the bed-center value obtained by numerical solution of the Brinkman–Forchheimer–Darcy (BFD) differential equation. In our implementation of the model, the interstitial velocity is used for consistency with Model 1 and Model 2, and the bed center velocity is replaced by the bed-average interstitial velocity, obtained from the CFD velocity profiles, to avoid having to solve the BFD equation. Parameter δ gives the thickness of the region next to the tube wall (in d_p) over which the dispersion coefficient rapidly decreases; n_m controls the shape of the decreasing function; while Pe_T represents the constant bed-center transverse dispersion.

3. Results and Discussion

Validation of the 3D CFD model was made against available experimental data in the literature. Figure 4a shows an example comparing void fractions from the CFD model as a function of radial coordinate to data from Giese et al. [38], for the case $N = 9.30$. The computed void fractions agree very well with the experimental ones, especially near the tube wall where the strong decrease occurs. The oscillatory nature of the void fraction in a bed of spheres is also well captured, with the locations and magnitudes of the peaks and troughs in the profile in good agreement. Only at the bed center, where the sampling isosurfaces become small, some differences were seen, which is typical in fixed beds. Figure 4b shows comparisons of the normalized superficial velocity profile obtained in [38] by laser-Doppler anemometry on a single plane to the bed-average normalized superficial velocities from CFD obtained by multiplying the interstitial velocity by the void fraction at each radial position and dividing by the inlet velocity. Detailed agreement is less good than for the bed structure, but the major features are reproduced to an acceptable degree. The sharp peak near the wall is predicted by CFD close to that seen in the experimental data, while the locations of maxima and minima in the velocity are in good agreement. Magnitudes of the peaks are somewhat under-predicted, and some deviation is seen at the bed center, which is usual in such studies. The differences between CFD and experiment could be attributed to the sampling of a single plane in the experiments versus averaged velocities over most of the packed bed length in the CFD simulations.

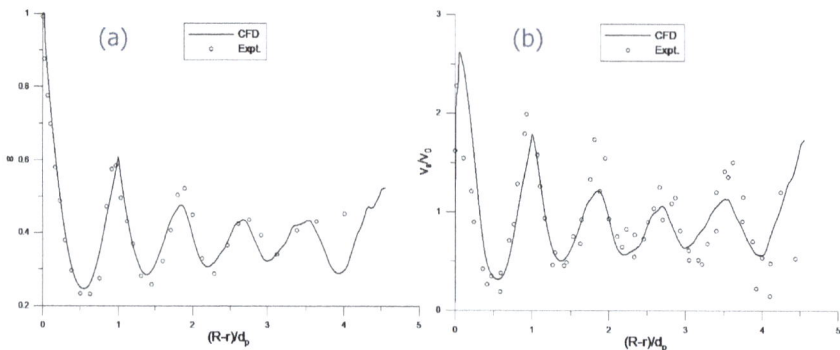

Figure 4. Validation of Computational Fluid Dynamics (CFD) simulations against data from [38] for (a) bed void fraction for $N = 9.30$; and (b) normalized axial superficial velocity profile for $Re = 532$ and $N = 9.30$.

3.1. 3D Computational Fluid Dynamics (CFD) Model

The images below in Figure 5 show contour plots of the concentration of methane taken from a center plane of the two beds, $N = 5.04$ and $N = 9.30$, for two selected values of particle Reynolds number Re, to illustrate the developing concentration fields obtained from the CFD simulations.

In the longer, narrower $N = 5.04$ tube, the full development of the concentration field can be seen. The local irregular penetration of the methane into the bed is evident, caused by the motion of the fluid between the wall particles. Regions of lower concentration near the wall can be seen caused by missing wall particles, which allow greater access of the fluid to the wall region. On average, however, a regular development is clearly seen. For the $N = 9.30$ tube, which due to computational constraints was necessarily shorter, the methane has only developed 2–3 particle diameters into the bed, and a large core of fluid free of the wall species persists the entire length of the bed. On exiting the bed, the fluid flow pattern exhibits regions of re-circulation and lateral motion, which are reflected in the irregular regions of high and low concentration.

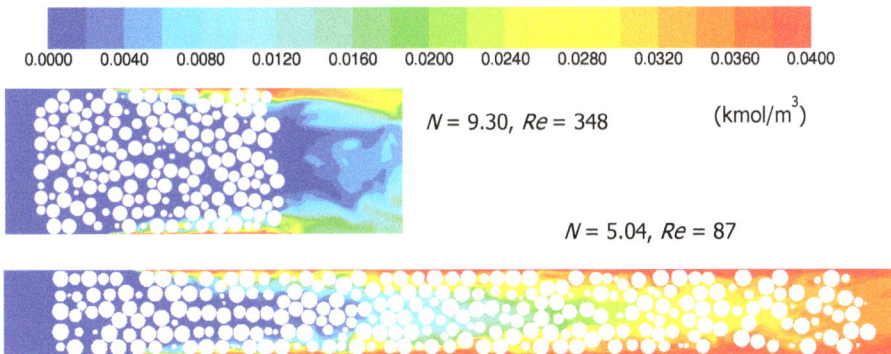

Figure 5. Full bed methane molar concentration (kmol/m³) contour plots on a slice through the midplane of two particle beds, for $N = 5.04$ and $N = 9.30$. The direction of flow for the two views is left to right.

Figure 6 displays the velocity profiles of three selected beds at all particle Reynolds numbers. Each plot contains axial interstitial velocity normalized by the superficial velocity set at the bed entrance. With each bed diameter, the velocity at the lowest Reynolds number ($Re = 87$) peaks higher than for the other three velocity profiles, which are essentially coincident. This is due to viscous forces in the flow becoming more significant compared to inertial forces at low Re. The velocities approach zero towards the tube wall, owing to the no-slip boundary condition. This low-velocity region is the reason for the separation of the diffusion-dominated wall boundary layer from the dispersion-dominated flow in the rest of the bed. From a comparison of the void fraction profiles, higher axial velocities appear in areas of highest radial void fraction (i.e., low velocities in areas of low bed porosity, except in the near-wall vicinity). Near the wall, when the void fraction is close to unity, a channeling effect occurs, in which flow becomes axial. In this region, resistance to radial mixing is strong, giving a decrease in transverse dispersion.

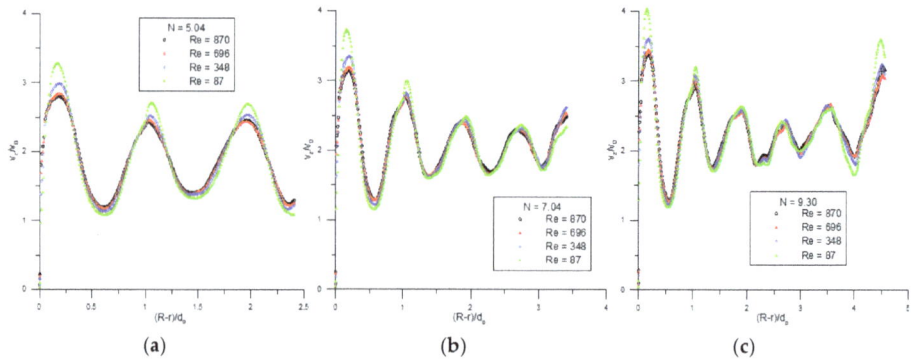

Figure 6. Comparison of all four interstitial velocity profiles for each of three values of (**a**) *N*: 5.04, (**b**) *N*: 7.04 and (**c**) *N*: 9.30.

3.2. One-Parameter Dispersion Model

Typical results for fits of the one-parameter model to single bed depth data are shown in Figure 7, for two values of *N* and *Re*. All combinations of *N* and *Re* were fitted and gave the same qualitative results. The one-parameter model is clearly unable to describe the developing concentration profiles, although literature results have indicated that it is sufficient for well-developed profiles. In this model, the concentration is forced to take the value c_w at the tube wall. This constrains the shape of the profile, which can reproduce neither the flatter profile region in the bed center, nor the steep increase near the tube wall. It was also found that there was a strong dependence of Pe_T on bed depth, *L*, and poor fits were obtained if the model was fit to all four bed depths simultaneously.

The dependence of Pe_T (and hence, D_T) on *L* can be understood by considering that in fact the dispersion must vary with radial position, due to the change in void fraction and velocity close to the tube wall. The use of a constant D_T means that this variation is being represented by an average value, and the value of this average depends on which parts of the radial concentration profile contribute most heavily to the least-squares objective function. That is, a profile that is just starting to develop will be influenced more by the lower values of D_T near the wall, than will a profile that is developed across the whole radius, which will be influenced also by higher dispersion values in the bed center.

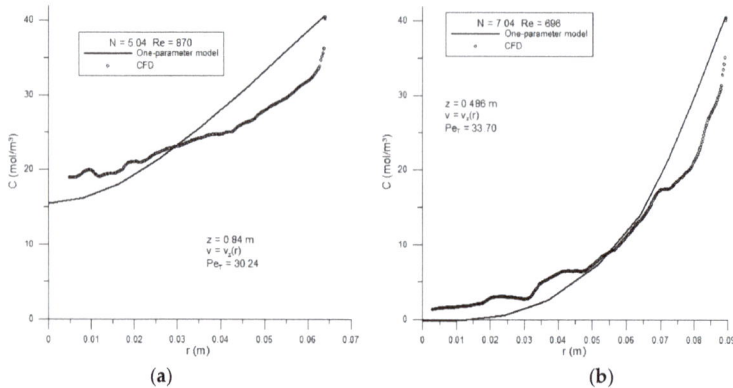

Figure 7. Representative comparisons of fits of the 1-parameter model using constant D_T to simulated data from Validation of Computational Fluid Dynamics (CFD), for (**a**) *N* = 5.04 and *Re* = 870 and for (**b**) *N* = 7.04 and *Re* = 696.

3.3. Two-Parameter Dispersion Model

From the one-parameter model results, the next idea is that since most of the change in D_T is anticipated to take place very close to the tube wall, it may be possible to account for it by introducing a wall mass transfer coefficient k_m and retaining the constant D_T. This essentially idealizes the extra resistance to mass transfer in the region of the tube wall, to a resistance located at the tube wall. Typical results are shown in Figure 8. In general, the fitted model predicts methane concentration away from the tube wall with both good qualitative and quantitative agreement. The model is unable to reproduce the increase in concentration in the region of the tube wall.

The wall concentration is better predicted for higher N, lower Re and lower bed depth combinations, as these correspond to situations in which the difference between the near-wall region and the bed center is reduced. In beds of lower N and higher Re and higher bed depth (e.g., $N = 5.04$, $Re = 870$), small gradients in methane concentration characterize the radial concentration profiles. In these cases, enough methane has dispersed laterally from the wall and been transported axially up the bed that a smaller drop in concentration from the wall value is seen at the longer bed depth.

The fitted values of Pe_T were somewhat higher than would be expected from literature studies of fully-developed profiles without wall effects [3,4]. This observation corresponds to lower values of the transverse dispersion coefficient. The values of Bi_m for the wall mass transfer coefficient were also quite high, which may be due to the same low D_T. A possible explanation is that the methane dispersion is controlled by slow diffusion through the wall boundary layer and then dispersed at a reduced rate of convective-dispersion at the interface between the diffusive-dispersion and convective-dispersion layers. Previous literature has shown also that 2D models predict higher dispersion coefficients to their 3D counterparts due to higher intercellular fluid motion in the lateral direction [30]. These coupled effects may serve as an explanation for the lower than expected dispersion coefficient values found in this research.

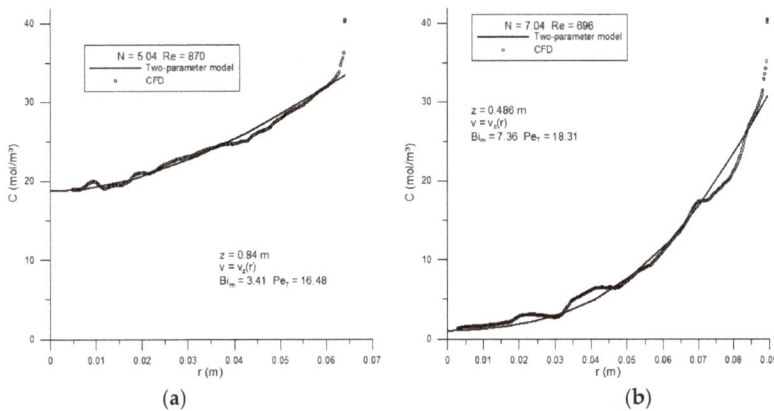

Figure 8. Representative comparisons of fits of the 2-parameter model using constant D_T and k_m to simulated data from Validation of Computational Fluid Dynamics (CFD), for (**a**) $N = 5.04$ and $Re = 870$, and for (**b**) $N = 7.04$ and $Re = 696$.

It appears that the mechanism of dispersion should be quantified by at least three types: slow, diffusive-dispersion in the tube wall boundary layer; then, an increasing rate of dispersion at the diffusive-dispersion and convective-dispersion interface and toward the tube center and, finally, a fully convective dispersion at the tube center. The use of the two-parameter model shown here reduces this mechanism to two parts: that of constant radial dispersion from the tube center and up to the wall, at which point a jump in concentration is implemented by the mass transfer coefficient. This limits the physical realization of the various dispersion mechanisms in the bed.

A separate study was conducted using $N = 5.04$, 6.40 and 9.3 in which the velocity profile was taken constant. The dispersion coefficients predicted with a constant velocity profile were 10%–20% higher than those obtained with a variable velocity. The unidirectional, constant velocity profile includes a nonzero velocity up to the wall, so that the axial convection of mass is higher in the near-wall region, which would reduce transverse mass transfer. To compensate for this, the transverse dispersion coefficient is increased in the fitting process, which would result in lower Pe_T values. The combination of fitting to developing profiles rather than fully-developed ones and using a radially-varying velocity profile rather than plug flow may account for the decreased transverse dispersion coefficients found for this model.

3.4. Three-Parameter Dispersion Model

The results from the two-parameter model suggested that more detail was required to represent the effects of the tube wall on D_T. In the three-parameter model, D_T varies with the radial coordinate controlled by three parameters, Pe_T, δ and n_m. A typical profile of $D_T(r)$ is shown in Figure 9.

The general form of the $D_T(r)$ profile is a constant through most of the bed, followed by a steep decrease in the vicinity of the wall, so that it may be termed a wall-function model. The magnitude of the bed center dispersion coefficient value is controlled by Pe_T, which changes in inverse proportion to D_T. The width of the region over which D_T decreases is given by δ as a fraction of a particle diameter, so that in the example provided in Figure 9, the near-wall region is of thickness $0.32d_p$. The form of the decrease is governed by the parameter n_m, which results in a convex function as shown for values less than unity and a concave function for values greater than unity. The convex function gives a decrease, which becomes steeper as the radial coordinate approaches the tube wall, as opposed to the concave function, which is steepest at the initial point of transition from a constant to a variable D_T. The profiles in the present study were always represented better with the convex shape with $n_m < 1$, a different result than that found for heat transfer in [32], where $n_m \approx 2$.

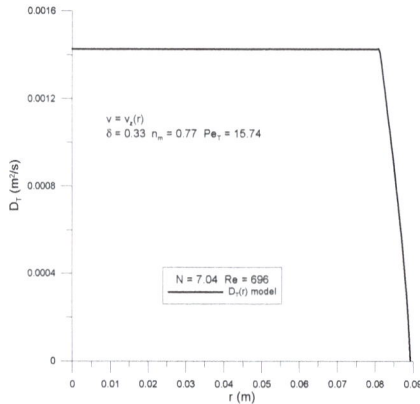

Figure 9. Typical form of $D_T(r)$, shown for the case $N = 7.04$ and $Re = 696$, with parameters $\delta = 0.33$, $n_m = 0.77$ and $Pe_T = 15.74$.

The results of fitting the radial methane concentration profiles one depth at a time using the three-parameter model gave a great improvement, as shown in Figure 10 for the same two cases that were presented previously in Figures 7 and 8, for ease of comparison. The fitted profiles are able to follow both the flat regions in the bed center, as well as the steep gradients very close to the tube wall. Again, similar results were seen for all eight values of N, four values of Re and four bed depths. The only aspects of the profiles that were not reproduced were the slight oscillations, familiar in heat transfer studies, which measure radial temperature profiles and which are attributable to the discrete nature of the ordered layers of the spheres adjacent to the wall. To reproduce these oscillations or "humps" would require a function for the radial dispersion parameter that mirrored the particle level changes. This is a level of detail that may not be worth pursuing. The estimates of all three parameters did change, often quite drastically, with bed depth, but no systematic trends in the changes could be found. The results of the individual bed depth fitting show that the fit was improved by the inclusion of a third parameter, which allowed the inclusion of a more nuanced treatment of the radial variation of the dispersion coefficient.

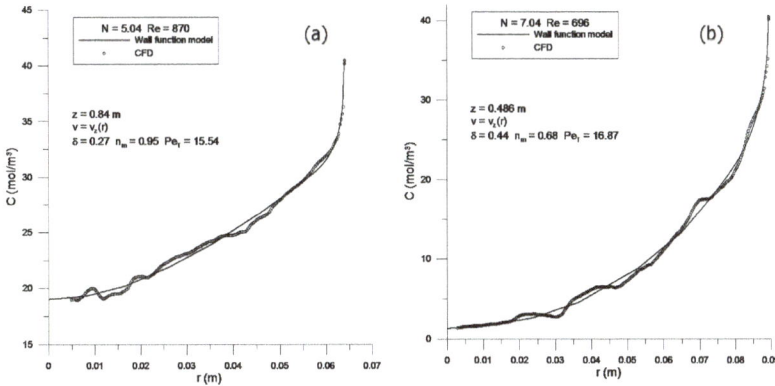

Figure 10. Results for three-parameter wall function model using $D_T(r)$: (**a**) $N = 5.04$, $Re = 870$; (**b**) $N = 7.04$, $Re = 696$. Single bed depth.

The next step was to see if the model could fit all bed depths simultaneously and whether depth-averaged parameter values would give acceptable fits. Results are presented in Figures 11–13. Figure 11 shows relatively well-developed methane radial concentration profiles for the $N = 5.04$ case. The methane has dispersed from the tube wall across the entire tube radius, so that the radial concentration gradients are relatively small, and at the longest bed depths, the profiles are quite flat. At all four values of Re, the model slightly over-predicts the concentrations in the bed center at the lowest bed depth and under-predicts the profile in the bed center at the highest bed depth. The two profiles at the middle depths are in good agreement, but even there, a slight trend can be seen suggesting that the predicted profiles show lower spread with bed depth than the CFD-generated ones. This disagreement occurs only for the profiles for $N = 5.04$; even for $N = 5.45$ and $N = 5.96$, the axial spread of CFD profiles and fitted profiles is in good agreement (see Figures S1–S5 in the Supplementary Material). Near the tube wall, the profiles all come closer together, as the boundary condition forces $c_A \rightarrow c_w$ for all. As for the individual bed depths, the steeper near-wall gradients are well captured. The fits to the set of four bed depths are reasonable with a single set of parameters.

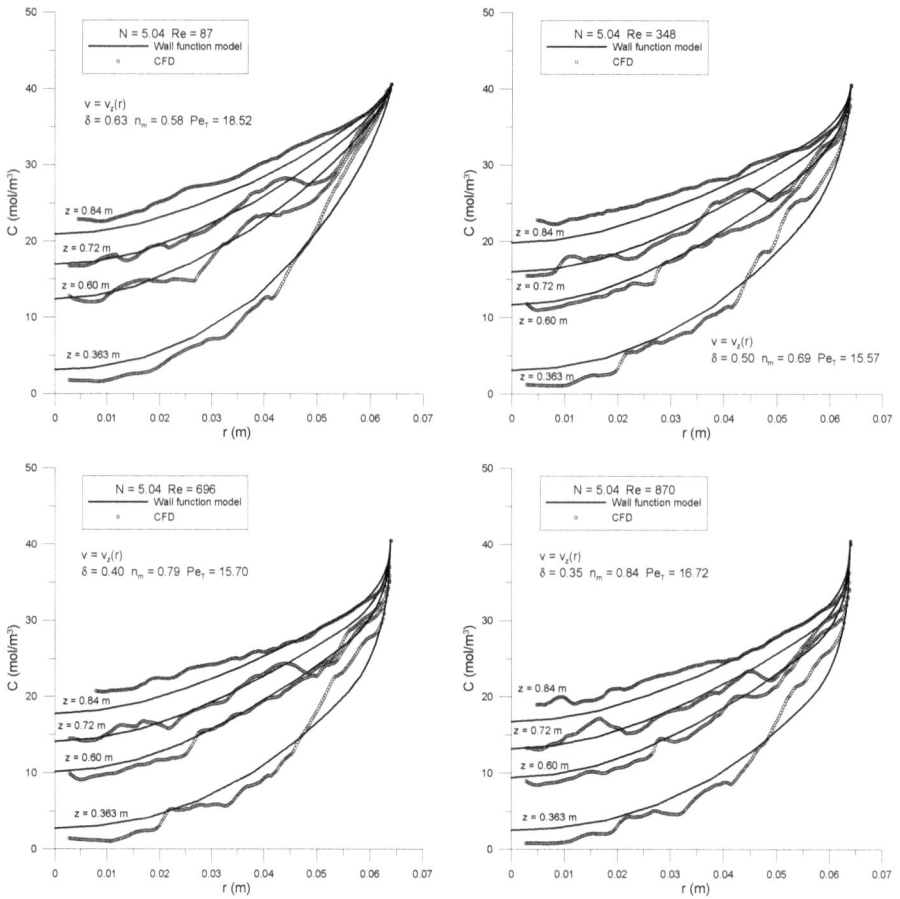

Figure 11. Results for three-parameter wall function model using $D_T(r)$ and $v_z(r)$ for the case $N = 5.04$, all four values of Re, each fitted to all four bed depths simultaneously.

Steeper radial concentration profiles are shown in Figure 12, for $N = 7.04$, in which the methane dispersing from the tube wall has only just reached the tube center at $r = 0$. The spread in the profiles in the center of the bed is much less than in the $N = 5.04$ case and is quite well captured by the model. Again, the profiles near the tube wall are close together, and the model does well in reproducing them. In particular, the increasingly steep gradient of concentration adjacent to the tube wall with increasing Re is followed, even with the fairly simple treatment of decreasing radial dispersion used in the model. In all cases, there is clearly considerable local variation in the shapes and details of each profile, but the overall trend, magnitudes and gradients of the profiles are fitted well by the three-parameter wall function model.

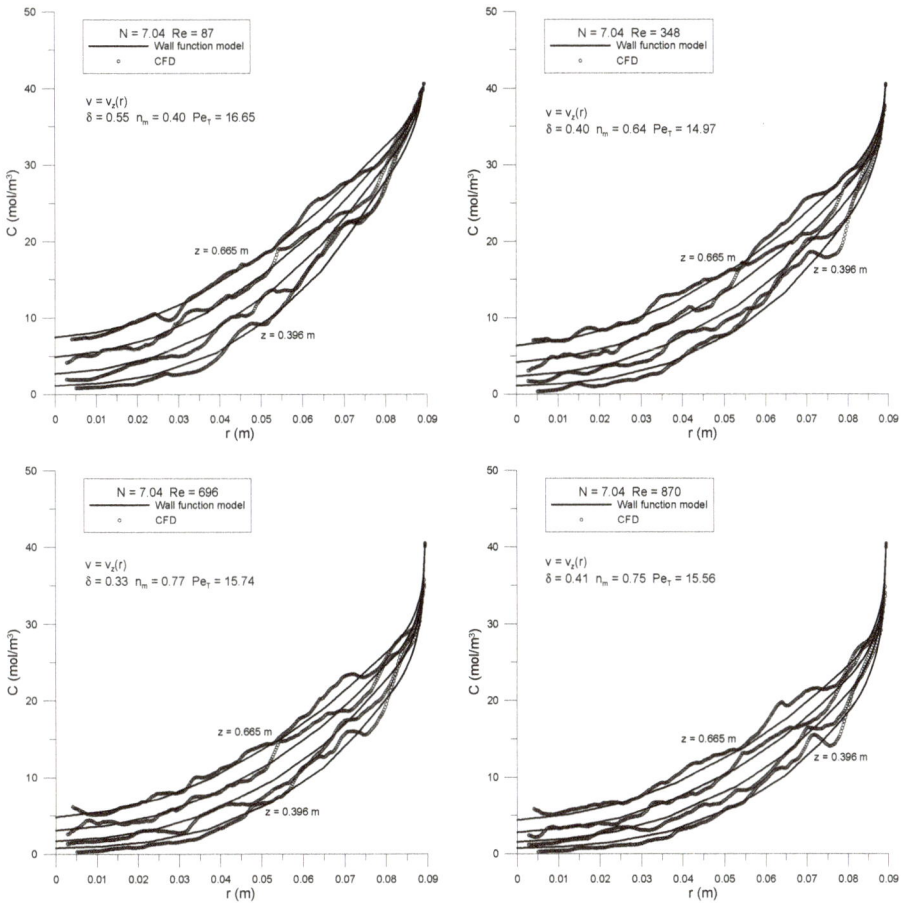

Figure 12. Results for three-parameter wall function model using $D_T(r)$ and $v_z(r)$ for the case $N = 7.04$, all four values of Re, each fitted to all four bed depths simultaneously.

In Figure 13, the fits for the non-developed profiles of the highest tube-to-particle diameter ratio $N = 9.30$ case are presented. The CFD-generated radial methane concentration profiles here are flat in the center of the tube, as the methane has not been able to disperse across the full tube radius. This is due to the wider tube, which increases the distance for material to disperse, and also the shorter tube length, which was constrained by the computational restrictions on the total number of particles that could be simulated. This shape of profile, which combines extremely sharp gradients near the tube wall and flat profiles with no gradients at all in the tube center, provides a severe test of the wall-function model's ability to reproduce a profile in the early stages of development. As can again be seen, the model is in excellent agreement with the CFD profiles over the entire radius, even as the gradients increase with higher Re. The spread of the four profiles with bed depth is less in this case than for the other two shown in Figures 11 and 12, and even that reduction is followed by the model.

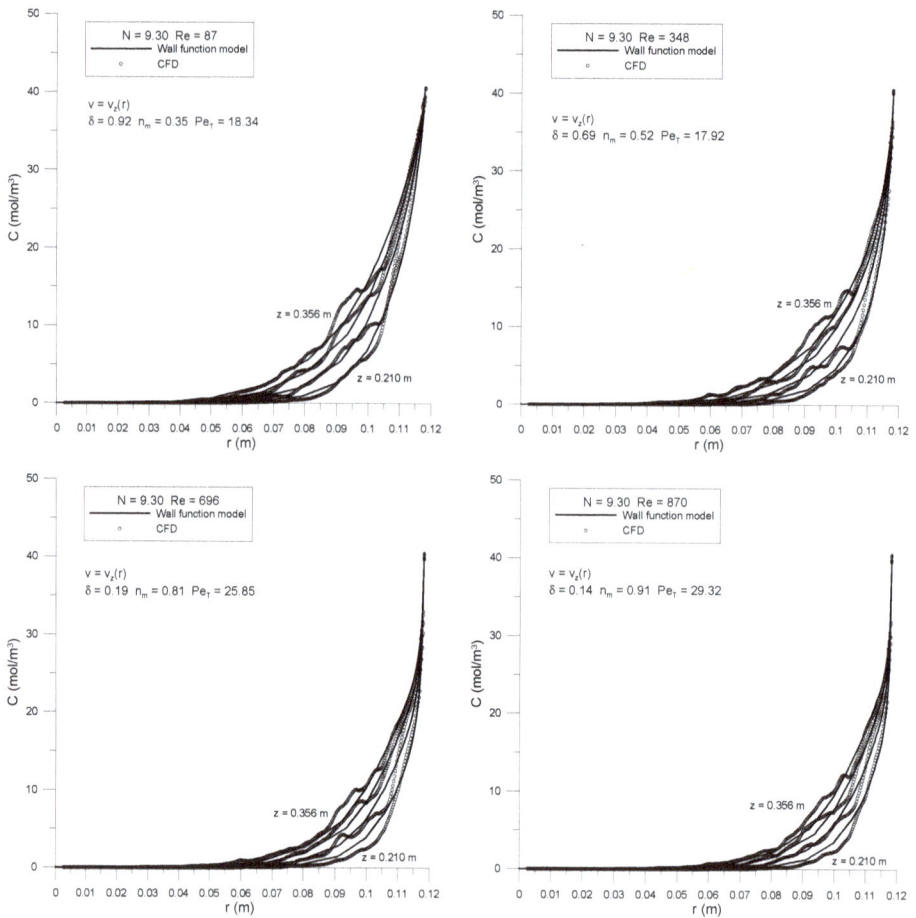

Figure 13. Results for three-parameter wall function model using $D_T(r)$ and $v_z(r)$ for the case $N = 9.30$, all four values of Re, each fitted to all four bed depths simultaneously.

The results shown for the three N values in Figures 11–13, as well as the five other N values presented in the Supplementary Material confirm that a model that incorporates the reduction in transverse dispersion over a region near the tube wall can fit a wide variety of profiles over the ranges of N and Re studied. This type of model improves on the representation of the complex radial variation of transverse dispersion by distributing a decrease over a finite region, instead of neglecting it or idealizing it to a point. Even a simple mixing-length approach is sufficient to accommodate both sharp and flat gradients, in profiles that range from ones starting to develop to others that are almost fully developed. The greatly improved ability of this three-parameter model to fit these profiles, compared to the one-parameter and two-parameter models, counter balances the disadvantage of having to estimate and correlate an extra parameter.

Having demonstrated that the three-parameter model can provide excellent agreement with CFD-generated developing radial concentration profiles, the next step is to determine whether reasonable correlations for the three parameters can be obtained. The parameters are presented in Figures 14–16 as functions of the main variables, N and Re. In each case, there is a reasonable amount of variation in the parameter, as well as one or two points that represent outliers from the main trends.

The runs corresponding to these outliers were re-examined for anomalies, but in each case, the results were confirmed. The fitted values of the parameters in these cases may be due to the randomly-generated bed structures, leading to the occasional anomalous feature in the concentration profiles.

From Figure 14, the parameter δ, which represents the width of the near-wall region scaled by particle diameter, can be seen to have a weak dependence on N and to decrease with increasing Re. These trends may be explained by considering that for any N, the arrangement of spheres next to the tube wall is similar, as has been suggested by radial void fraction profiles [37] and velocity profiles (Figure 6 above), which almost (with the exception of $Re = 87$) fall onto a single curve when properly scaled. As Re increases, also, the convective contribution to transverse dispersion will increase, and the diffusive layer next to the tube wall will be expected to decrease in thickness.

The shape of the decrease in D_T is represented by the parameter n_m, which is seen in Figure 15 to have a strong dependence on Re, increasing from approximately 0.2 up to nearly 1.0 over the range studied here. It is possible that at higher Re, the shape of the function may change and the value may approach 2.0, as found for heat transfer by Winterberg et al. [32]. The reason may be that at higher flow rates, the differences between the two approaches in use of the velocity profile may become less important. The parameter n_m may be taken as independent of N, as shown in Figure 15b.

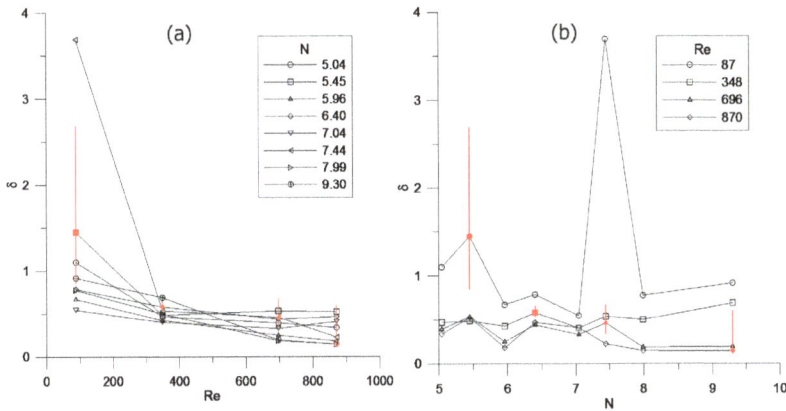

Figure 14. Variation of parameter δ against (**a**) Re for all values of N; and (**b**) N for all values of Re.

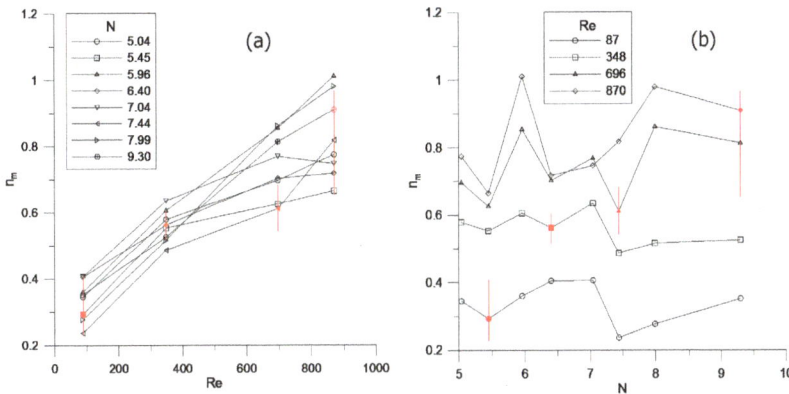

Figure 15. Variation of parameter n_m against (**a**) Re for all values of N; and (**b**) N for all values of Re.

Finally, the parameter Pe_T as shown in Figure 16 is independent of Re except that it is possibly a little higher for $Re = 87$. This is consistent with values of transverse dispersion for gases [1–4] in which the convective contribution dominates and the asymptotic value of Pe_T is reached for $Re > 100$–200, approximately. Pe_T depends only weakly on N, showing a slight increase at higher N values. The range for Pe_T is 16–20, which is again a little higher than for the cases in which only fully-developed profiles are considered, without wall effects.

For a chosen set of runs, the variance inherent in the CFD data (no replicates in computer simulation) was estimated by forming the sum of squares of the deviations of the CFD points around the fitted curve and dividing by the number of degrees of freedom. Two alternative cases were made, one by adding the square root of this variance (or a sample estimate of standard deviation) to each point of the original CFD data and one by subtracting the standard deviation from each point. The model was then re-fitted to each of the two new datasets, and the range of parameter estimates obtained was taken as an indication of the uncertainty in the parameter estimates. These ranges were then plotted in Figures 13–16 as red, vertical lines, and the original CFD data points were also plotted in red so that they would be clearly visible. Note that these "confidence intervals" are not centered on the original data, which is normal for nonlinear parameter estimation. The uncertainty ranges depend on both N and Re and can vary considerably in magnitude. Despite that, the observed trends in the parameters are seen to be significant even in comparison to the uncertainties involved.

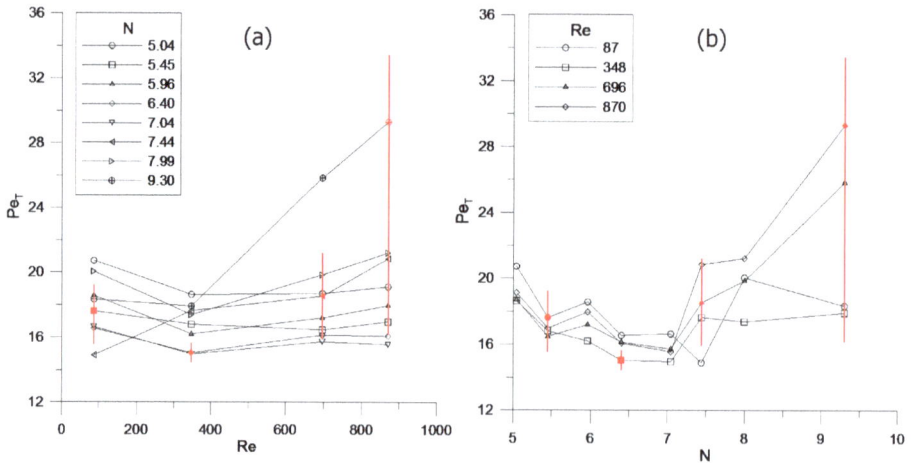

Figure 16. Variation of parameter Pe_T against (**a**) Re for all values of N; and (**b**) N for all values of Re.

4. Conclusions

Literature studies [1–4] have shown that in the absence of wall effects, i.e., for well-developed concentration profiles, radial dispersion in a packed tube can be described satisfactorily through the use of a single transverse dispersion coefficient. The present study has shown that to describe the development of radial concentration profiles in the presence of wall effects, neither a one-parameter model, nor a two-parameter model that incorporates a wall mass transfer coefficient are adequate. It is necessary to allow for a region of primarily diffusive mass transfer adjacent to the tube wall, a transition region where both diffusive and convective transfer are important, and the bed center region, which is dominated by convective dispersion. Our work has shown that a three-parameter model based on a mixing length concept [31,32] in which the transverse dispersion coefficient varies with radial position can represent the developing profiles even with a relatively simple functional dependence on tube radius. Further work should investigate the development of an equation for the velocity profile in the bed that can be used in engineering (as opposed to CFD) models and the provision of quantitative and reliable correlations for the three model parameters.

Supplementary Materials: The following are available online at http://www.mdpi.com/2311-5521/2/4/56/s1, Figure S1: Results for the three-parameter wall function model using $D_T(r)$ and $v_z(r)$ for the case $N = 5.45$; Figure S2: Results for the three-parameter wall function model using $D_T(r)$ and $v_z(r)$ for the case $N = 5.96$; Figure S3: Results for the three-parameter wall function model using $D_T(r)$ and $v_z(r)$ for the case $N = 6.40$; Figure S4: Results for the three-parameter wall function model using $D_T(r)$ and $v_z(r)$ for the case $N = 7.44$; Figure S5: Results for the three-parameter wall function model using $D_T(r)$ and $v_z(r)$ for the case $N = 7.99$.

Author Contributions: Anthony G. Dixon developed the initial fixed bed CFD model geometries and supervised Nicholas J. Medeiros, who developed the full CFD models and dispersion Models I and II, ran the simulations and performed analysis of the results. Anthony G. Dixon ran further simulations for Model III and performed the analysis in that section. Both authors contributed to the discussion of the results. The majority of the manuscript was based on the MS thesis of Nicholas J. Medeiros, which was re-written and extended for journal publication by Anthony G. Dixon.

Conflicts of Interest: The authors declare no conflict of interest.

References

1. Gunn, D.J. Theory of axial and radial dispersion in packed beds. *Trans. Inst. Chem. Eng. Chem. Eng.* **1969**, *47*, T351–T359.

2. Gunn, D.J. Axial and radial dispersion in fixed beds. *Chem. Eng. Sci.* **1987**, *42*, 363–373. [CrossRef]

3. Delgado, J.M.P.Q. A critical review of dispersion in packed beds. *Heat Mass Transf.* **2006**, *42*, 279–310. [CrossRef]

4. Delgado, J.M.P.Q. Longitudinal and transverse dispersion in porous media. *Chem. Eng. Res. Des.* **2007**, *85*, 1245–1252. [CrossRef]

5. Bernard, R.A.; Wilhelm, R.H. Turbulent diffusion in fixed beds of packed solids. *Chem. Eng. Prog.* **1950**, *46*, 233–244.

6. Fahien, R.W.; Smith, J.M. Mass transfer in packed beds. *AIChE J.* **1955**, *1*, 28–37. [CrossRef]

7. Hiby, J.W. Longitudinal and transverse mixing during single-phase flow through granular beds. *Interact. Fluid. Part.* **1962**, 312–325.

8. Gunn, D.J.; Pryce, C. Dispersion in packed beds. *Trans. Inst. Chem. Eng.* **1969**, *47*, T341–T350.

9. Han, N.-W.; Bhakta, J.; Carbonell, R. Longitudinal and lateral dispersion in packed beds: Effect of column length and particle size distribution. *AIChE J.* **1985**, *31*, 277–287. [CrossRef]

10. Schertz, W.W.; Bischoff, K.B. Thermal and material transport in nonisothermal packed beds. *AIChE J.* **1969**, *15*, 597–604. [CrossRef]

11. Foumeny, E.A.; Chowdhury, M.A.; McGreavy, C.; Castro, J.A.A. Estimation of dispersion coefficients in packed beds. *Chem. Eng. Technol.* **1992**, *15*, 168–181. [CrossRef]

12. Yagi, S.; Wakao, N. Heat and mass transfer from wall to fluid in packed beds. *AIChE J.* **1959**, *5*, 79–85. [CrossRef]

13. Kunii, D.; Suzuki, M. Heat and mass transfer from wall surface to packed beds. *J. Fac. Eng. Univ. Tokyo B* **1969**, *30*, 1–15.

14. Storck, A.; Coeuret, F. Mass transfer between a flowing liquid and a wall or an immersed surface in fixed and fluidised beds. *Chem. Eng. J.* **1980**, *20*, 149–156. [CrossRef]

15. Dixon, A.G.; DiCostanzo, M.A.; Soucy, B.A. Fluid-phase radial transport in packed beds of low tube-to-particle diameter ratio. *Int. J. Heat Mass Transf.* **1984**, *27*, 1701–1713. [CrossRef]

16. Colledge, R.A.; Paterson, W.R. Heat transfer at the wall of a packed bed: A j-factor analogy established. In Proceedings of the 11th Annual Research Meeting, Bath, UK, 9–10 April 1984; Collected Papers; Institution of Chemical Engineers: Rugby, UK, 1984; pp. 103–108.

17. Dixon, A.G.; LaBua, L.A. Wall-to-fluid coefficients for fixed bed heat and mass transfer. *Int. J. Heat Mass Transf.* **1985**, *28*, 879–881. [CrossRef]

18. Latifi, M.A.; Laurent, A.; Wild, G.; Storck, A. Wall-to-liquid mass transfer in a packed-bed reactor: Influence of Schmidt number. *Chem. Eng. Process Intensif.* **1994**, *33*, 189–192. [CrossRef]

19. Dixon, A.G.; Arias, J.; Willey, J. Wall-to-liquid mass transfer in fixed beds at low flow rates. *Chem. Eng. Sci.* **2003**, *58*, 1847–1857. [CrossRef]

20. Coelho, M.A.N.; Guedes de Carvalho, J.R.F. Transverse dispersion in granular beds. Part I Mass transfer from a wall and the dispersion coefficient in packed beds. *Chem. Eng. Res. Des.* **1988**, *66*, 165–177.

21. Guedes de Carvalho, J.R.F.; Delgado, J.M.P.Q. Lateral dispersion in liquid flow through packed beds at $Pe_m < 1400$. *AIChE J.* **2000**, *46*, 1089–1095.

22. Guedes de Carvalho, J.R.F.; Delgado, J.M.P.Q. Overall map and correlation of dispersion data for flow through granular packed beds. *Chem. Eng. Sci.* **2005**, *60*, 365–375. [CrossRef]

23. Schnitzlein, K. Modelling radial dispersion in terms of the local structure of packed beds. *Chem. Eng. Sci.* **2001**, *56*, 579–585. [CrossRef]

24. Schnitzlein, K. Modeling radial dispersion in terms of the local structure of packed beds. Part II: Discrete modeling approach. *Chem. Eng. Sci.* **2007**, *62*, 4944–4947.

25. Magnico, P. Hydrodynamic and transport properties of packed beds in small tube-to-sphere diameter ratio; pore scale simulation using an Eulerian and a Lagrangian approach. *Chem. Eng. Sci.* **2003**, *58*, 5005–5024. [CrossRef]

26. Freund, H.; Bauer, J.; Zeiser, T.; Emig, G. Detailed simulation of transport processes in fixed beds. *Ind. Eng. Chem. Res.* **2005**, *44*, 6423–6434. [CrossRef]

27. Soleymani, A.; Turunen, I.; Yousefi, H.; Bastani, D. Numerical investigations of fluid flow and lateral fluid dispersion in bounded granular beds in a cylindrical coordinates system. *Chem. Eng. Technol.* **2007**, *30*, 1369–1375. [CrossRef]

28. Jafari, A.; Zamankhan, P.; Mousavi, S.M.; Pietarinen, K. Modeling and CFD simulation of flow behavior and dispersivity through randomly packed bed reactors. *Chem. Eng. J.* **2008**, *144*, 476–482. [CrossRef]

29. Augier, F.; Idoux, F.; Delenne, J.Y. Numerical simulations of transfer and transport properties inside packed beds of spherical particles. *Chem. Eng. Sci.* **2010**, *65*, 1055–1064. [CrossRef]

30. Jourak, A.; Hellström, J.G.I.; Lundström, T.S.; Frishfelds, V. Numerical derivation of dispersion coefficients for flow through three-dimensional randomly packed beds of monodisperse spheres. *AIChE J.* **2014**, *60*, 749–761. [CrossRef]

31. Cheng, P.; Vortmeyer, D. Transverse thermal dispersion and wall channeling in a packed bed with forced convective flow. *Chem. Eng. Sci.* **1988**, *43*, 2523–2532. [CrossRef]

32. Winterberg, M.; Tsotsas, E.; Krischke, A.; Vortmeyer, D. A simple and coherent set of coefficients for modelling of heat and mass transport with and without chemical reaction in tubes filled with spheres. *Chem. Eng. Sci.* **2000**, *55*, 967–979. [CrossRef]

33. *Fluent UNS Version 14.5 User's Guide*; ANSYS, Inc.: Canonsburg, PA, USA, 2011.

34. Salvat, W.I.; Mariani, N.J.; Barreto, G.F.; Martinez, O.M. An algorithm to simulate packing structure in cylindrical containers. *Catal. Today* **2005**, *107–108*, 513–519. [CrossRef]

35. Mueller, G.E. Numerical simulation of packed beds with monosized spheres in cylindrical containers. *Powder Technol.* **1997**, *92*, 179–183. [CrossRef]

36. Dixon, A.G.; Nijemiesland, M.; Stitt, E.H. Systematic mesh development for 3D CFD simulation of fixed beds: Contact point study. *Comp. Chem. Eng.* **2013**, *48*, 135–153. [CrossRef]

37. Eppinger, T.; Seidler, K.; Kraume, M. DEM-CFD simulations of fixed bed reactors with small tube to particle diameter ratios. *Chem. Eng. J.* **2011**, *166*, 324–331. [CrossRef]

38. Giese, M.; Rottschafer, K.; Vortmeyer, D. Measured and modeled superficial flow profiles in packed beds with liquid flow. *AIChE J.* **1998**, *44*, 484–490. [CrossRef]

Article

Characterization of Bubble Size Distributions within a Bubble Column

Shahrouz Mohagheghian and Brian R. Elbing *

Mechanical and Aerospace Engineering, Oklahoma State University, Stillwater, OK 74078, USA;
mohaghe@okstate.edu
* Correspondence: elbing@okstate.edu; Tel.: +1-405-744-5897

Received: 15 December 2017; Accepted: 4 February 2018; Published: 7 February 2018

Abstract: The current study experimentally examines bubble size distribution (BSD) within a bubble column and the associated characteristic length scales. Air was injected into a column of water via a single injection tube. The column diameter (63–102 mm), injection tube diameter (0.8–1.6 mm) and superficial gas velocity (1.4–55 mm/s) were varied. Large samples (up to 54,000 bubbles) of bubble sizes measured via 2D imaging were used to produce probability density functions (PDFs). The PDFs were used to identify an alternative length scale termed the most frequent bubble size (d_{mf}) and defined as the peak in the PDF. This length scale as well as the traditional Sauter mean diameter were used to assess the sensitivity of the BSD to gas injection rate, injector tube diameter, injection tube angle and column diameter. The d_{mf} was relatively insensitive to most variation, which indicates these bubbles are produced by the turbulent wakes. In addition, the current work examines higher order statistics (standard deviation, skewness and kurtosis) and notes that there is evidence in support of using these statistics to quantify the influence of specific parameters on the flow-field as well as a potential indicator of regime transitions.

Keywords: bubble column; bubble size distribution; Sauter mean diameter; probability density function; skewness; kurtosis

---

1. Introduction

Bubble columns are frequently used as contact reactors in chemical processing, bio-chemical applications and metallurgical applications due to their simplicity (e.g., no moving parts), low operation cost and high efficiency at heat and mass transfer. Design and scale up of a bubble column relies on characterization of transport coefficients, which are sensitive to the bubble size and spatial distribution (local void fraction). Relative velocity between phases coupled with nonhomogeneous distributions has significantly limited the ability to apply laboratory insights to industrial applications. This is due in part to the fact that bubble size is frequently characterized with a single length scale (commonly the Sauter mean diameter, d_{32}), which fails to capture details of the size distribution. Thus, the current work aims to characterize the bubble size distribution (BSD) and its dependence on bubble column conditions via examination of the probability density function (PDF) and higher order statistics.

BSDs are heavily dependent on the operating regime, which determines dominant fluid mechanisms active within the multiphase system [1]. The current work does not aim to provide an analysis of characteristic length scales over a range of flow regimes, but rather focuses on relatively low volumetric injection fluxes to assess the sensitivity of the distribution to a range of parameters. However, it is hypothesized that the size distribution characteristics can provide a robust means of identifying regime transitions.

Use of a single length scale would be appropriate for characterizing the bubble size if the bubble size/shape is readily represented with a single length (e.g., spherical bubbles) and the shape of the size distribution was constant. Many researchers implicitly make this assumption without examining the higher order statistics, primarily due to the challenge of generating a sufficiently large sample size to accurately estimate the higher order statistics. Sauter mean diameter (d_{32}) is the most widely used characteristic length in bubble column studies (e.g., [2–5]). Sauter mean diameter,

$$d_{32} = \frac{\sum_{i=1}^{n} n_i d_i^3}{\sum_{i=1}^{n} n_i d_i^2} \tag{1}$$

is the ratio of the representative bubble volume to the bubble surface area, which is a weighted average. Here n_i is the number of bubbles with size d_i. Sauter mean diameter is frequently used when the sizes are acquired using optical photography techniques. Here the bubble cross sectional area (A_{proj}) is determined from the projected image. Then assuming that the bubbles are well approximated as ellipsoids (or more specifically an oblate spheroid), an equivalent bubble chord length,

$$d_b = \sqrt{\frac{4b A_{proj}}{\pi}} \tag{2}$$

can be computed. Here b is the ratio of the large diameter to the small diameter (i.e., aspect ratio). This equivalent bubble diameter is then used for d_i in Equation (1).

A common alternative to d_{32} is a probabilistic approach, which uses the mean of the PDF of the bubble chord length [5–9]. This method is most common when the measurements are acquired with electrical impedance/resistivity [10–12], wire mesh [13–15] or optical [16] point probes, which can only provide a single length scale but a relatively large sample size. These measurements are sensitive to the bubble size, velocity, shape and orientation as well as the sensor design (e.g., response from optical/impedance probes are unique to the sensor design and fluid properties). Consequently, these measurements are unable to provide details about the shape due to the required ad hoc assumptions to relate the signals to a bubble size. The current work uses bubble imaging of a large bubble population to produce PDFs that are not dependent on the assumption that the bubbles are spherical. These PDFs are then analyzed to identify an alternative length scale based on the peak in the PDF, which is then used along with the Sauter mean diameter to test sensitivity of the scales to operation conditions. In addition, higher order statistics from the PDFs are reported.

2. Experimental Methods

2.1. Test Facility and Air Injection

The bubble columns used in this work were originally designed as part of a vibrating bubble column facility that examined the influence of unsteady loading on bubble dynamics [17–19]. The current work utilizes the facility's bubble columns and a modified compressed air injection setup. Here we provide details of the components used in the current study and the interested reader is directed to Still [17] for additional details on the test facility and its design. The current work used two columns with inner diameters of 63 mm and 102 mm with corresponding lengths of 610 mm and 1220 mm, respectively.

The columns were fabricated from cast acrylic for optical access for bubble imaging. The ends of the columns were capped with flanges, which were used for mounting as well as a pressure seal. Both columns were filled with water and the ratio of the bubble column height (H) to the column diameter (D) was held constant at nine throughout testing. This aspect ratio was selected following the recommendations of Besagni et al. [20] to mitigate the effect of column height on the physical behavior of the system.

Fluid properties are known to affect the size distribution with bubble size decreasing with increasing liquid density [21] and decreasing surface tension [22]. Viscosity plays a more complex role [23] as illustrated with reports of increasing [24] and decreasing [25] bubble size with increasing viscosity. However, fluid properties had minimal variation in the current work with filtered air injected into filtered water at nearly constant temperature. The columns were filled with tap water that was passed through a cartridge filter (W10-BC, American Plumber, Pentair Residential Filtration, LLC, Brookfield, WI, USA) that had 5 μm nominal filtration. Surface tension of the filtered water supply was measured with a force tensiometer (K6, Krüss GmbH, Hamburg, Germany) and platinum ring (RI0111-28438, Krüss GmbH, Hamburg, Germany). The surface tension of the filtered water supply over several days was 72.6 ± 0.4 mN/m, which is comparable to the nominal surface tension of pure water (~72.8 mN/m). The column pressure for all data in the current study was at atmospheric pressure (plus hydrostatic pressure).

The injection method is known to impact BSDs [26]. The current study injected compressed air into the column via a single gas injector tube mounted near the column base as shown schematically in Figure 1a. After the stainless steel tube passed through the column wall, it was smoothly curved to produce either a 45° or 90° bend. The tube outlet was centered in the cylinder and pointed vertically upward. The injector tube had an inner diameter of 0.8 or 1.6 mm, which, for reference, produce initial bubble sizes of 3.4 and 4.3 mm, respectively, when surface tension dominates detachment [27]. Thus the dimensionless injector (sparger) opening parameter for the 0.8 and 1.6 mm injector tubes are 4.3 and 2.7 [28], respectively. The injection tube was supplied with compressed air from the building that was filtered to 5 μm (SGY-AIR9JH, Kobalt, Lowe's Companies, Inc., Mooresville, NC, USA). Immediately upstream of the injection tube was a check valve to prevent water from back filling the air supply line between test runs.

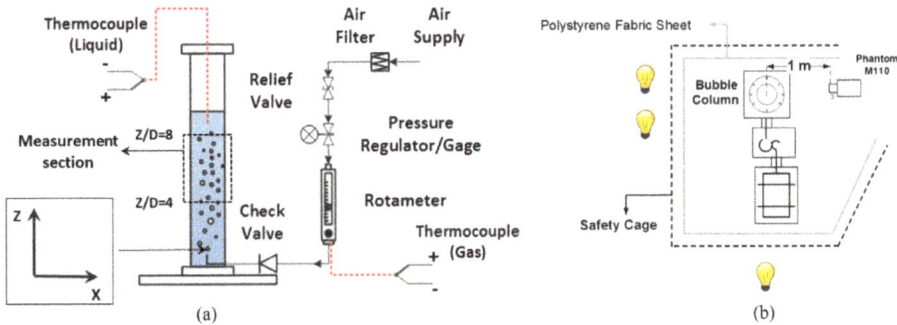

Figure 1. (**a**) Schematic of the experimental setup including the bubble column, gas injection system and instrumentation to monitor and control flowrates. (**b**) Top view of the column showing the camera and lighting configuration for bubble imaging.

The mass airflow rate into the bubble column was controlled and monitored with a combination of a rotameter (EW-32461-50, Cole-Parmer, Vernon Hills, IL, USA), thermocouple (5SC-TT-K-40-36, Omega Engineering, Norwalk, CT, USA) and pressure regulator (Spectra Gases, Inc., Stewartsville, NJ, USA). The rotameter had a manufacturer specified accuracy of 2% full scale and a measurement range of 0.4 to 5 liters per minute (lpm). The thermocouple measured the air temperature immediately upstream of the rotameter with an accuracy of ±0.1 °C. The pressure regulator was used to both measure the pressure at the rotameter as well as control the mass injection rate. The ideal gas law was used to convert the volumetric flowrate, air temperature and pressure to a mass airflow rate. Note that the air temperature difference between the regulator and the rotameter was ≤0.3 °C over all test conditions. The mass airflow rate was converted back to a local volumetric flow rate within the bubble column using hydrostatic pressure and the water temperature, which was measured with a separate thermocouple (HSTC-TT-K-20S-120-SMPW-CC, Omega). All tests were conducted with the air temperature between 20 °C and 22 °C, and temperature difference between the air and water was less than ±2 °C.

2.2. Bubble Size Measurement

The BSDs were determined from 2D optical imaging with a high-speed complementary metal–oxide–semiconductor (CMOS) camera (Phantom Miro 110, Vision Research, Wayne, NJ, USA), which has a resolution of 1280 × 800 pixels. The camera pixel size was 20 μm × 20 μm with a 12-bit depth. For the current work, the sample rate was 400 Hz with a reduced resolution of 1280 × 400 pixels, which the on-board memory (12 GB) allows ~38 s of recording with these settings. A 60 mm diameter, f/2.8D lens (AF-Micro-NIKKOR, Nikon Corporation, Tokyo, Japan) was used with the camera, which produced a field-of-view of 470 mm × 150 mm. The exposure time was 600 μs to provide maximum illumination without bubble blurring. The column was backlit with four 500 W halogen lights and twelve 45 W fluorescent lights. The light was uniformly diffused using several 2.3 m × 2.6 m solid white microfiber fabric sheets. Consistent and uniform backlighting simplifies image-processing and decreases uncertainty. The final lighting configuration (shown in Figure 1b) produced a homogenous light intensity distribution.

Imaging through a round cylindrical column produces optical distortions, especially near the column edges. A spatial calibration was performed with a particle-image-velocimetry (PIV) calibration target (Type 58-5, LaVision, Göttingen, Germany), which also quantified the impact of these distortions. Figure 2a illustrates use of the target to identify the distortions, and Figure 2b shows the spatial variation of the calibration coefficient for each column. Cropping the images at the lines shown (11 mm and 14 mm from the wall for the 63 and 102 mm columns, respectively) results in a maximum size variation of ±0.4 mm due to the calibration variation, which is below the minimum bubble size (1.6 mm). Since this variation is comparable to the variation associated with out-of-plane motion, an average mid-plane spatial calibration was used for the entire image.

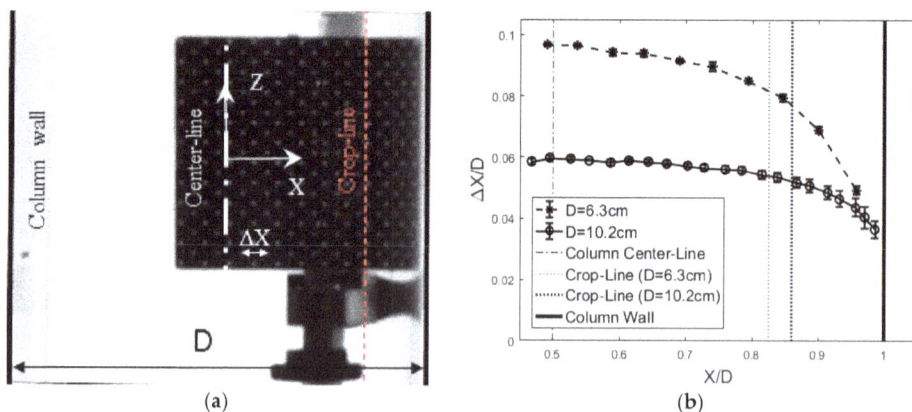

Figure 2. (a) Effect of column curvature on spatial calibration, ΔX = 5 mm, D = 102 mm. (b) Spatial variation of the calibration coefficient across the column mid-plane.

Bubble images were acquired with commercial data acquisition software (2.5.744.0v, Phantom Camera Control, Vision Research, Wayne, NJ, USA) and then post-processed using ImageJ (1.49v, National Institutes of Health (NIH), Bethesda, MD, USA) [29–31], a common open access image-processing program. Within ImageJ, an edge detection algorithm was used to sharpen the bubble edges, the background was subtracted and then a grayscale threshold was used to convert the 12-bit images to binary images. A subset of images from each condition were manually processed and then used to determine the appropriate grayscale threshold. Note that a range of acceptable threshold values were explored and had a 2% variation on measured bubble size. Figure 3 provides an example of a raw image with the identified bubbles using the appropriate threshold outlined. This illustrates that the processing algorithm can identify in-focus bubbles and exclude out-of-focus bubbles, which minimizes the impact of out-of-plane bubble locations on the spatial calibration. Note that for the current study in-focus bubbles were limited to ±12 mm of the focal plane. Figure 3 was also selected to show that, even with a proper threshold, overlapping and defective bubbles (e.g., defected bubble outlined at bottom left of Figure 3) can contaminate the size distributions. Consequently, each image was manually inspected for the aforementioned problems and impacted bubbles were removed from the population sample. These manual inspections were also used to confirm that the grayscale threshold was not impacted by changes in void fraction between conditions.

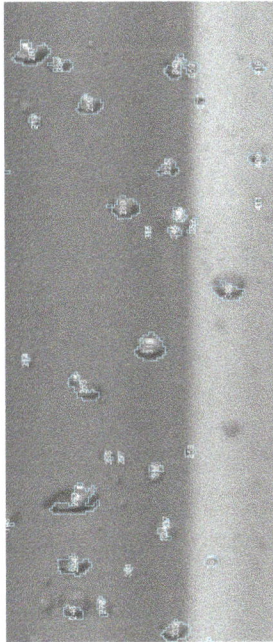

Figure 3. Example image of bubble identification (identified bubbles are outlined). Note that out-of-focus bubbles are not identified due to the blurred edges.

The cross-sectional area, bubble centroid location and the aspect ratio were then calculated for identified bubbles. Note that any deviation in orientation perpendicular to the visualization plane when the aspect ratio is greater than 1 (i.e., bubbles larger than ~2 mm) results in an overestimate of the bubble projected area. A high-pass filter with a cutoff area of $A_{proj} = 2$ mm^2 was used to remove noise contamination from BSD and consequently the PDFs. Given the cross-sectional area, a nominal bubble size was determined using Equation (2). Note that not every image was processed because the sample rate (400 Hz) did not produce a sufficient duration for a new bubble population in each image. Consequently, the period between processed images was increased such that each processed image contained a new bubble population to ensure statistically independent bubble samples.

2.3. Repeatability

A subset of tests were conducted to evaluate the repeatability of the experiment, which also provided insight into the target measurement location. Three air volumetric flowrates (Q_m) were selected that produced superficial gas velocities (volume averaged phase velocity; $U_{sg} = Q_c/A_{cs}$) of 6.9 mm/s, 27.6 mm/s and 55.1 mm/s. Under these conditions, the bubble column was operating within the poly-dispersed homogenous regime [18], which is true throughout the current study. Each condition was repeated at least ten times with a minimum of 3000 bubbles sampled per condition. Results from these tests are shown in Figure 4 with the Sauter mean diameter (d_{32}) plotted versus the vertical distance above the injection location (Z) scaled with the column diameter (D). Error bars represent the standard deviation of the mean for each condition. Similar to Akita and Yoshida [22], these results exhibit a decrease in d_{32} with increasing gas flux for locations sufficiently far from the injection location. Note that increasing superficial velocity is known to increase or decrease [32,33] bubble size due to its complex role modifying bubble formation processes and liquid circulation. Figure 4 also indicates that beyond $Z \sim 4D$ the bubble size remains constant within the measurement uncertainty. Consequently, the current work focuses on bubble measurements in the range of $4 < Z/D < 6$ to minimize the influence of the injection method. It is noteworthy that the minimum height above the injector will be sensitive to the injection condition, which will be discussed subsequently. Furthermore, inspection of the images within the target height range showed minimal influence of bubble breakup and/or coalescence.

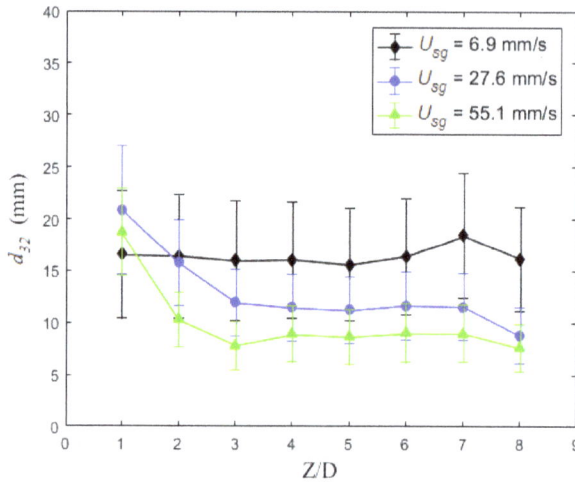

Figure 4. Sauter mean diameter (d_{32}) plotted versus the scaled vertical distance above the injection location. Each data point is the average of 10 repetitions, and the error bars are the standard deviation of the mean ($P_m = 600$ kPa, $T_c = 21 \pm 1\,°C$, $D = 102$ mm, $d_{inj} = 1.6$ mm).

3. Bubble Size Length Scales

While Sauter mean diameter (d_{32}) is widely used as the characteristic bubble length scale, bubble size distributions are often poly-dispersed, which makes a single length scale insufficient to characterize the distribution. Consequently, in the current work the PDF was examined to identify a length scale(s) that represents the size distribution. The PDFs generated from counting at least 10,000 bubbles per condition is provided in Figure 5a (PDF of conditions shown in Figure 4, though limited to $4 < Z/D < 6$). Here there is a noticeable shift between the PDF peaks and d_{32}. Consequently, the most frequent bubble size (d_{mf}) was defined as the size corresponding to the peak in the PDF (mode). These representative conditions illustrate the different behavior between d_{32} and d_{mf}, with d_{mf} being significantly smaller than d_{32} over the range tested. In addition, while there is a noticeable dependence between d_{32} and the volumetric gas flux, d_{mf} appears to have negligible variation. It is worth mentioning that the high-pass filter forces the left leg of PDFs to be zero when $A_{proj} < 2$ mm^2. For a spherical bubble ($b = 1$; aspect ratios are nominally one for smallest bubbles), this minimum area translates into a minimum bubble size of $d_b < 1.6$ mm.

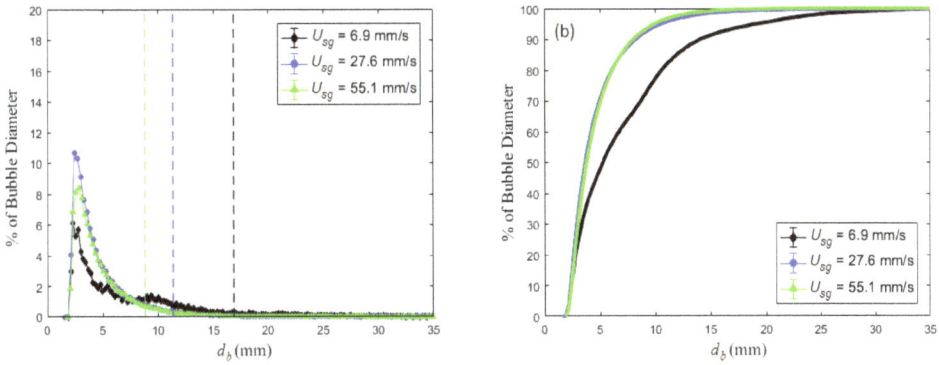

Figure 5. (**a**) Probability density functions (PDF) and (**b**) cumulative density function (CDF) of bubble size (d_b) for the same conditions shown in Figure 4. The PDF/CDF for each U_{sg} was determined from counting at least 10,000 bubbles. Dashed lines in (**a**) correspond to the d_{32} values for each condition ($P_m = 600$ kPa, $T_c = 21 \pm 1\,°C$, $D = 102$ mm, $d_{inj} = 1.6$ mm).

The obvious question is what accounts for the difference between d_{32} and d_{mf}. As seen in Equation (1), d_{32} is a weighted average that is biased towards the largest bubbles generated due to the diameters being raised to powers before summing. Consequently, the influence of a large quantity of small bubbles has a weaker impact on d_{32} than a few large bubbles. This can be seen in the cumulative density function (CDF) for these conditions provided in Figure 5b. The lowest flow rate exhibits significantly more large bubbles (e.g., 23% of bubbles are larger than 10 mm) than the highest injection flux (<5% of bubbles are larger than 10 mm), thus illustrating how these three conditions with nearly identical d_{mf} values generate measurable deviations in d_{32}.

A comprehensive examination of the variation between d_{32} and d_{mf} is provided in Figure 6 with the most frequent bubble size plotted versus Sauter mean diameter for all test conditions. For reference, a dashed line corresponding to $d_{mf} = d_{32}$ has been included, which shows that for all conditions d_{mf} is smaller than d_{32}. The majority of the data points collapse on a curve that appears to asymptote to $d_{mf} \approx 2$ mm. The uniformity of these bubbles and insensitivity to the injection condition suggests that they are being generated by the flow-field, which the most likely mechanism would be the turbulent motions generated by the bubble wakes. This would suggest that d_{mf} is a length scale associated with the velocity fluctuations within the flow-field. This conjecture is supported by the known Reynolds number dependence of bubble wakes. Bubble diameter (d_{mf}) based Reynolds numbers ($Re = V_b \cdot d_{mf}/\nu$, where V_b is the mean bubble rise velocity that is nominally U_{sg}/α, α is the void fraction and ν is the kinematic viscosity) tested ranged between 590 and 11,000. Starting at a Reynolds number of ~500, vortices begin to be shed from bubbles and the flow-field becomes quite unsteady until ~1000. Starting at Re ~1000, a boundary layer forms on the bubble with a laminar near-wake region. However, the shear layer spreads resulting in a turbulent far-wake region. This behavior exits until Re ~3×10^5, which is beyond the range of bubbles observed in the current study. Of note, a bimodal distribution is observed at lower Reynolds numbers ($590 < Re < 2300$), which is shown in Figure 7. This is a curious observation given that in this range the bubble wakes are unsteady with periodic shedding of vortex rings. The Strouhal number for Re ~1000 is ~0.3 [34], which the shedding from a 2.5 mm diameter bubble (nominal d_{mf} for conditions in Figure 7) would produce an 8.3 mm long wavelength. This is comparable to the size of the second peak in the distribution.

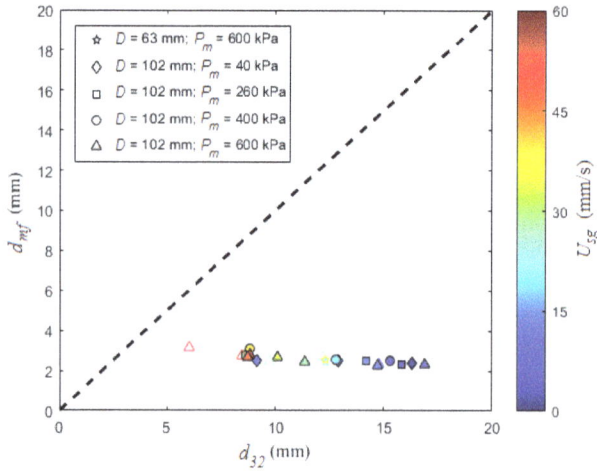

Figure 6. Comparison between the most frequent bubble size (d_{mf}) and the Sauter mean diameter (d_{32}). The dashed line corresponds to $d_{mf} = d_{32}$. Open and closed symbols correspond $d_{inj} = 0.8$ and 1.6 mm, respectively.

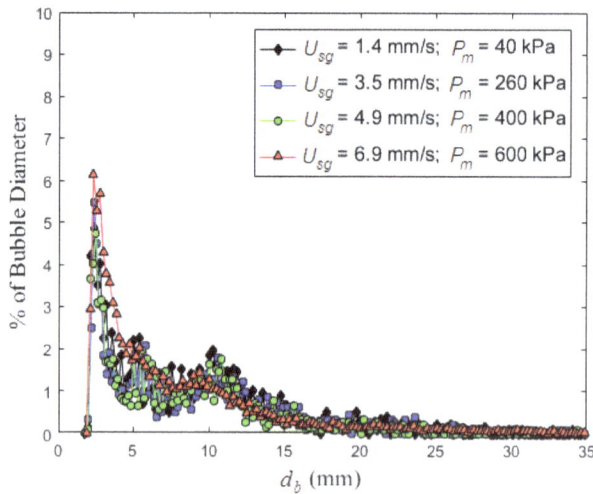

Figure 7. PDFs from bimodal conditions (U_{sg} = 1.4, 3.5, 4.9 and 6.9 mm/s). While the d_{mf} is still determined from the smaller bubbles, there is a second weaker peak near 10 mm (D = 102 mm; d_{inj} = 1.6 mm).

Assuming that the PDF shape changes are related to regime transitions, higher order statistics (i.e., standard deviation, skewness and kurtosis) for a subset of conditions are presented in the conclusions of the parametric study. The use of both d_{mf} and d_{32} are explored in more detail in the following section with a parametric study to assess the sensitivity to individual control parameters. Of note, over the conditions explored d_{mf} (mode of PDF) is similar to d_{10} (mean of PDF; defined using Equation (1) with the powers changed from 3.2 to 1.0). Given that the PDFs are skewed to larger bubbles, d_{10} is generally larger than d_{mf} and smaller than d_{32}. While the behaviors are similar, they carry distinctly different physical information. While not explored in the current study, if the Reynolds number based on bubble diameter decreased below ~500, it is expected that $d_{mf} > d_{10}$. This is contrary to the current work where $d_{mf} < d_{10}$ for all conditions.

4. Parametric Studies

4.1. Gas Injection Rate

The volumetric flowrate of gas within the column (Q_c) is determined from the mass flowrate into the column ($\dot{m}$), column pressure (P_c) and column temperature (T_c). In the current experiment, the column temperature and pressure were held nearly constant at $T_c = 21 \pm 1\,°C$ and atmospheric pressure (plus hydrostatic pressure), respectively. Consequently, the mass flow rate was the only parameter varied, which was controlled with a combination of meter pressure (P_m) and metered volumetric flow rate (Q_m). Figure 8 compares d_{32} and d_{mf} dependence on the superficial velocity (U_{sg}). Four different meter gauge pressures ($P_m = 40, 260, 400$ and 600 kPa) were used to achieve $1.4 \leq U_{sg} \leq 55$ mm/s. Sauter mean diameter shows good collapse over most of the test conditions, but there is some deviation observed with the $P_m = 40$ kPa condition. Conversely, d_{mf} collapses at lower superficial velocities but show some deviation at higher fluxes with $P_m = 600$ kPa.

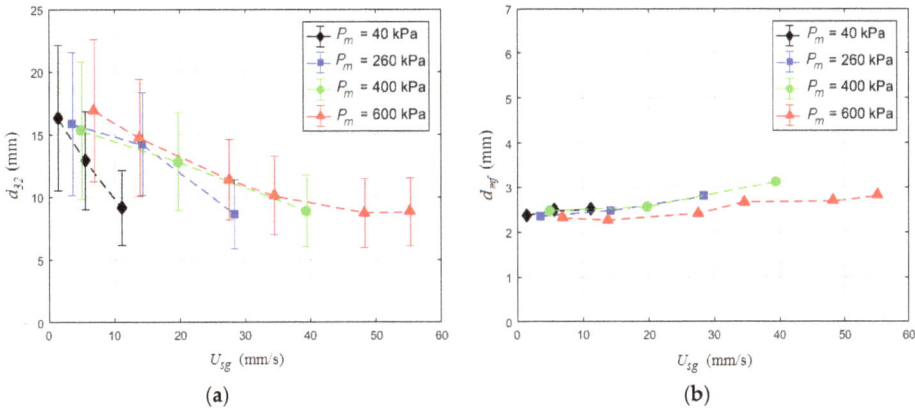

Figure 8. (a) Sauter mean diameter and (b) most frequent bubble size plotted versus the superficial gas velocity. Error bars represent the standard deviation for the given condition ($D = 102$ mm; $d_{inj} = 1.6$ mm; $T_c = 21 \pm 1\,°C$).

The only significant outlier condition from Figure 8a is the $P_m = 40$ kPa with $U_{sg} = 11.1$ mm/s condition. Images at the injection location compare this condition with other low mass flux conditions in Figure 9. Here it is apparent that the initial bubble size distribution is significantly different compared to the other low mass flux conditions. The Reynolds number based on the injector tube diameter for the outlier condition is 4800, which is at the transition between laminar and turbulent flow in a pipe. This makes the airflow at this superficial gas velocity transitional, which transitional flows are extremely sensitive to the operating condition. The data suggests that the lower metering pressure makes the initial bubble formation more sensitive to the inlet airflow condition. The metering pressure could impact bubble detachment from the injection tube since the upstream pressure could modify the bubble shape during expansion (especially with transitional flow). In addition, the initial bubble size distribution as well as breakup and coalescence behaviors are sensitive to the density of the gas [35].

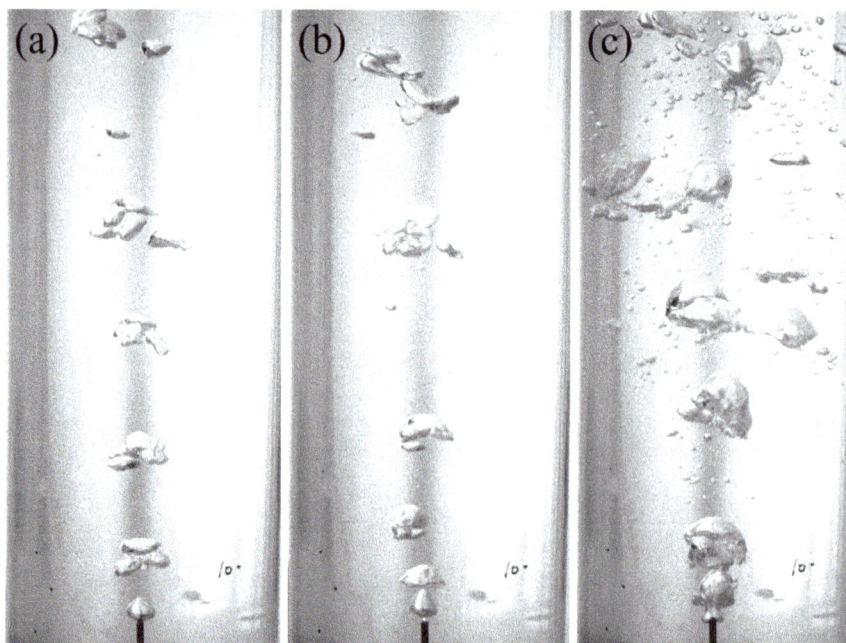

Figure 9. Still frames in the $D = 102$ mm column with $d_{inj} = 1.6$ mm with an injection condition of (**a**) $P_m = 260$ kPa, $U_{sg} = 3.5$ mm/s; (**b**) $P_m = 600$ kPa, $U_{sg} = 6.9$ mm/s and (**c**) $P_m = 40$ kPa, $U_{sg} = 11.1$ mm/s.

4.2. Injector Tube Angle

The experimental setup had the injector tube positioned such that it was pointed upward and aligned with gravity. However, the setup made fine adjustments to the injector tube orientation difficult once installed. Thus, testing was performed to assess the sensitivity of the BSD to injector orientation. Here, two different injector orientations were tested, 45° and 90° (vertical, design condition) measured from horizontal with $D = 102$ mm and $d_{inj} = 0.8$ mm. Results for both d_{mf} and d_{32} are provided in Figure 10 at each injector tube angle. These results show that d_{mf} has negligible variation even with the significant misalignment. Conversely, d_{32} has a measurable decrease at 45° relative to the 90° condition. There are two potential mechanisms responsible for this deviation; (i) the misalignment between gravity (buoyancy force) and the bubble wake where the turbulent production is located and/or (ii) increased influence of wall effects as the initial bubbles were directed into the column wall where the stress distribution will deviate from the core of the column. The wall effects are mostly likely for the current work since the decrease in bubble size suggests a higher shear stress.

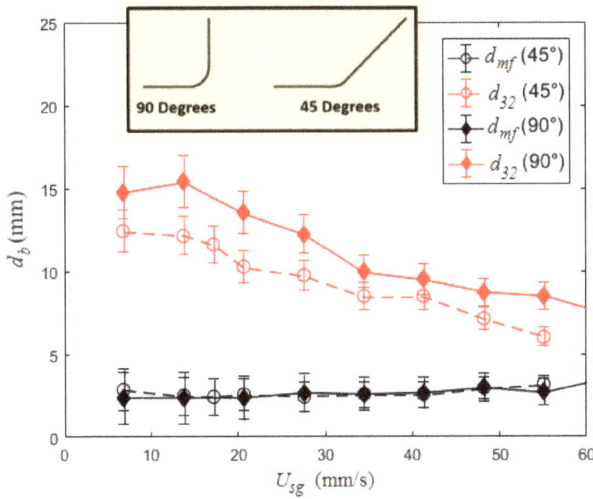

Figure 10. Bubble sizes (d_{mf} and d_{32}) plotted versus the superficial gas velocity with the injector tube angle either 45° or 90° from horizontal (see insert sketch) (D = 102 mm; d_{inj} = 0.8 mm; P_m = 600 kPa).

4.3. Injection Tube Diameter

The injection tube diameter is one of the key parameters that modifies the BSD, especially in the homogenous regime by effecting the bubble formation process. It is commonly accepted that bubble chord (vertical length from tube to top of bubble) at detachment is on the same order of magnitude as the injector tube diameter [36], which is supported with observations that decreasing orifice diameter decreases the bubble size [37]. This is because at the time of detachment the surface tension forces are balanced with hydrostatic pressure and buoyancy forces, where the outer diameter of the injector sets the contact angle [38]. Thus, the injector size has a significant impact on the initial bubble size, which is known to affect the flow pattern [39] and consequently the flow regime as discussed above.

Given the above observations, the current study examined the effect injector diameter has on the BSD with two injector sizes (d_{inj} = 0.8 and 1.6 mm). Based on past observations [40], it is expected that increasing the injector tube diameter will increase the bubble size. Results for both d_{mf} and d_{32} are provided in Figure 11. The most frequent bubble size shows negligible variation between the injector tube diameters. This is consistent with the turbulent scales within the wakes setting d_{mf}. The Sauter mean diameter trend is nearly identical between tube diameters, but the curve for the smaller tube is shifted downward slightly. This supports previous observations since it exhibits a dependence on the tube diameter, but the tube diameter was not varied by an order of magnitude resulting in the bubble size having a relatively small variation.

It is instructive to examine the PDFs from these conditions to determine how the tube diameter is modifying the BSD.

Figure 12 provides the PDF for two of the volumetric flow rates tested with each of the injector tube diameters. These two representative conditions (U_{sg} = 6.9 mm/s produced PDFs with and without an apparent second peak) demonstrate that the PDFs are nearly identical between the two injectors. This explains why d_{mf} is nearly identical between the two injector diameters, but not the shift in d_{32}. The difference between the PDFs is that the larger injector tube diameters produced larger maximum sized bubbles (i.e., larger tube diameter produces a longer tail in the PDFs). Maximum measured bubble sizes (d_{max}) for U_{sg} = 6.9, 27.6 and 55.1 mm/s are provided in Table 1. This shows that the smaller bubble tube diameter produces significantly smaller d_{max} (up to 40% smaller than the large tube). This supports the comments that both length scales are important since while d_{mf} is insensitive to these changes, d_{32} is modified because of these larger bubbles. While d_{32} is sensitive to these variations, higher order statistics (particularly skewness, a measure the asymmetry of a distribution) should be more sensitive to these variations.

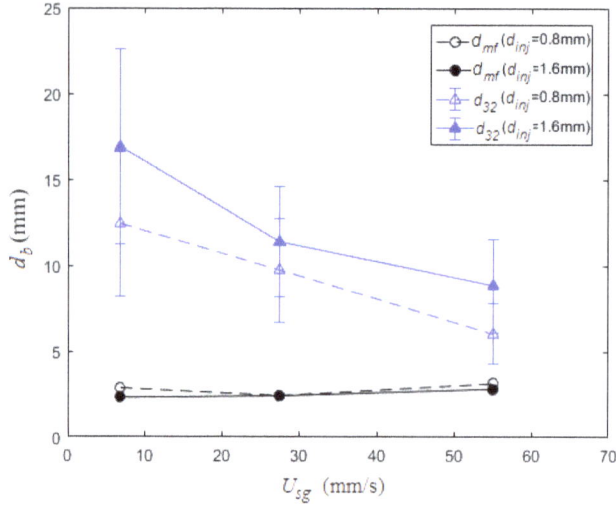

Figure 11. Bubble sizes (d_{mf} and d_{32}) plotted versus the superficial gas velocity varying the injector tube diameter (D = 102 mm; P_m = 600 kPa).

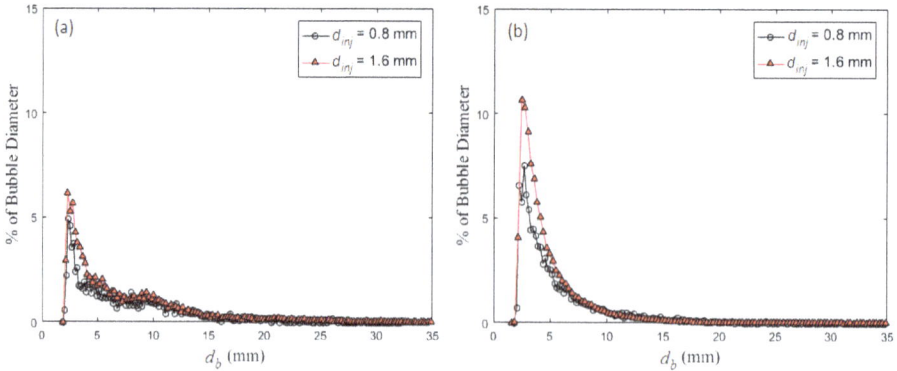

Figure 12. Bubble size PDF for the two injector tube diameters (d_{inj} = 0.8 or 1.6 mm) tested at (a) U_{sg} = 6.9 mm/s and (b) U_{sg} = 27.6 mm/s (D = 102 mm; P_m = 600 kPa; T_c = 21 ± 1 °C).

Table 1. The maximum measured bubbles size (d_{max}) spanning the flow rates tested with both injector tube diameters.

U_{sg} (mm/s)	Maximum Measured Bubble Size (mm)	
	d_{inj} = 0.8 mm	d_{inj} = 1.6 mm
6.9	10.2	11.7
27.6	9.9	16.7
55.1	9.4	15.8

4.4. Column Diameter

Wall effects play a significant role when the column diameter is below 0.15 m [40]. This explains the contradictory trends between bubble size and column diameter in the literature [41–43]. These contradictory observations are the product of operation within different flow regimes or transitioning between regimes. In particular, there are a number of studies [44–46] that indicate column diameter has an impact on the transition superficial gas velocity, but currently there is no comprehensive understanding of the influence of column diameter. The current study does not aim to assess the overall impact of column diameter, but does examine variation of the BSD with two different column diameters (D = 63 and 102 mm). Results in Figure 13 show no significant deviation for either bubble size measurements between the two column diameters.

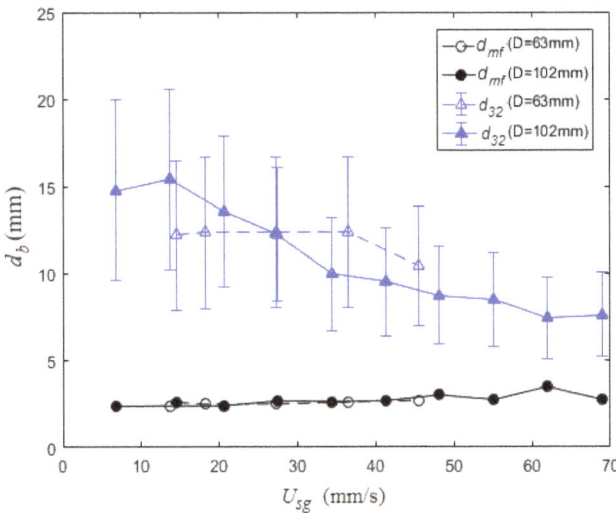

Figure 13. Bubble sizes (d_{mf} and d_{32}) plotted versus the superficial gas velocity with different column diameters (d_{inj} = 0.8 mm; P_m = 600 kPa).

5. Higher Order Statistics

While the parameter space of the current study is insufficient to provide a detailed analysis of higher order statistics (i.e., standard deviation σ, skewness S and kurtosis κ), the available results are provided in Figure 14 given the dearth of available data in the literature. Based on the previous observations/discussion, there are a few expected trends in the higher order statistics. In particular, increasing the injector diameter is expected to increase the skewness given that larger injection tubes generate larger maximum bubbles, which will result in a longer tail in the PDF. This is observed in Figure 14c, noting that the open symbols are d_{inj} = 0.8 mm and the closed symbols are d_{inj} = 1.6 mm. Thus focusing on the large column (D = 102 mm) and P_m = 600 kPa, the smaller injector tube diameter results in a smaller skewness at a given U_{sg}. The kurtosis (a measure of "tailedness" of a distribution) is provided in Figure 14d, which for all conditions the kurtosis is greater than that of a normal distribution (κ = 3). The relatively high kurtosis values indicate the presence of infrequent excessive deviations from the mean. Furthermore, use of the skewness and kurtosis can provide a quantitative measure of the bimodality of the distribution (e.g., Sarle's bimodality coefficient). There is a peak in this bimodality coefficient at a Reynolds number based on the d_{mf} at ~1000. This supports the previous observations that the bimodality could be the product of the transition from the unsteady flow-field between 500 < Re < 1000 and the turbulent far-wake with Strouhal shedding above 1000. Thus, the higher order statistics are a potential means for identifying regime transitions within the column.

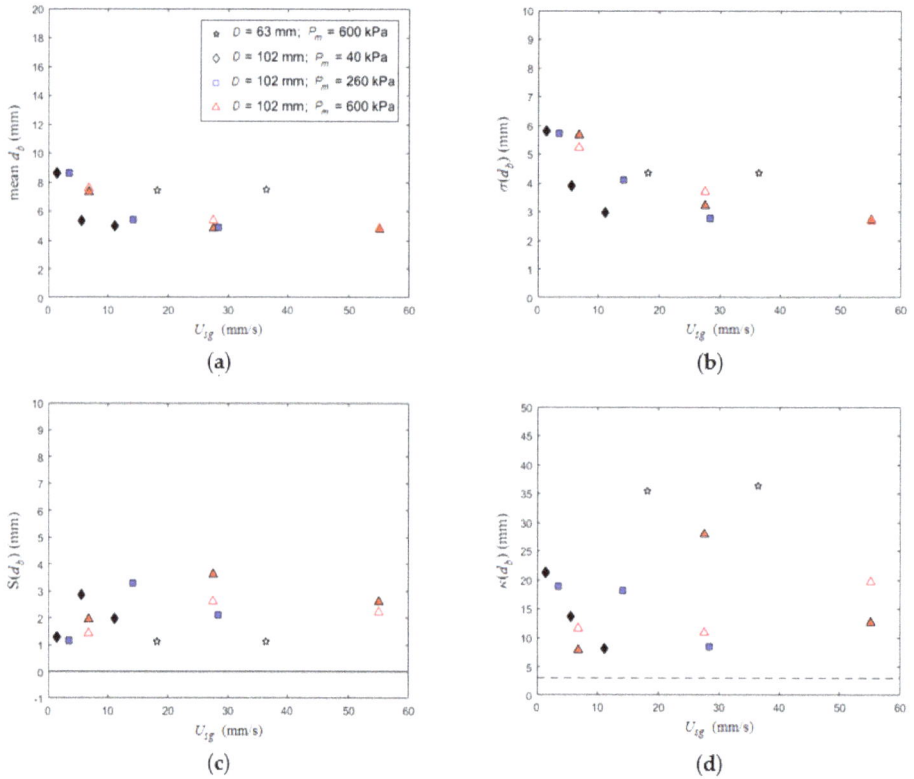

Figure 14. Higher order statistics from the PDFs including (**a**) unweighted mean, (**b**) standard deviation σ, (**c**) skewness S and (**d**) kurtosis κ of the bubble diameter. Dashed line on the kurtosis plot at $\kappa(d_b) = 3$ corresponds to the kurtosis value of a normal distribution. The same legend is used for all plots. Open and closed symbols correspond to $d_{inj} = 0.8$ and 1.6 mm, respectively.

6. Conclusions

The current study analyzed the BSD within a bubble column using high-speed imaging of a large population of bubbles. The gas phase (air) was injected into the liquid phase (water) near the base of the column via a single injection tube. The column diameter (D = 63 or 102 mm), injection tube diameter (d_{inj} = 0.8 or 1.6 mm) and the superficial gas flux (1.4 < U_{sg} < 55 mm/s) were varied during testing. The range of superficial gas fluxes was controlled via a combination of pressure and volumetric flux measured/controlled upstream of the injection tube. However, the temperature both at the metering location as well as within the column were held nearly constant throughout testing. The large sampling of bubbles were used to generate PDFs for each test condition. The maximum peak in the PDFs was used to identify a new bubble length scale, which was termed the most frequent bubble size (d_{mf}). This bubble length scale was compared with the traditional Sauter mean diameter (d_{32}), which is a weighted average. Both were applied to a parametric study to determine the information that each length scale provides. In general, Sauter mean diameter is more sensitive to the largest bubbles within the flow while d_{mf} is related to the turbulent structures created in the bubble wakes. Consequently, the difference between d_{32} and d_{mf} is a nominal range of bubble sizes expected within a given flow.

The parametric study examined the dependence of each of the bubble length scales on the gas injection rate, injector tube diameter, injection tube angle and column diameter. These individual studies are not meant to be an exhaustive assessment of these dependences, but rather a case study to assess how these length scales can be utilized to characterize multiphase flow behavior. The Sauter mean diameter was sensitive to the tube angle and injection tube diameter, but was relatively insensitive to the gas injection rate (except when the flow within the injection tube was transitional) and column diameter. Sauter mean diameter did exhibit a sensitivity to tube angle with misalignment between the tube and gravity resulting in a ~25% decrease in bubble size. The two most likely mechanisms for the sensitivity are the buoyancy force not aligning with the initial bubble wakes (i.e., turbulent regions) and the gas being directed towards the walls where the shear stress distribution is higher. Since the Sauter mean diameter decreased, the increase in shear stress at the wall is the likely cause for the sensitivity since the misalignment would decrease the turbulence level the bubbles experience at low void fractions. Conversely, doubling the injector tube diameter produced a ~33% increase in the Sauter mean diameter. This dependence is expected based on previous work that noted that the detachment bubble size is of the same order of magnitude as the injector tube, and the detachment bubble size is directly related to the largest bubbles. Since d_{32} is biased towards the largest bubbles within the flow, it is expected that there would be a dependence of d_{32} on injector tube diameter.

Conversely, the most frequent bubble size was relatively insensitive to gas injection rate, injection tube diameter, tube angle and column diameter. The insensitivity to most parameters is due to the minimum bubble diameter based Reynolds number tested being greater than 500 and most bubbles greater than 1000. In this range, coherent structures shed from the bubbles produce turbulent far-wakes. It is known that turbulent flow-fields produce relatively uniform bubble distributions, which is consistent with the observations of d_{mf}.

Higher order statistics (standard deviation, skewness and kurtosis) were also reported for the test conditions. While the range of test conditions limited the insights from these results, they were reported due to dearth of available data in the literature. The limited data were consistent with some expected behavior given conclusions drawn from the parametric study assessing the behavior of the Sauter mean diameter and the most frequent bubble diameter. These conclusions are (i) skewness increases with increasing injection tube diameter due to the longer tail in the PDF, (ii) high kurtosis values indicate the presence of infrequent excessive deviations from the mean and (iii) higher order statistics could be used as an indicator for a regime change since a bimodal coefficient peaked at *Re* ~1000. The overall evaluation is that the combination of Sauter mean diameter and most frequent bubble diameter provides a more comprehensive assessment of the flow behavior.

Acknowledgments: The authors would like to thank Afshin Ghajar for the use of his vibrating bubble column facility as well as Adam Still's assistance with familiarizing our team with the use of the facility. We would also like to thank Trevor Wilson, Bret Valenzuela and Tyler Hinds for their assistance counting bubbles and execution of the experiment as well as Amir Erfani for his assistance with surface tension measurement. This research did not receive any specific grant from funding agencies in the public, commercial, or not-for-profit sectors.

Author Contributions: Shahrouz Mohagheghian and Brian R. Elbing conceived and designed the experiments; Shahrouz Mohagheghian performed the experiments; Shahrouz Mohagheghian and Brian R. Elbing analyzed the data; Shahrouz Mohagheghian and Brian R. Elbing wrote the paper.

Conflicts of Interest: The authors declare no conflict of interest.

References

1. Kantarci, N.; Borak, F.; Ulgen, K.O. Bubble column reactors. *Process Biochem.* **2005**, *40*, 2263–2283. [CrossRef]
2. Krishna, R.; Ellenberger, J.; Urseanu, M.I.; Keil, F.J. Utilisation of bubble resonance phenomena to improve gas-liquid contact. *Naturwissenschaften* **2000**, *87*, 455–459. [CrossRef] [PubMed]
3. Oliveira, M.S.N.; Ni, X. Gas hold-up and bubble diameters in a gassed oscillatory baffled column. *Chem. Eng. Sci.* **2001**, *56*, 6143–6148. [CrossRef]
4. Waghmare, Y.G.; Rice, R.G.; Knopf, F.C. Mass transfer in a viscous bubble column with forced oscillations. *Ind. Eng. Chem. Res.* **2008**, *47*, 5386–5394. [CrossRef]

5. Hur, Y.G.; Yang, J.H.; Jung, H.; Park, S.B. Origin of regime transition to turbulent flow in bubble column: Orifice- and column-induced transitions. *Int. J. Multiph. Flow* **2013**, *50*, 89–97. [CrossRef]

6. Clark, N.N.; Turton, R. Chord length distributions related to bubble size distributions in multiphase flows. *Int. J. Multiph. Flow* **1988**, *14*, 413–424. [CrossRef]

7. Wu, C.; Suddard, K.; Al-Dahhan, M.H. Bubble dynamics investigation in a slurry bubble column. *AIChE J.* **2008**, *54*, 1203–1212. [CrossRef]

8. Xue, J.; Al-Dahhan, M.; Dudukovic, M.P.; Mudde, R.F. Four-point optical probe for measurement of bubble dynamics: Validation of the technique. *Flow Meas. Instrum.* **2008**, *19*, 293–300. [CrossRef]

9. Xue, J.; Al-Dahhan, M.; Dudukovic, M.P.; Mudde, R.F. Bubble velocity, size, and interfacial area measurements in a bubble column by four-point optical probe. *AIChE J.* **2008**, *54*, 350–363. [CrossRef]

10. Van Der Welle, R. Void fraction, bubble velocity and bubble size in two-phase flow. *Int. J. Multiph. Flow* **1985**, *11*, 317–345. [CrossRef]

11. George, D.L.; Torczynski, J.R.; Shollenberger, K.A.; O'Hern, T.J.; Ceccio, S.L. Validation of electrical-impedance tomography for measurements of material distribution in two-phase flows. *Int. J. Multiph. Flow* **2000**, *26*, 549–581. [CrossRef]

12. Mäkiharju, S.; Elbing, B.R.; Wiggins, A.; Schinasi, S.; Vanden-Broeck, J.M.; Perlin, M.; Dowling, D.R.; Ceccio, S.L. On the scaling of air entrainment from a ventilated partial cavity. *J. Fluid Mech.* **2013**, *732*, 47–76. [CrossRef]

13. Manera, A.; Prasser, H.M.; Lucas, D.; van der Hagen, T.H.J.J. Three-dimensional flow pattern visualization and bubble size distributions in stationary and transient upward flashing flow. *Int. J. Multiph. Flow* **2006**, *32*, 996–1016. [CrossRef]

14. Manera, A.; Prasser, H.M.; Lucas, D. Experimental investigation on bubble turbulent diffusion in a vertical large-diameter pipe by wire-mesh sensors and correlation techniques. *Nuclear Technol.* **2007**, *158*, 275–290. [CrossRef]

15. Omebere-Iyari, N.K.; Azzopardi, B.J.; Lucas, D.; Beyer, M.; Prasser, H.-M. The characteristics of gas/liquid flow in large risers at high pressures. *Int. J. Multiph. Flow* **2008**, *34*, 461–476. [CrossRef]

16. Youssef, A.A.; Al-Dahhan, M.H. Impact of internals on the gas holdup and bubble properties of a bubble column. *Ind. Eng. Chem. Res.* **2009**, *48*, 8007–8013. [CrossRef]

17. Still, A.L. Multiphase Phenomena in a Vibrating Bubble Column Reactor. Master's Thesis, Mechanical & Aerospace Engineering, Oklahoma State University, Stillwater, OK, USA, 2012.

18. Still, A.L.; Ghajar, A.J.; O'Hern, T.J. Effect of amplitude on mass transport, void fraction and bubble size in a vertically vibrating liquid-gas bubble column reactor. In Proceedings of the ASME Fluids Engineering Summer Meeting-FEDSM2013-16116, Incline Village, NV, USA, 7–11 July 2013.

19. Mohagheghian, S.; Elbing, B.R. Study of bubble size and velocity in a vibrating bubble column. In Proceedings of the ASME 2016 14th International Conference on Nanochannels, Microchannels, and Minichannels collocated with the ASME 2016 Heat Transfer Summer Conference and the ASME 2016 Fluids Engineering Division Summer Meeting, FEDSM2016-1056, V01BT33A012, Washington, DC, USA, 10–14 July 2016.

20. Besagni, G.; Pasquali, A.D.; Gallazzini, L.; Gottardi, E.; Colombo, L.P.M.; Inzoli, F. The effect of aspect ratio in counter-current gas-liquid bubble columns: Experimental results and gas holdup correlations. *Int. J. Multiph. Flow* **2017**, *94*, 53–78. [CrossRef]

21. Luo, X.; Lee, D.J.; Lau, R.; Yang, G.; Fan, L.-S. Maximum stable bubble size and gas holdup in high-pressure slurry bubble columns. *AIChE J.* **1999**, *45*, 665–680. [CrossRef]

22. Akita, K.; Yoshida, F. Bubble size, interfacial area, and liquid-phase mass-transfer coefficient in bubble columns. *Ind. Eng. Chem. Process Des. Dev.* **1974**, *13*, 84–91. [CrossRef]

23. Besagni, G.; Inzoli, F.; De Guido, G.; Pellegrini, L.A. The dual effect of viscosity on bubble column hydrodynamics. *Chem. Eng. Sci.* **2017**, *158*, 509–538. [CrossRef]

24. Li, H.; Prakash, A. Heat transfer and hydrodynamics in a three-phase slurry bubble column. *Ind. Eng. Chem. Res.* **1997**, *36*, 4688–4694. [CrossRef]

25. Schafer, R.; Merten, C.; Eigenberger, G. Bubble size distributions in a bubble column reactor under industrial conditions. *Exp. Therm. Fluid Sci.* **2002**, *26*, 595–604. [CrossRef]

26. Ohnuki, A.; Akimoto, H. An experimental study on developing air-water two-phase flow along a large vertical pipe: Effect of air injection method. *Int. J. Multiph. Flow* **1996**, *22*, 1143–1154. [CrossRef]

27. Gaddis, E.S.; Vogelpohl, A. Bubble formation in quiescent liquids under constant flow conditions. *Chem. Eng. Sci.* **1986**, *41*, 97–105. [CrossRef]

28. Besagni, G.; Inzoli, F. Novel gas holdup and regime transition correlation for two-phase bubble columns. In *Journal of Physics: Conference Series*; IOP Publishing: Bristol, UK, 2017.

29. Abràmoff, M.D.; Magalhães, P.J.; Ram, S.J. Image processing with ImageJ. *Biophotonics Int.* **2004**, *11*, 36–42.

30. Schneider, C.A.; Rasband, W.S.; Eliceiri, K.W. NIH Image to ImageJ: 25 years of image analysis. *Nat. Methods* **2012**, *9*, 671–675. [CrossRef] [PubMed]

31. Rasband, W.S. ImageJ. U.S. National Institutes of Health, Bethesda, MD, USA, 1997–2016. Available online: http://imagej.nih.gov/ij (accessed on 16 April 2013).

32. Fukuma, M.; Muroyama, K.; Yasunishi, A. Properties of bubble swarm in a slurry bubble column. *J. Chem. Eng. Jpn.* **1987**, *20*, 28–33. [CrossRef]

33. Saxena, S.C.; Rao, N.S.; Saxena, A.C. Heat-transfer and gas-holdup studies in a bubble column: Air-water-glass bead system. *Chem. Eng. Commun.* **1990**, *96*, 31–55. [CrossRef]

34. Brennen, C.E. *Fundamental of Multiphase Flow*, 1st ed.; Cambridge University Press: New York, NY, USA, 2005; pp. 60–66. ISBN 0521139988.

35. Hecht, K.; Bey, O.; Ettmüller, J.; Graefen, P.; Friehmelt, R.; Nilles, M. Effect of gas density on gas holdup in bubble columns. *Chem. Ing. Tech.* **2015**, *87*, 762–772. [CrossRef]

36. Kulkarni, A.A.; Joshi, J.B. Bubble formation and bubble rise velocity in gas-liquid systems: A review. *Ind. Eng. Chem. Res.* **2005**, *44*, 5873–5931. [CrossRef]

37. Basha, O.M.; Sehabiague, L.; Abdel-Wahab, A.; Morsi, B.I. Fischer-Tropsch synthesis in slurry bubble column reactors: Experimental investigations and modeling—A review. *Int. J. Chem. React. Eng.* **2015**, *13*, 201–288. [CrossRef]

38. Liow, J.-L. Quasi-equilibrium bubble formation during top-submerged gas injection. *Chem. Eng. Sci.* **2000**, *55*, 4515–4524. [CrossRef]

39. Cheng, H.; Hills, J.H.; Azzopardi, B.J. Effects of initial bubble size on flow pattern transition in a 28.9 mm diameter column. *Int. J. Multiph. Flow* **2002**, *28*, 1047–1062. [CrossRef]

40. Wilkinson, P.M.; Spek, A.P.; van Dierendonck, L.L. Design parameters estimation for scale-up of high-pressure bubble columns. *AIChE J.* **1992**, *38*, 544–554. [CrossRef]

41. Daly, J.G.; Patel, S.A.; Bukur, D.B. Measurement of gas holdups and Sauter mean bubble diameters in bubble column reactors by dynamic gas disengagement method. *Chem. Eng. Sci.* **1992**, *47*, 3647–3654. [CrossRef]

42. Koide, K.; Morooka, S.; Ueyama, K.; Matsuura, A.; Yamashita, F.; Iwamoto, S.; Kato, Y.; Inoue, H.; Shigeta, M.; Suzuki, S.; et al. Behavior of bubbles in large-scale bubble column. *J. Chem. Eng. Jpn.* **1979**, *12*, 98–104. [CrossRef]

43. Sasaki, S.; Uchida, K.; Hayashi, K.; Tomiyama, A. Effects of column diameter and liquid height on gas holdup in air-water bubble columns. *Exp. Therm. Fluid Sci.* **2017**, *82*, 359–366. [CrossRef]

44. Zahradnik, J.; Fialová, M.; Ruzicka, M.; Drahos, J.; Kastanek, F.; Thomas, N.H. Duality of the gas-liquid flow regimes in bubble column reactors. *Chem. Eng. Sci.* **1997**, *52*, 3811–3826. [CrossRef]

45. Sarrafi, A.; Müller-Steinhagen, H.; Smith, J.M.; Jamialahmadi, M. Gas holdup in homogeneous and heterogeneous gas—liquid bubble column reactors. *Can. J. Chem. Eng.* **1999**, *77*, 11–21. [CrossRef]

46. Ruzicka, M.C.; Drahos, J.; Fialova, M.; Thomas, N.H. Effect of bubble column dimensions on flow regime transition. *Chem. Eng. Sci.* **2001**, *56*, 6117–6124. [CrossRef]

fluids

MDPI

Article

Experimental Analysis of a Bubble Wake Influenced by a Vortex Street

Sophie Rüttinger, Marko Hoffmann and Michael Schlüter *

Institute of Multiphase Flows, Hamburg University of Technology, Eissendorfer Str. 38,
D-21073 Hamburg, Germany; sophie.ruettinger@tuhh.de (S.R.); marko.hoffmann@tuhh.de (M.H.)
* Correspondence: michael.schlueter@tuhh.de; Tel.: +49-404-2878-3252

Received: 29 November 2017; Accepted: 18 January 2018; Published: 20 January 2018

Abstract: Bubble column reactors are ubiquitous in engineering processes. They are used in waste water treatment, as well as in the chemical, pharmaceutical, biological and food industry. Mass transfer and mixing, as well as biochemical or chemical reactions in such reactors are determined by the hydrodynamics of the bubbly flow. The hydrodynamics of bubbly flows is dominated by bubble wake interactions. Despite the fact that bubble wakes have been investigated intensively in the past, there is still a lack of knowledge about how mass transfer from bubbles is influenced by bubble wake interactions in detail. The scientific scope of this work is to answer the question how bubble wakes are influenced by external flow structures like a vortex street behind a cylinder. For this purpose, the flow field in the vicinity of a single bubble is investigated systematically with high spatial and temporal resolution. High-speed Particle Image Velocimetry (PIV) measurements are conducted monitoring the flow structure in the equatorial plane of the single bubble. It is shown that the root mean square (RMS) velocity profiles downstream the bubble are influenced significantly by the interaction of vortices. In the presence of a vortex street, the deceleration of the fluid behind the bubble is compensated earlier than in the absence of a vortex street. This happens due to momentum transfer by cross-mixing. Both effects indicate that the interaction of vortices enhances the cross-mixing close to the bubble. Time series of instantaneous velocity fields show the formation of an inner shear layer and coupled vortices. In conclusion, this study shows in detail how the bubble wake is influenced by a vortex street and gives deep insights into possible effects on mixing and mass transfer in bubbly flows.

Keywords: particle image velocimetry; flow structure; single bubbles; convective transfer; mixing

---

1. Introduction

The efficiency of process engineering operations strongly depends on the transport of heat, mass, and momentum. As the transport from a dispersed phase into a bulk phase is often the limiting step, fundamental knowledge of phenomena close to fluidic interfaces is crucial to selectively improve chemical reactors like, e.g., bubble columns, stirred tanks or jet loop reactors. The hydrodynamics of bubbly flows is dominated by bubble wake interactions. There is a large number of experimental, as well as numerical studies dealing with the characterization of bubble wakes. While there are good standard references on mass transfer and multiphase flows [1,2], Fan and Tsuchiya were the first who collected the state of the art for a book dealing with bubble wakes [3]. There are many experimental studies dealing with single rising bubbles, e.g., [4–8] or bubble swarms, e.g., [9]. It is known that bubbles are able to induce turbulence [10] and that they can enhance mixing processes [11]. Furthermore, the primary bubble wake is known to dictate transport phenomena in multiphase systems [3]. There are also recent numerical studies on reactive mass transfer from single bubbles which are much elaborated [12–14]. Despite the great amount of carefully performed studies, the question of how bubble wakes are influenced by external flow structures is still not answered completely. Due to the wide knowledge of flows around cylinders [15–18], an experiment has been designed with the aim to investigate the influence of a vortex street generated within a cylinder wake on the hydrodynamics of a single bubble, which is fixed in place.

There are studies on fixed bubbles presented in literature: No et al. [19] and Tokuhiro et al. [20] compared the flow around a bubble, which is kept in place by a cap, to the flow around a solid ellipsoid. They found a major difference in turbulent production terms. Tokuhiro et al. [21] investigated the hydrodynamics around two confined bubbles. They observed wake interaction and a jet-like flow between the bubbles. When comparing a confined ellipsoidal bubble to a solid ellipsoid, Tokuhiro et al. [22] found the turbulent kinetic energy to be more uniformly distributed directly behind the bubble than it was directly behind the solid body.

The flow behind one or two cylinders has been studied extensively in the literature for many years [23,24]. It is well known how Strouhal numbers, i.e., dimensionless frequencies, depend on Reynolds numbers. There are correlations for different Reynolds numbers. Roshko [23] proposed the following correlations:

$$Sr = 0.212\left(1 - \frac{21.2}{Re}\right), \quad 50 < Re < 150$$

$$Sr = 0.212\left(1 - \frac{12.7}{Re}\right), \quad 150 < Re < 500$$

Due to the wide knowledge of von Karman vortex streets, a cylinder can be used to produce well-defined vortices with a certain frequency. There are two flow instability points that describe the bifurcation process in the subcritical Reynolds number regime: The first flow instability that occurs in the cylinder wake is at $Re = 49$. From this point, it is unsteady, but two-dimensional. At $Re = 190$, the second instability occurs, and the cylinder wake becomes three-dimensional and is no longer laminar [25]. The wake transition regime occurs at $Re = 190$ [25,26]. The irregular regime of the von Karman vortex street is found for Reynolds numbers higher than $Re = 300$ [23,25].

A thorough understanding of processes at fluidic interfaces and how they are influenced by external structures is crucial to specifically influence multiphase processes in the future. This study is a first step towards this aim. It deals with the interaction of a cylinder wake and a fixed bubble held in place by a cap. The cap ensures a nearly spherical bubble shape. Strouhal numbers of the bubble wake and the cylinder wake are calculated on the basis of Particle Image Velocimetry (PIV) data. The fluctuating motions behind the bubble are monitored, and the velocity field of the bubble is characterized for different Reynolds numbers. This paper is organized as follows: In Section 2, the experimental setup is depicted, and the data processing is described. Section 3 consists of three parts. The first part gives information about the cylinder wake, which is used to create vortical structures within the fluid flow that approaches the bubble. The second part shows the analysis of the flow situation around the single bubble without an upstream cylinder wake. Finally, the third part illustrates the flow structure of the flow around the single bubble influenced by cylinder wakes.

2. Experimental Set-up and Data Processing

2D2C high-speed PIV measurements are conducted in a duct made from acrylic glass with a cross-section of 0.1×0.1 m^2. Demineralized water with PIV seeding particles (details are listed in Table 1), which have a density very close to water, is supplied continuously through the duct with adjustable volumetric flow rates (250–875 L/h). The single bubble is produced by a hypodermic needle, which is pushed through a septum into the fluid flow. The bubble (CO_2 or air) is kept in place using a spherical cap. A cylinder is brought into the duct to produce vortices of different frequencies by means of a von Karman vortex street. The streamwise distance between the cylinder and the bubble is varied while the transverse distance is held constant. In Figure 1, the staggered configurations of bubble and cylinder within the duct are sketched in detail. The first configuration (Figure 1a) has the cylinder diameter as the distance between the bubble and cylinder; the second (Figure 1b) has several cylinder diameters as the distance. This is chosen in order to observe different wake interactions.

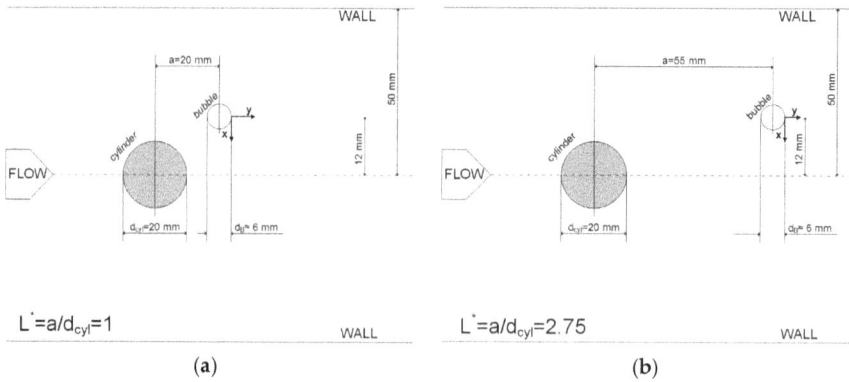

Figure 1. Staggered configurations of the cylinder and bubble.

PIV experiments are conducted to characterize the flow field around the bubble and to monitor the influence of the cylinder wake. For this purpose, the laser light sheet (thickness ≈ 1 mm), which is produced using a light sheet optics with a $50°$ rod lens (ILA_5150 GmbH, Aachen, Germany), is adjusted at the equatorial plane of the bubble (see Figure 2). In Table 1, the experimental parameters are summarized. The temperature of the fluid is held nearly constant, and the experiments are conducted under atmospheric pressure.

Table 1. Parameters of 2D high-speed Particle Image Velocimetry (PIV) experiments.

Parameters	Settings
Camera	PCO dimax HS2 (PCO AG, Kelheim, Germany), 1400×1000 Px^2, 12 bit
Objective	Zeiss macro planar 2/50 mm
Laser	Quantronix Darwin-Duo-100M, Nd:YLF (Quantronix Inc., Hamden, CT, USA), total energy > 60 mL, average power at 3 kHz > 90 W
Seeding Particles	PS-FluoRed-Fi203, monodisperse 3.16 μm, abs/em = 530/607 nm (MicroParticles GmbH, Berlin, Germany)
Frame Rate	500 fps
Acquisition Time	20 s
Number of Images Processed	10,000
Spatial Resolution (vector-to-vector spacing)	0.36 ... 0.69 mm (24 Px)
Temperature	$20 \pm 1.5\,°C$
PIV Data Processing Software	PivView 3.60 (PivTec GmbH, ILA_5150 GmbH, Aachen, Germany)

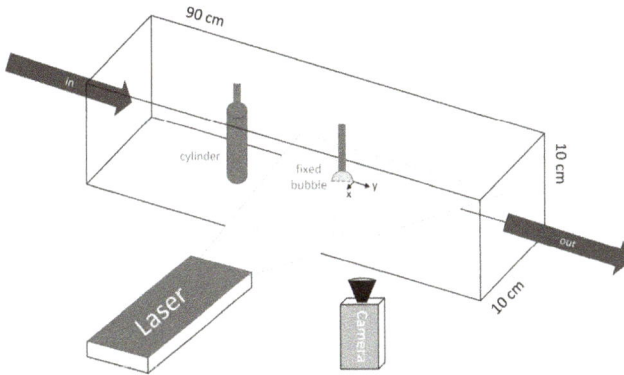

Figure 2. Schematic diagram of the experiment.

The high-speed images with an equidistant timing of 2 ms between the images are processed using commercial software (see Table 1). A Fast Fourier Transformation (FFT) correlation with a multi-grid refinement algorithm is used. Peak search is conducted using the least squares Gauss algorithm.

In Figure 3, a raw PIV image and the related instantaneous velocity field is shown. The axis coordinates, as well as the velocity magnitude values are given in physical coordinates to obtain a first overview of the flow case. The direction of flow is from left to right. Later on in this study, velocity fields are shown turned 90° counterclockwise. This is done for an easier readability when comparing to velocity profiles. Velocity vector arrows are included, as well as stream traces. By following the stream traces, one can easily observe the vortices generated within the cylinder wake approaching the bubble.

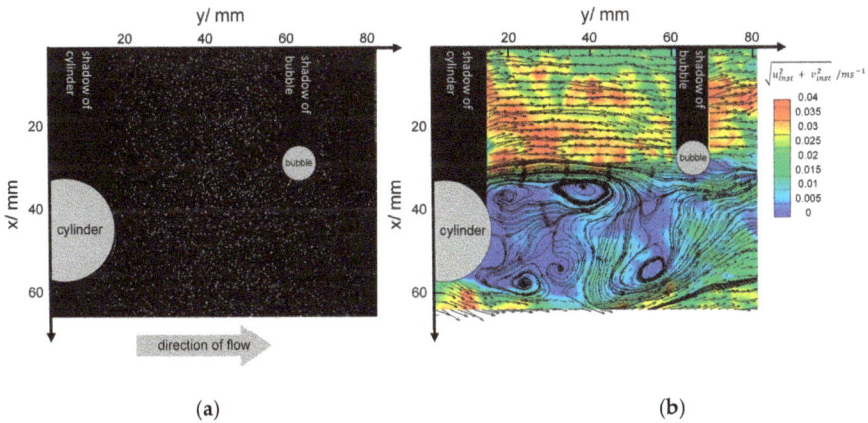

Figure 3. PIV processing: raw image (**a**) and instantaneous velocity field with physical coordinates (**b**).

Via MATLAB® (The MathWorks, Natick, MA, USA), the PIV data are processed further. Fast Fourier Transformation (FFT) is used to obtain frequencies dominating the bubble wake or the cylinder wake, respectively. Strouhal numbers are calculated using the frequencies obtained by FFT. The frequency analysis is carried out 1 cm downstream of the bubble or the cylinder (see Figure 4).

(a)

(b)

(c)

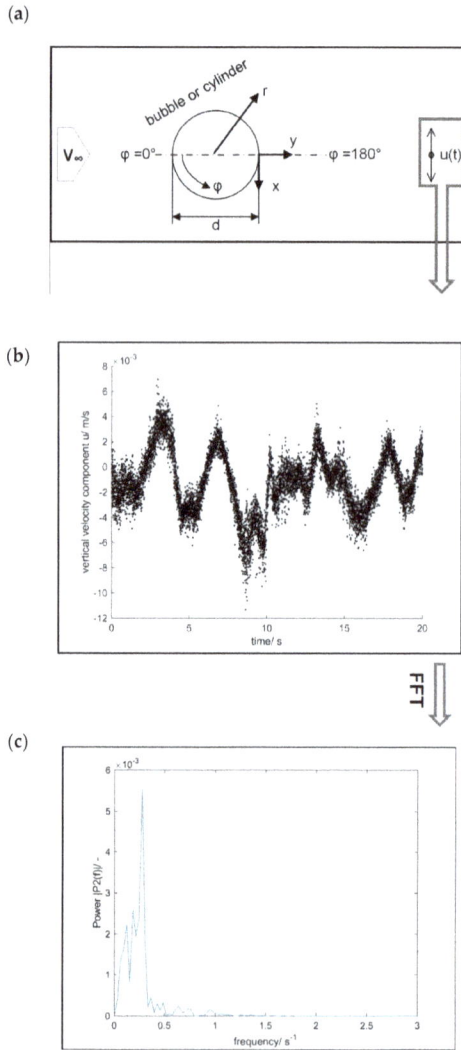

Figure 4. Nomenclature (**a**); procedure of velocity analysis (**b**) and frequency analysis (power spectrum) (**c**).

At this point, the velocity fluctuations over time perpendicular to the flow are monitored (see Figure 4). An FFT leads to a spectrum of frequencies with a sharp peak. This frequency f is the frequency of the wake and can therefore be used to characterize it. This is taken to calculate the Strouhal number of the bubble:

$$Sr_b = \frac{f \cdot d_b}{v_\infty}$$

with the frequency f, the bubble diameter d_b and the mean velocity that approaches the bubble v_∞. For v_∞, which is calculated from the volumetric flow rate $\dot{V}$ and the cross-section of the duct A_{duct}:

$$v_\infty = \frac{\dot{V}}{A_{duct}}$$

the conservation of mass is used. This is done because due to the field of view, a complete flow profile of the duct flow is not obtained. Due to a preferably high spatial resolution (see Table 1), the whole duct was not recorded. With the so calculated velocity, also the Reynolds numbers:

$$Re_b = \frac{v_\infty \cdot d_b}{\nu}$$

are calculated (kinematic viscosity ν). If the cylinder is characterized (Re_{cyl}, Sr_{cyl}), the diameter of the cylinder is used as a characteristic length.

3. Results and Discussion

3.1. Characterization of the Vortex Street

The vortex street arising from the flow around the circular cylinder is analyzed via the calculation of Strouhal numbers for several Reynolds numbers. Table 2 shows the results in combination with well-established correlations from the literature [23], which are depicted in Section 1. The flow regimes are also added to the table so that it becomes clear that the laminar and irregular vortex street with three-dimensional effects can be produced within the experimental setup. It is visible that the Strouhal numbers achieved in the experimental setup deviate from the theoretical values. Slightly different geometric relationships may be the reason for this fact: The cylinder diameter is 2 cm, and the hydraulic diameter of the duct is 10 cm (see Figure 1). Therefore, the vortex street can be influenced by the proximity of the walls. The geometry of the duct is chosen this way to obtain the best optical access to investigate the bubble. The optical access can be limited by choosing, e.g., a large cross-sectional area.

From the comparably high Strouhal numbers achieved in this work, it can be deduced that the wake is stable and remains more two-dimensional than an ideal von Karman vortex street would. This is found when comparing the results to data from numerical simulation in the transition regime [25]. Furthermore, the cylinder has a mean averaged surface roughness of 8.8 µm. In many literature references, also the one used for Table 2 [23], there is no information about surface roughness. The cylinder used may therefore be rougher than the ones used by other authors. This can lead to an enhanced detachment of the boundary layer and, thus, to enhanced vortex shedding. However, Chen [24] summarizes that the roughness does not have a great influence on the subcritical Reynolds number regime ($Re_{cyl} < 3 \times 10^4$). Nevertheless, there is a noticeable frequency of velocity fluctuation (see Table 2). In particular, the vortices generated behind the cylinder are very well visible when having a look at the streamlines in Figure 3. Therefore, it can be concluded the von Karman vortex street is generated successfully.

Table 2. Details of frequency analysis behind the cylinder.

Reynolds Number Re_{cyl}	Flow Regime [25]	Frequency of $u(t)$ Fluctuation/s^{-1}	Strouhal Number Sr_{cyl}	Calculated Strouhal Number [23] Sr_{calc}
100	2D (laminar)	0.06	0.24	0.17
418	3D (irregular)	0.24	0.23	0.21
418	3D (irregular)	0.27	0.26	0.21
478	3D (irregular)	0.31	0.25	0.21

3.2. Characterization of the Single Bubble

In the following, the fluid flow around the single bubble (no cylinder) is characterized. This is done by a frequency analysis and by the analysis of the streamwise velocity component v. Since only the equatorial plane of the bubble is illuminated by the laser light sheet (2D experiments), the velocity is evaluated for three azimuthal angles φ. By monitoring several Reynolds numbers, an overview of the flow structure depending on the relative velocity between the bubble and the liquid is obtained.

A dependency of the frequencies behind the bubble on the Reynolds number is clearly visible in Figure 5. There, the Strouhal number of the single bubbles (right red y axis) and the frequency of velocity fluctuations (left blue y axis) is plotted against the Reynolds number of the bubble. For the flow regime investigated in this work, no asymptotic value for the Strouhal number is achieved. For the Reynolds numbers in this regime, Strouhal numbers between 0.1 and 0.18 are expected, similar to the cylinder Strouhal number. This can be estimated from the first equation given in Section 1. While the Strouhal numbers for the lower three Reynolds numbers are slightly below the values estimated from the equation, the Strouhal number for $Re = 148$ is higher. The differences of the values that are presented in Figure 5 may arise due to the fluidic interface; they can also be caused by three-dimensional effects. In Section 1, critical points for the cylinder wake are mentioned. For the bubble, the three-dimensional effects are different from the ones of the cylinder wake. For spheres, a large amount of information is available from the literature [27]. In contrast to flow past cylinders, the wake structure behind spheres looks different and stays planar up to $Re = 350$. However, there is still a lack of information regarding the flow past spheres with the fluidic interface, as is the case for a bubble in three-dimensional flow. For this reason, a clear statement concerning three-dimensional effects in the bubble wake cannot be made, and further research is needed. Furthermore, it is also likely that the differences arise due to experimental limits. For example, there is only a given amount of frequencies that can be calculated during the FFT. When the flow is non-linear and the data are noisy, higher order dynamic mode decomposition can be used to calculate dominant frequencies [28]. However, with the given temporal resolution (see Table 1), the data basis in this study is considered to be large enough. It is worth noting that the Reynolds numbers in this work are in a low-to-medium regime. For bubbles, there are only Strouhal number correlations for much higher Reynolds numbers [2,29]. The duct Reynolds numbers are added to an upper x axis. The fact that the data points at the highest Reynolds number are already in a transitional regime ($Re_d = 2421 > 2300$) can be an explanation for the remarkably higher Strouhal number.

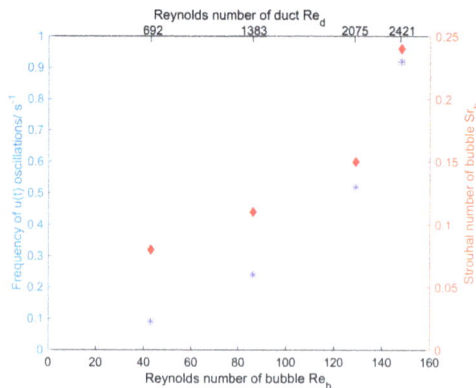

Figure 5. Strouhal numbers of the bubble. Red diamond items belong to the red (right) axis, and blue star items belong to the blue (left) axis.

To characterize the flow field in the vicinity of the bubble, the mean velocity v_{mean} in the direction of flow is used. The non-dimensional velocities v_{mean}/v_∞ of the single bubble depend strongly on the azimuthal angle φ and scale with the Reynolds number of the bubble. In Figure 6, velocities of the bubble are plotted for two different Reynolds numbers. They are plotted over the distance from the bubble center. Utilizing v_∞ and d_B, the axes are made non-dimensional. The horizontal axis starts with 0.5, since PIV only yields information regarding the flow of the continuous phase outside the bubble. For detailed information on the coordinate system, a schematic overview is added to Figure 6. Data points are shown for three azimuthal angles of $0°$, $90°$ and $180°$.

At the angle $\varphi = 0°$ (empty icons), the velocity decreases with decreasing distance from the bubble until the stagnation point is nearly reached. Further away from the bubble, at $r/d_B < 1.5$, it is well visible that v/v_∞ is nearly equal to one. This means that the velocity calculated from the continuity equation dominates there. At the very surface of the bubble, no velocity field information is available since a mask has to be used to cover the bubble during PIV data processing.

At $\varphi = 90°$ (light-colored icons), the fluid is accelerated near the bubble and then approaches to a profile which is the undisturbed flow profile in the duct. Due to the position of the bubble in the center of the duct, the undisturbed flow profile does not equal one. In the literature, there is a theoretical curve given by Oellrich et al. [29] (semi-analytical solution for a single bubble of constant spherical shape). This curve goes at $\varphi = 90$ straight to $v/v_\infty = 1$. The reason why the experimental results from this study differ slightly from [29] is that there is a flow profile approaching the bubble and not one single velocity v_∞. Close to the interface, the experimental data from Figure 6 decrease significantly. This leads to the assumption that the single bubble in the experiments has only a little inner circulation. This can be a result of the bubble fixation.

Behind the bubble, which is $\varphi = 180°$ (dark-colored icons), the tangential velocity is very low and also can be slightly negative due to back flow (Figure 6b). Here, the fluid is decelerated due to the bubble. However, there is no back flow for the higher Reynolds number (Figure 6a). The wake at $Re = 31$ is very steady, and vortex shedding is very low or does not occur at all. According to Komasawa et al. [30], $Re = 31$ lies in the regime of a laminar wake, while $Re = 120$ lies in the regime of a transitional wake. When comparing the findings of Komasawa et al. [30] to the findings of this work, it has to be kept in mind that this study deals with fixed bubbles. Due to the fixation, the relative velocity between bubble and liquid is determined by the duct flow. The shape of the bubble is spherical due to the fixation cap, although free rising bubbles with a comparable diameter may often be already of an ellipsoidal or cap shape. Clift et al. [2] present cases of wake shapes for several Reynolds numbers. They describe a change from a convex to a concave shape at $Re = 35$.

Finally, some considerations concerning the hydrodynamic boundary layer can be made by further analyzing the $\varphi = 90°$ curve. The hydrodynamic boundary layer is defined as the thickness that is needed until 99% of the velocity far away from the interface is reached. This is highlighted in Figure 6a,b for the location $\varphi = 90°$. It is well visible that this boundary layer is thicker for the lower Reynolds number and thinner for the higher one. This meets the expectations. There is a rule of thumb to estimate the boundary layer thickness of spheres [31]:

$$\delta_{hydr,\,90°} = 1.7 d_b / \sqrt{Re_b}$$

Utilizing this equation leads to theoretical values of $\delta_{hydr,\,90°} = 1.89$ mm for the lower Reynolds number and $\delta_{hydr,\,90°} = 0.88$ mm for the higher Reynolds number. The values match the experimental results well.

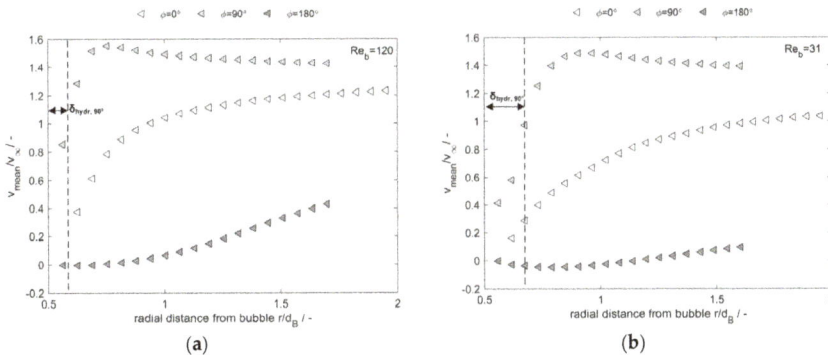

Figure 6. *Cont.*

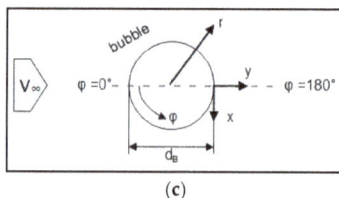

(c)

Figure 6. Non-dimensional streamwise velocity component v/v_φ of the single bubble for two different Reynolds numbers (**a**,**b**); and nomenclature (**c**). r/d_B denotes the surface of the single bubble.

3.3. Characterization of the Single Bubble Wake Influenced by the Vortex Street

The observation of velocity fluctuations is a helpful and widely-used tool to describe unsteady flow structures. By decomposing the velocity vector into a temporal mean (highlighted by a bar over the symbol or a mean in the lower index) and a fluctuating part (Reynolds decomposition), statements on the turbulence and unsteadiness of a fluid flow can be made. In the Reynolds number regime investigated in this work, the fluid flow around the bubble cannot be described as turbulent. Nevertheless, there are unsteady motions in the wake of the bubble, and therefore, it is reasonable to use root mean square (RMS) velocities. They are calculated by the following:

$$u_{RMS} = \sqrt{(u - \bar{u})^2}$$

In this work, u_{RMS} depicts the fluctuations perpendicular to the flow direction. This is shown in Figure 7a,b along the y coordinates along a line downstream the bubble (see Figure 4 or Figure 6 for orientation). Utilizing v_∞ and d_B, the axes are made non-dimensional. Expectedly, u_{RMS} is dependent on the Reynolds number and increases with increasing distance from the bubble. It is expected to again decrease further downstream. The dependency on the Reynolds number is weak for the lower three Reynolds numbers. The profiles for $Re = 43$, $Re = 86$ and $Re = 129$ do not differ much; they even intersect. However, what is remarkable is that for similar Reynolds numbers of about $Re = 150$ (circular symbols), u_{RMS} is influenced by the vortex street (filled icons). There are two different staggered configurations presented in Figure 7: A cylinder with 2 cm in diameter, 5.5 cm from the bubble ($L^* = 5.5$ cm/2 cm $= 2.75$), and a cylinder with 2 cm in diameter, 2 cm from the bubble ($L^* = 2$ cm/2 cm $= 1$) (see Figure 1). Both cylinders enhance the velocity fluctuations downstream the bubble and, thus, the cross-mixing within the bubble wake. The cylinder further away ($L^* = 2.75$) increases u_{RMS} to a higher extent (dark-colored icons) than the cylinder which is closer to the bubble ($L^* = 1$) (light-colored icons). To further discuss this point, the fluctuations for the two configurations are plotted for two different Reynolds numbers in Figure 7b. There, it is visible that the type of configuration, i.e., the value of L^*, affects the fluctuations to a greater extent than the Reynolds number. The development of u_{RMS} is very similar for $Re = 126/L^* = 2.75$ and $Re = 163/L^* = 1$, even though the Reynolds numbers differ. The stronger influence of the $L^* = 2.75$ configuration of the u_{RMS} profile can be explained as follows. Due to the larger distance between the bubble and the cylinder, the vortex street is further developed for $L^* = 2.75$. With increasing distance from the cylinder, this momentum affects a larger area of the flow field (like in a free jet) and therefore leads to higher values for u_{RMS} and to a wake that remains for a longer distance because even the dissipation of energy in the larger area needs a longer time (respectively longer way) compared to the case for the $L^* = 1$ configuration. The vortices are detached from the cylinder over a longer distance and are accelerated by the surrounding fluid flow. This development leads to a higher momentum within the wake.

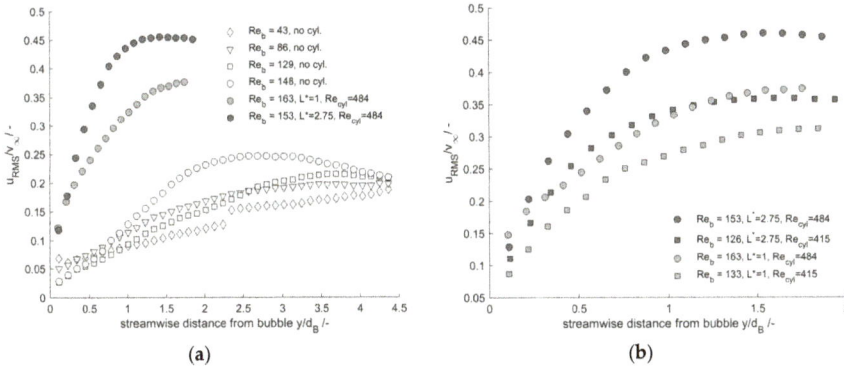

(a)

(b)

Figure 7. Vertical root mean square velocities downstream the bubble at the rear stagnation point, $\varphi = 180°$ ($y/d_B = 0$: surface of the bubble). (**a**) Comparison with and without the vortex street; (**b**) comparison of different configurations and Reynolds numbers.

To get a deeper insight into this finding, velocity profiles of mean velocities in the flow direction (y direction) are plotted in Figures 8–10 at several locations downstream the bubble. The distances from the bubble are given in bubble diameters (d). Velocity fields for the whole viewing field including streamlines are added in small pictures to the profiles. The bubble, as well as the shadow of the bubble, which occurs as a result of the laser light sheet illumination, are marked. Due to the bubble, the mean velocity in the y direction v_{mean} decreases dramatically. Further away from the bubble, v_{mean} again increases. The width of the primary bubble wake is in the order of the bubble diameter. The sharp changes in v_{mean} depict the spatial extension of the primary bubble wake. In Figure 8, it can be seen that in absence of a cylinder wake, these sharp changes blur with increasing distance from the bubble until they are no longer sharp for a distance three bubble diameters from the rear stagnation point (180°). In the presence of cylinder wakes (Figures 9 and 10), the velocity profiles look different. Due to the cylinder, which is located at positive x values, upstream the bubble, the velocity increases to a lesser extent to the right from the bubble (increasing x values). The influence of the different cylinder setups is well visible: for the $L^* = 1$ configuration, the primary cylinder wake is still visible. This leads to a more rapid blurring of the bubble wake: already one bubble diameter behind the rear stagnation point, the bubble wake is no longer sharply recognizable. For the $L^* = 2.75$ configuration, the bubble wake is blurred 1.5 diameters behind the rear stagnation point. This means that the presence of the cylinder wake shortens the primary bubble wake length. The three-dimensional effects found in the cylinder wake may enhance this effect. The higher value of L^* leads to a stronger increase of the vertical RMS velocities, whereas the bubble wake length remains longer. Between the bubble and the cylinder, an inner shear layer is formed, which is described later on when Figure 11 is discussed. This shear layer transfers momentum from the cylinder wake into the bubble wake. Since the momentum within the cylinder wake for the $L^* = 2.75$ configuration is considered to have a larger momentum (see the discussion of Figure 7b) close to the bubble, the inner shear layer can transfer the momentum to the bubble wake. Thus, the bubble wake remains longer for this configuration. Furthermore, it is visible that for the $L^* = 1$ configuration, the bubble wake is attracted strongly to the cylinder wake. The bubble wake is therefore pulled away. It does not remain downstream the bubble, but it is drawn in the positive x/d_B direction. As mentioned earlier, the cylinders enhance the cross-mixing (higher u_{RMS} velocities, Figure 7) and, thus, the interaction of the wake with the external flow. The observation of the shortened primary wake leads to the assumption that not only the cross-mixing, but also the downstream mixing is enhanced. This downstream mixing is enhanced more strongly for the $L^* = 1$ configuration due to the shortened wake. Transferred mass from the bubble into the liquid may therefore be mixed into the liquid remarkably faster.

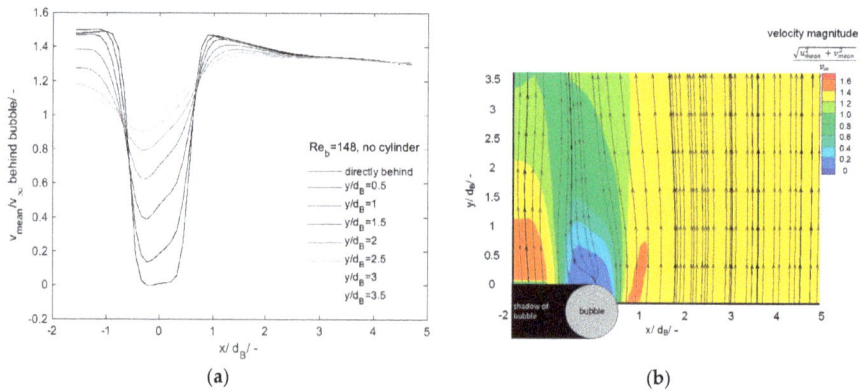

Figure 8. Horizontal velocity profiles (**a**) and velocity fields with stream traces (**b**) around a single bubble (no cylinder).

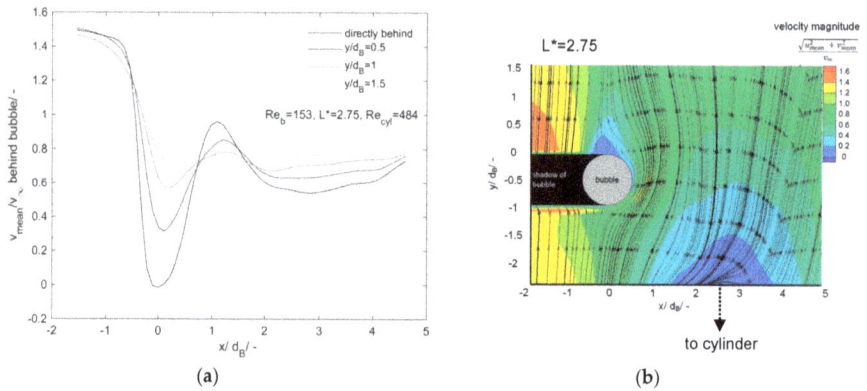

Figure 9. Horizontal velocity profiles (**a**) and velocity fields with stream traces (**b**) around a staggered configuration of a single bubble and a cylinder (*L** = 2.75).

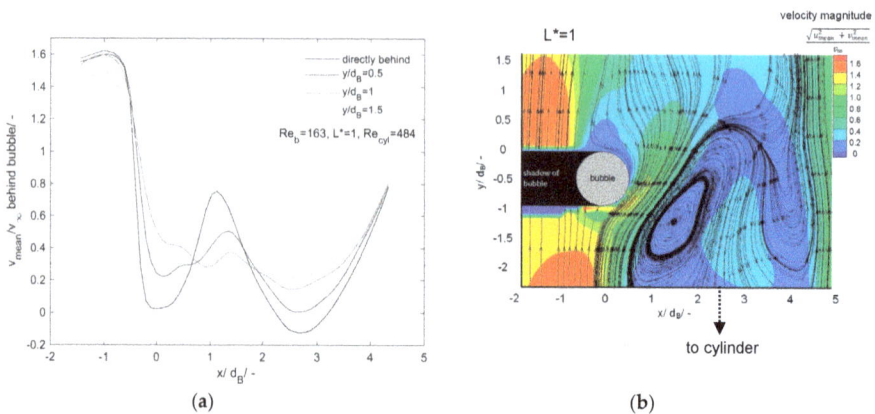

Figure 10. Horizontal velocity profiles (**a**) and velocity fields with stream traces (**b**) around a staggered configuration of a single bubble and a cylinder (*L** = 1).

Due to the fact that the centers of the cylinder and the bubble are at different x coordinates, the cylinder wake attracts the bubble wake, and therefore, it is slightly shifted to the right (higher x values). This is more clearly visible for the $L^* = 1$ configuration. The formation of an asymmetric wake structure of both the bubble and the cylinder is a convincing explanation of why the $L^* = 2.75$ configuration leads to higher velocity fluctuations directly behind the bubble in Figure 7.

For both configurations of bubble and cylinder, no dominating frequency could be found at the observation point 1 cm behind the bubble. Thus, no Strouhal numbers are calculated. From Figure 7, it could be learned, however, that the vertical velocity fluctuations increase behind the bubble. By observation of the temporal vertical velocity profiles, it could be deduced that the fluctuations increase in an irregular manner, which explains the fact that no sharp peak in the power spectrum is found.

The interaction of the bubble wake and the cylinder wake is strongly dependent on the distance between them. Strong wake interaction is well visible for $L^* = 1$. The formation of an inner shear layer resulting from proximity and wake interference can be observed by following the stream traces. In this study, for the sake of simplicity, it is called the inner shear layer. Strictly speaking, according to Sumner [17], it consists of the upstream inner shear layer (cylinder), the downstream inner shear layer (bubble), as well as the gap between them. Tokuhiro et al. [21] observed a jet-like behavior when investigating two bubbles in a parallel configuration. Due to the staggered configuration of this work, a jet-like behavior cannot arise; however, the complex interaction of shear layers is visible. Along the inner shear layer, vortices and even coupled vortices occur. It has to be pointed out that wake interaction can be investigated better by analyzing instantaneous velocity fields. This is depicted in Figure 11. Instantaneous velocity fields of the velocity magnitude are presented for three time steps. Again, the bubble and the bubble shadow are marked. Observation of the stream traces gives insight into the vortex structures, which approach the interface. Many vortices are formed along the inner shear layer, which is highlighted white in Figure 11. A good way to analyze flow structures is to use flow topology [32–34]. For the flow around cylinders, proper orthogonal decomposition, dynamic mode decomposition and critical point theory have been applied to experimental and numerical data to analyze the flow structure [35,36]. This gives the opportunity to make statements concerning the kinetic energy distribution within nonlinear and unsteady flow [35]. Vortices can be detected furthermore via vortex detection criteria [37–39]. Following the streamlines can give a first visualization of vortices within instantaneous flow fields. A node (critical point) shows the presence of a vortex, while the limiting streamlines follow the trajectory of the inner shear layer. The vortex is separated from the bubble by the inner shear layer. Jeong and Hussain [38] suggest a criterion that identifies local pressure minima by making a critical point analysis of the Hessian of pressure. Starting with the Navier–Stokes equation for incompressible flows and neglecting unsteady irrotational straining and viscous effects, they find that a vortex core is a connected region with two negative eigenvalues of $S^2 + \Omega^2$. This condition is equivalent to the condition that after ordering the eigenvalues according to size ($\lambda_1 \geq \lambda_2 \geq \lambda_3$), the second eigenvalue has to be negative: $\lambda_2 < 0$. For this reason, it is called the λ_2 criterion. At the point where the inner shear layer touches the bubble, the application of the λ_2 criterion yields a minimum (see Figure 11, marked purple). This location can therefore be interpreted as an important point from which the separation of vortices from the bubble starts. Only the vortices that are formed downstream the bubble move visibly in the direction of flow. Interestingly, it can be observed how a coupled vortex is formed, which is also located close to the inner shear layer: in Figure 11c, two vortices rotating in opposite directions are visible at the upper part of the image. This is also found by Hu and Zhou [40], who call this phenomenon vortex pairing and enveloping. In studies dealing with flow around circular cylinders [15,16], the flow is classified based on interference. This classification says for the flow case in this work that there should be proximity and wake interference, which matches the findings of this work. This accordance is remarkable since most studies are carried out at much higher Reynolds numbers and with cylindrical bodies with equal diameters.

As is the case for all experimental investigations, there are several shortcomings of this study, which will be discussed in the following. It has to be mentioned that the measurements take place in a two-dimensional plane. Therefore, three-dimensional effects cannot be observed. Vortices moving out of the laser light sheet plane cannot be tracked once they have left the illuminated area. Trajectories of vortices are therefore not identified in this work. This can be overcome in the future by carrying out three-dimensional measurement methods, e.g., tomographic PIV [41]. As was already discussed earlier, events that take place very close to the interface cannot be tracked via PIV. This is observable in Figure 6. Due to a mask, which has to be put over the bubble during PIV processing, the velocities close to the bubble are low. It is not clear whether there is in fact only very little internal circulation within the bubble or whether the velocities close to the bubble are underestimated due to the data processing. Eventually, the fixation of the bubble reduces the degrees of freedom of the bubble motion. Free rising bubbles undergo different shape regimes as a function of the three non-dimensional numbers, Reynolds, Eotvos and Morton [2]. In contrast to free rising bubbles, the shape of the bubble in this study is constantly spherical. This slightly idealized system has been chosen deliberately since only with a fixed bubble, these deep insights into bubble and cylinder interactions could be carried out.

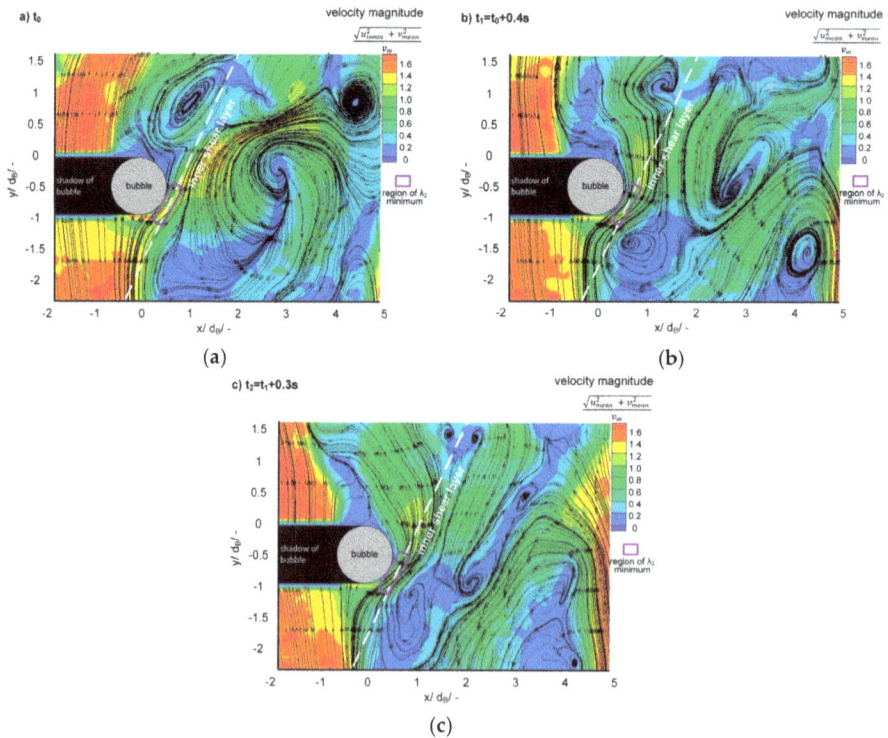

Figure 11. Instantaneous velocity fields and stream traces of a staggered configuration of a single bubble and a cylinder ($L^* = 1$); $Re_b = 163$, $Re_{cyl} = 484$.

4. Conclusions

This study presents the detailed experimental investigation of a single bubble wake in a vortex street behind a cylinder. 2D2C high-speed PIV experiments are conducted to obtain information about velocity profiles and flow structures close to the single bubble. After characterizing the cylinder wake, the flow field around the single bubble is analyzed. For this purpose, streamwise velocity components

Fluids **2018**, *3*, 8

are determined for three azimuthal angles and discussed. Perpendicular to the direction of flow, at $\varphi = 90°$, the thickness of the hydrodynamic boundary layer is estimated.

The streamwise velocity profiles of the single bubble are highly dependent on the azimuthal angle and on the Reynolds number of the bubble. The hydrodynamic boundary layer scales with the Reynolds number. Due to the flow profile within the duct, the velocity towards the bubble is not uniformly v_∞. This leads to the question of how the boundary layer can be estimated even if there is no uniform velocity. This is also the case for the flow field of the interacting bubble and cylinder. From our point of view, it is necessary to have a more general definition of the boundary layer than the 99% criterion.

By analyzing the root mean square velocities behind the bubble, the influence of the cylinder wake is clearly visible. Velocity profiles of the mean velocity in the direction of flow show the different effects of the two configurations of a bubble with and without a cylinder. The type of configuration (L^*) influences the fluctuations to a greater extent than the Reynolds number. Both configurations show that the deceleration of the fluid behind the bubble is compensated earlier than in absence of a cylinder due to momentum transfer by cross-mixing. In combination with the higher RMS velocity profiles, this finding indicates that the cylinder wake enhances the cross-mixing close to the bubble.

For the $L^* = 1$ configuration, direct wake interactions are apparent. The evaluation of instantaneous velocity fields shows the formation of an inner shear layer, which is comprised of the shear layers of the bubble and the cylinder and a gap between them. Along this inner shear layer, vortices and even coupled vortices are formed. It furthermore turns out that the cylinder wake attracts the bubble wake and a joint wake is formed. This interaction of shear layers is considered to be a dominant mechanism for mixing and mass transfer in bubble swarms. Furthermore, in the case of fast parallel or consecutive reactions, the contact time between gaseous and liquid educts might be influenced by the interaction of shear layers, leading to different yield and selectivity.

Future studies are intended to study how mass transfer from a single bubble is influenced by the interaction of shear layers in a vortex street without and with chemical reaction. In combination with the results of this study, a deeper understanding of the influence of vortex structures on the transport of momentum and mass at fluidic interfaces will be achieved.

Acknowledgments: The authors gratefully acknowledge the support that was given by the German Research Foundation (Deutsch Forschungsgemeinschaft (DFG)) within the priority program SPP1740 "Reactive bubbly flows" under Grant No. SCHL 617/12-2.

Author Contributions: Sophie Rüttinger, Marko Hoffmann, and Michael Schlüter conceived and designed the experiments; Sophie Rüttinger performed the experiments; Sophie Rüttinger and Michael Schlüter analyzed the data; Marko Hoffmann contributed reagents/materials/analysis tools; Sophie Rüttinger wrote the paper.

Conflicts of Interest: The authors declare no conflict of interest.

References

1. Brauer, H. *Grundlagen der Einphasen-und Mehrphasenstroemungen*, 1st ed.; Sauerlaender AG: Aarau, Switzerland, 1971.
2. Clift, R.; Grace, J.R.; Weber, M.E. *Bubbles, Drops, and Particles*; Dover Publication, Inc.: Mineola, NY, USA, 1978.
3. Fan, L.S.; Tsuchiya, K. *Bubble Wake Dynamics in Liquids and Liquid-Solid Suspensions*; Butterworth-Heinemann: Quebec City, QC, Canada, 1990.
4. Tsuchiya, K.; Mikasa, H.; Saito, T. Absorption dynamics of CO_2 bubbles in a pressurized liquid flowing downward and its simulation in seawater. *Chem. Eng. Sci.* **1997**, *52*, 4119–4126. [CrossRef]
5. Tsuchiya, K.; Ishida, T.; Saito, T.; Kajishima, T. Dynamics of interfacial mass transfer in a gas-dispersed system. *Can. J. Chem. Eng.* **2003**, *81*, 647–654. [CrossRef]
6. Dani, A.; Guiraud, P.; Cockx, A. Local measurement of oxygen transfer around a single bubble by planar laser-induced fluorescence. *Chem. Eng. Sci.* **2007**, *62*, 7245–7252. [CrossRef]
7. Hanyu, K.; Saito, T. Dynamical mass-transfer process of a CO_2 bubble measured by using LIF/HPTS visualisation and photoelectric probing. *Can. J. Chem. Eng.* **2010**, *139*, 551–560. [CrossRef]

8. Saito, T.; Toriu, M. Effects of a bubble and the surrounding liquid motions on the instantaneous mass transfer across the gas-liquid interface. *Chem. Eng. J.* **2015**, *265*, 164–175. [CrossRef]

9. Bork, O.; Schlueter, M.; Raebiger, N. The impact of local phenomena on mass transfer in gas-liquid systems. *Can. J. Chem. Eng.* **2005**, *83*, 658–666. [CrossRef]

10. Joshi, J.B.; Nandakumar, K.; Evans, G.M.; Pareek, V.K.; Gumulya, M.M.; Sathe, M.J.; Khanwale, M.A. Bubble generated turbulence and direct numerical simulations. *Chem. Eng. Sci.* **2017**, *157*, 26–75. [CrossRef]

11. Alméras, E.; Cazin, S.; Roig, V.; Risso, F.; Augier, F.; Plais, C. Time-resolved measurement of concentration fluctuations in a confined bubbly flow by LIF. *Int. J. Multiph. Flow* **2016**, *83*, 153–161. [CrossRef]

12. Falcone, M.; Bothe, D.; Marschall, H. 3D direct numerical simulations of reactive mass transfer from deformable single bubbles: An analysis of mass transfer coefficients and reaction selectivities. *Chem. Eng. Sci.* **2018**, *177*, 523–536. [CrossRef]

13. Weber, P.S.; Marschall, H.; Bothe, D. Highly accurate two-phase species transfer based on ALE Interface Tracking. *Int. J. Heat Mass Transf.* **2017**, *104*, 759–773. [CrossRef]

14. Krauß, M.; Rzehak, R. Reactive absorption of CO_2 in NaOH: Detailed study of enhancement factor models. *Chem. Eng. Sci.* **2017**, *166*, 193–209. [CrossRef]

15. Zdravkovich, M.M. Review of flow interference between two circular cylinders in various arrangements. *J. Fluids Struct.* **1977**, *1*, 239–261. [CrossRef]

16. Zdravkovich, M.M. The effects of interference between circular cylinders in cross flow. *J. Fluid. Struct.* **1987**, *1*, 239–261. [CrossRef]

17. Sumner, D. Two circular cylinders in cross-flow: A review. *J. Fluid. Struct.* **2010**, *26*, 849–899. [CrossRef]

18. Zhou, Y.; Alam, M.M. Wake of two interacting circular cylinders: A review. *Int. J. Heat Fluid Flow* **2016**, *62*, 510–537. [CrossRef]

19. No, H.; Call, M.; Tokuhiro, A.T. Comparison of Near Wake-Flow Structure Behind a Solid Cap with an Attached Bubble and a Solid Counterpart. In Proceedings of the 4th Joint Fluids Summer Engineering Conference, Honolulu, HI, USA, 6–10 July 2003; pp. 1721–1725.

20. Tokuhiro, A.T.; No, H.; Call, M.; Hishida, K. Comparison of near wake-flow structure behind a solid cap with an attached bubble and a solid counterpart. *JSME Int. J. Ser. B* **2006**, *49*, 737–747. [CrossRef]

21. Tokuhiro, A.; Fujiwara, A.; Hishida, K.; Maeda, M. Measurement in the wake region of two bubbles in close proximity by combined shadow-image and PIV techniques. *J. Fluid. Eng.* **1999**, *121*, 191–197. [CrossRef]

22. Tokuhiro, A.; Maekawa, M.; Iizuka, K.; Hishida, K.; Maeda, M. Turbulent flow past a bubble and an ellipsoid using shadow-image and PIV techniques. *Int. J. Multiph. Flow* **1998**, *24*, 1383–1406. [CrossRef]

23. Roshko, A. On the Development of Turbulent Wakes from Vortex Streets. *Tech. Rep. Arch. Image Libr.* **1954**, 1–28.

24. Chen, Y.N. Jahre Forschung über die Kàrmànschen Wirbelstrassen-Ein Rückblick. *Schweiz. Bauz.* **1973**, *44*, 1079–1096.

25. Williamson, C.H.K. Three-Dimensional Wake Transition. In *Advances in Turbulence VI*; Moreau, R., Gavrilakis, S., Machiels, L., Monkewitz, P.A., Eds.; Springer: Dordrecht, The Netherlands, 1996; pp. 399–402.

26. Barkley, D.; Henderson, R.D. Three-dimensional Floquet stability analysis of the wake of a circular cylinder. *J. Fluid Mech.* **1996**, *322*, 215–241. [CrossRef]

27. Tomboulides, A.G.; Orszag, S.A. Numerical investigation of transitional and weak turbulent flow past a sphere. *J. Fluid Mech.* **2000**, *416*, 45–73. [CrossRef]

28. Le Clainche, S.; Vega, J.M. Higher order dynamic mode decomposition. *SIAM J. Appl. Dyn. Syst.* **2017**, *16*, 882–925. [CrossRef]

29. Oellrich, L.; Schmidt-Traub, H.; Brauer, H. Theoretische berechnung des stofftransports in der umgebung einer einzelblase. *Chem. Eng. Sci.* **1973**, *28*, 711–721. [CrossRef]

30. Komasawa, I.; Otake, T.; Kamojima, M. Wake behavior and its effect on interaction between spherical-cap bubbles. *J. Chem. Eng. Jpn.* **1980**, *13*, 103–109. [CrossRef]

31. Böswirth, L.; Bschorer, S. *Technische Strömungslehre*, 10th ed.; Springer: Berlin, Germany, 2014.

32. Perry, A.E.; Fairlie, B.D. Critical points in flow patterns. *Adv. Geophys.* **1975**, *18*, 299–315.

33. Dallmann, U. Topological structures of three-dimensional vortex flow separation. In Proceedings of the AIAAA 16th Fluid and Plasma Dynamics Conference, Danvers, MA, USA, 12–14 July 1983.

34. Chong, M.S.; Perry, A.E.; Cantwell, B.J. A general classification of three-dimensional flow fields. *Phys. Fluids A Fluid Dyn.* **1990**, *2*, 765–777. [CrossRef]

35. Le Clainche, S.; Li, J.I.; Theofilis, V.; Soria, J. Flow around a hemisphere-cylinder at high angle of attack and low Reynolds number. Part I: Experimental and numerical investigation. *Aerosp. Sci. Technol.* **2015**, *44*, 77–87. [CrossRef]

36. Le Clainche, S.; Rodríguez, D.; Theofilis, V.; Soria, J. Flow around a hemisphere-cylinder at high angle of attack and low Reynolds number. Part II: POD and DMD applied to reduced domains. *Aerosp. Sci. Technol.* **2015**, *44*, 88–100. [CrossRef]

37. Hunt, J.C.R.; Wray, A.A.; Moin, P. Eddies, Streams, and Convergence Zones in Turbulent Flows. Center for Turbulence Research. In Proceedings of the Summer Program, Stanford, CA, USA, 27 June–22 July 1988; pp. 193–208.

38. Jeong, J.; Hussain, F. On the identification of a vortex. *J. Fluid Mech.* **1995**, *285*, 69–94. [CrossRef]

39. Kolář, V. Vortex identification: New requirements and limitations. *Int. J. Heat Fluid Flow* **2007**, *28*, 638–652. [CrossRef]

40. Hu, J.C.; Zhou, Y. Flow structure behind two staggered circular cylinders. Part 1. Downstream evolution and classification. *J. Fluid Mech.* **2008**, *607*, 51–80. [CrossRef]

41. Elsinga, G.E.; Scarano, F.; Wieneke, B.; van Oudheusden, B.W. Tomographic particle image velocimetry. *Exp. Fluids* **2006**, *41*, 933–947. [CrossRef]

fluids

MDPI

Article

On the Bias in the Danckwerts' Plot Method for the Determination of the Gas–Liquid Mass-Transfer Coefficient and Interfacial Area

German E. Cortes Garcia [1], Kevin M. P. van Eeten [1], Michiel M. de Beer [2], Jaap C. Schouten [1] and John van der Schaaf [1,*]

[1] Laboratory of Chemical Reactor Engineering, Department of Chemical Engineering & Chemistry, Eindhoven University of Technology, P.O. Box 513, 5600 MB Eindhoven, The Netherlands; g.e.cortes.garcia@tue.nl (G.E.C.G.); k.m.p.v.eeten@tue.nl (K.M.P.v.E.); j.c.schouten@tue.nl (J.C.S.)

[2] AkzoNobel Chemicals B.V., RD&I, Expert Capability Group-Process Technology, Zutphenseweg 10, 7418 AJ Deventer, The Netherlands; michiel.debeer@akzonobel.com

* Correspondence: j.vanderschaaf@tue.nl; Tel.: +31-40-247-2850; Fax: +31-40-244-6653

Received: 14 December 2017; Accepted: 11 February 2018; Published: 20 February 2018

Abstract: The Danckwerts' plot method is a commonly used graphical technique to independently determine the interfacial area and mass-transfer coefficient in gas–liquid contactors. The method was derived in 1963 when computational capabilities were limited and intensified process equipment did not exist. A numerical analysis of the underlying assumptions of the method in this paper has shown a bias in the technique, especially for situations where mass-transfer rates are intensified, or where there is limited liquid holdup in the bulk compared to the film layers. In fact, systematic errors of up to 50% in the interfacial area, and as high as 90% in the mass-transfer coefficients, can be expected for modern, intensified gas–liquid contactors, even within the commonly accepted validity limits of a pseudo-first-order reaction and Hatta numbers in the range of $0.3 < Ha < 3$. Given the current computational capabilities and the intensified mass-transfer rates in modern gas–liquid contactors, it is therefore imperative that the equations for reaction and diffusion in the liquid films are numerically solved and subsequently used to fit the interfacial area and mass-transfer coefficient to experimental data, which would traditionally be used in the graphical Danckwerts' method.

Keywords: gas–liquid mass transfer; Danckwerts' plot method; numerical simulation; mass-transfer coefficient; interfacial area

1. Introduction

Novel types of process equipment have recently been developed in which the mixing and hydrodynamics of gasses and liquids directly lead to better gas–liquid mass-transfer performance [1–5]. By reducing limitations in heat and mass transfer, chemical processes can be performed at their intrinsic kinetic conditions. These intensified processes will help meet the societal demand for safer, more efficient, and economical production of chemicals.

Accurate measurements of the interfacial area for gas–liquid mass transfer (a_{GL}), and of the gas- and liquid-phase mass-transfer coefficients (k_G and k_L, respectively), are thus required in order to understand the effect of different design and operating parameters on the mass-transfer performance of gas–liquid contactors and reactors.

Numerous methods have been used to measure these parameters and several reviews about them are available in the literature [6–10]. The most commonly used methods rely on chemical and physical absorption experiments and, in most cases, they allow for the measurement of the volumetric mass-transfer coefficients (i.e., the products $k_G \cdot a_{GL}$ and $k_L \cdot a_{GL}$). However, it is often desired to measure a_{GL} independently, since design and operating conditions affect each of these parameters differently.

One of the methods most often used for the independent determination of k_L and a_{GL} is the Danckwerts' plot method. This method was proposed by Danckwerts et al. in 1963 [11] and it has been used ever since to characterize the mass-transfer performance of several gas–liquid contactors and reactors, such as stirred gas–liquid reactors [12], three-phase fluidized beds [13], packed columns [6], bubble columns [14], Venturi contactors [15], and even for gas–liquid–liquid systems [16]. The method relies on the absorption of gaseous A into a liquid, in which B is pre-dissolved. A and B react inside the liquid, increasing the concentration gradient of A near the interface, effectively enhancing the mass-transfer rate. The absorption rate can then be expressed as:

$$Ra_{GL} = k_L a_{GL} C_A^* E \tag{1}$$

where E represents the enhancement factor, which is defined as the factor with which the transport of A through the interface is increased due to the effect of reaction.

According to Hatta's theory [17], the enhancement factor for an irreversible first-order reaction can be expressed in terms of Ha as:

$$E = \frac{1}{C_A^*}\left(C_A^* - \frac{C_A^\infty}{\cosh(Ha)}\right)\frac{Ha}{\tanh(Ha)} \tag{2}$$

where Ha is the Hatta number, defined by:

$$Ha = \frac{\sqrt{k_2 D_A C_B^\infty}}{k_L} \tag{3}$$

Based on surface renewal theory, Danckwerts further derived an approximation for the enhancement factor for the case where the bulk concentration of dissolved gas (A^∞) is zero [6]:

$$E = \sqrt{1 + Ha^2} \tag{4}$$

which holds for $Ha > 0.3$.

If the reaction is fast enough, i.e., $Ha > 3$, Equation (4) can be approximated as

$$E = Ha \tag{5}$$

and the rate of absorption becomes independent of the mass-transfer coefficient

$$Ra_{GL} = a_{GL} C_A^* \sqrt{k_2 D_A C_B^\infty} \tag{6}$$

Equations (4) and (5) can also be applied to irreversible second-order reactions as long as the reactions can be considered pseudo-first-order. It should then hold that:

$$Ha << k_L^2 \cdot E_i \tag{7}$$

Combining Equations (1) and (4) yields the following relationship for the absorption rate:

$$Ra_{GL} = a_{GL} C_A^* \sqrt{D_A k_2 C_B^\infty + k_L^2} \tag{8}$$

which can be rearranged to give:

$$\left(\frac{Ra_{GL}}{C_A^*}\right)^2 = a_{GL}^2 \cdot D_A \cdot k_{app} + (k_L a_{GL})^2 \tag{9}$$

in which the apparent first-order rate constant is

$$k_{app} = k_2 C_B^\infty \tag{10}$$

Chemical absorption experiments with a pseudo-first-order reaction can then be used to characterize the mass-transfer performance of gas–liquid contactors and reactors in two different ways: (a) by measuring a_{GL} in the fast reaction regime (i.e., $3 < Ha < 0.5E_i$) based on Equation (6), and (b) by measuring both a_{GL} and k_L in the intermediate reaction regime (i.e., $0.3 < Ha < 3$) based on Equation (9). Method (b) is known as the Danckwerts' plot method. This method uses a pseudo-first-order reaction between a liquid and an absorbed gas to measure the absorption rate at different apparent rate constants (k_{app}). The data thus obtained is used to construct a Danckwerts' plot, as schematically depicted in Figure 1, from which a_{GL} and k_L can be found from the slope and the intercept, respectively. The apparent first-order rate constant can be varied by either changing the concentration of reactant B or by modifying the reaction rate constant with the help of a catalyst [6].

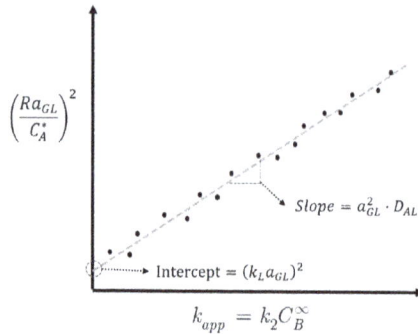

Figure 1. Schematic representation of the Danckwerts' plot method.

The method was developed in 1963 when there was no global trend in process intensification for the development of novel gas–liquid contactors. Although the method still has its merit, it has an important limitation: the existence of a systematic error, or bias, in the estimation of the mass-transfer coefficient (k_L). This paper discusses this bias and its dependence on the Hatta number and on the relative liquid volume in the bulk versus that in the film layers and how this limits its use for novel, intensified gas–liquid contactors.

2. Numerical Methods

Every experimental method is only as accurate as the underlying data. In order to present the bias in the Danckwerts' plot method in the most accurate way, data is generated by solving the continuum equations for diffusion and reaction in the liquid-film layer near the interface. Since this simplified, one-dimensional (1D)-diffusion model is the basis for the original method, the same system is solved here, but in a more rigorous way. The numerical data generated with the following method will then be subjected to the Danckwerts' plot. The mass-transfer parameters will be extracted from the slope and the intercept and will be compared to parameters initially fixed in the model in order to analyze the accuracy of the method.

2.1. Diffusion-Reaction Model

The numerical method is based on a Two-Film model [18]. In this model, reactant A is in the gas phase and reactant B is in the liquid as shown in Figure 2. When gas-side mass transfer is neglected, the concentration of A on the interface is determined by the partial pressure of A in the gas and by

Henry's solubility constant (H). Component A diffuses through the film of thickness δ into the bulk while simultaneously reacting with B, which diffuses in the opposite direction.

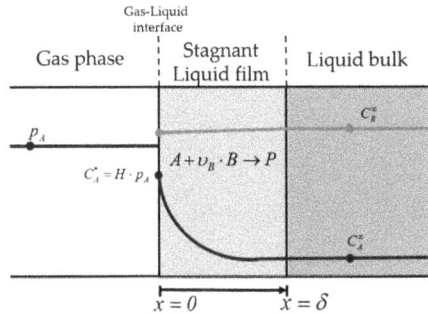

Figure 2. Concentration profiles of A and B within stagnant liquid film (Film Theory).

The liquid-side mass-transfer coefficient is determined by the thickness of this stagnant liquid film and the diffusivity of the species (D_A), as shown in Equation (11).

$$k_L = \frac{D_A}{\delta} \tag{11}$$

A mass balance over the liquid film results in the following two equations:

$$D_A \frac{d^2 C_A}{dx^2} = k_2 C_A C_B \tag{12}$$

$$D_B \frac{d^2 C_B}{dx^2} = v_B \cdot k_2 C_A C_B \tag{13}$$

These two equations can be solved simultaneously in MATLAB (R2017a, MathWorks, Inc., Natick, MA, USA) using the bvp4c solver subject to boundary conditions (14)–(17) to obtain the concentration profiles of A and B within the liquid film.

$$C_A|_{x=0} = C_A^* \tag{14}$$

$$\left(-D_A \frac{dC_A}{dx} \right)\bigg|_{x=\delta} \cdot a_{GL} = k_2 C_A^\infty C_B^\infty \cdot (\varepsilon_l - \delta \cdot a_{GL}) \tag{15}$$

$$\frac{dC_B}{dx}\bigg|_{x=0} = 0 \tag{16}$$

$$C_B|_{x=\delta} = C_B^\infty \tag{17}$$

Boundary condition (15) states that at $x = \delta$ the total flux of A equals the rate of reaction in the liquid bulk. Consequently, $(\varepsilon_l - \delta \cdot a_{GL})$ equals the bulk volume, excluding the liquid film.

After solving Equations (12) and (13), the overall rate of absorption can be calculated from the slope of the concentration profile at the gas–liquid interface, given by Equation (18).

$$Ra_{GL} = -D_A \cdot \left(\frac{dC_A}{dx} \right)_{x=0} \cdot a_{GL} \tag{18}$$

This model was used to simulate a bubble column, a packed bed, and a rotating packed bed so that the effect of the holdup ε_l and of the magnitude of the interfacial area a_{GL} can be accounted for. The model requires input for the liquid holdup and interfacial area. Typical values occurring in bubble columns, packed beds, and rotating packed beds are chosen as represented in Table 1. Please note that k_L is fixed in this model by choosing a value for δ in boundary conditions (15) and (17), while a_{GL} is set via Equation (18). Other parameters used in the simulation are shown in Table 2.

Table 1. Parameters used in the simulations.

Equipment	ε_l ($m_L^3 \cdot m_R^{-3}$)	a_{GL} ($m_i^2 \cdot m_L^{-3}$)	k_L ($m_L^3 \cdot m_i^{-2} \cdot s^{-1}$)	($\varepsilon_l / a_{GL} \cdot \delta$)	References
Bubble Column	0.90	50	1.0×10^{-3}	1.0×10^4	[19–21]
			1.0×10^{-4}	1.0×10^3	
Packed Bed	0.15	150	1.0×10^{-3}	5.6×10^2	[22–24]
			1.0×10^{-4}	5.6×10^1	
Rotating Packed Bed	0.03	700	1.0×10^{-3}	2.4×10^1	[25–27]
			1.0×10^{-4}	2.4×10	

Table 2. Other parameters used in the simulations (based on properties of the system $CO_2/NaOH$ [6]).

Parameter	Value
D_A ($m_L^4 \cdot m_i^{-2} \cdot s^{-1}$)	1.80×10^{-9}
D_B ($m_L^4 \cdot m_i^{-2} \cdot s^{-1}$)	3.10×10^{-9}
C_B^∞ ($mol \cdot m_L^{-3}$)	1.00×10^3
C_A^* ($mol \cdot m_L^{-3}$)	3.90×10^{-1}
E_i	2.18×10^2

From the results following Equation (18), a Danckwerts' plot can be finally made from which the regressed values of a_{GL} and k_L are compared with their counterparts in Table 1.

3. Results and Discussion

3.1. Accuracy of the Danckwerts' Plot Method

The model described in Section 2.1 was used to obtain the concentration profiles for a bubble column using its respective values from Tables 1 and 2, with a k_L value of 1.0×10^4 $m_L^3 \cdot m_i^{-2} \cdot s^{-1}$. Figure 3a shows typical concentration profiles for a bubble column at Hatta numbers between 0.3 and 3, with a spacing between them of $\Delta Ha = 0.15$. The concentration of B is nearly constant and the reaction can therefore be considered pseudo-first-order in A. Since Equation (4) holds, the graphical method should, in principle, be able to be applied to obtain a_{GL} and k_L.

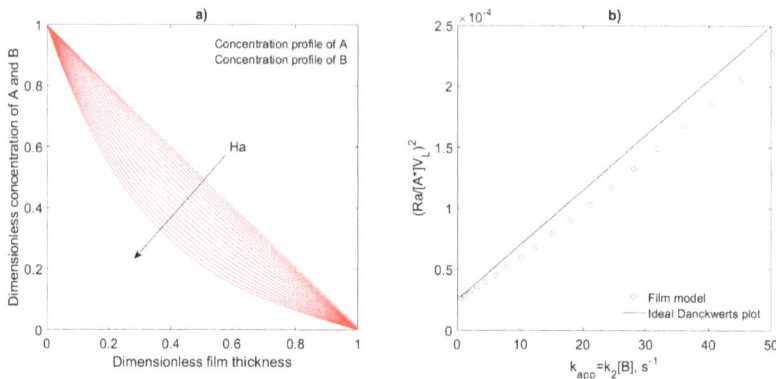

Figure 3. Concentration profiles (**a**) and Danckwerts' plot (**b**) for $\varepsilon_l / (a_{GL} \cdot \delta) = 1.0 \times 10^3$ (bubble column).

In Figure 3b, the results from the simulation data are plotted alongside the theoretical line expected from the Danckwerts' method using a_{GL} and k_L as reported in Table 1 for a bubble column. If the method were accurate enough, regressing the simulation results would produce a straight line with a very similar slope and intercept to those of the "ideal" plot. Nevertheless, inspection of this figure shows that the slope of the numerical data approaches that of the theoretical line only for high values of the apparent rate constant. On the other hand, the Danckwerts' method results in the right value of the mass-transfer coefficient at low values of k_{app} at the cost of accuracy in a_{GL}. In addition to this effect, a small deviation in the slope of k_{app} further reduced the accuracy in the mass-transfer coefficient. It is thus clear that a trade-off exists between the accuracy at which each of these parameters can be estimated.

Since the slope and the intercept of a regressed plot change as k_{app} increases, the mass-transfer parameters thus calculated should vary with the Hatta number. As the Hatta number depends on the mass-transfer coefficient, it cannot be known a priori at which position in Figure 3b the experiments were performed, and thus to which accuracy a_{GL} and k_L were determined.

To study the effect of Hatta number on the accuracy of the method, regressions of the simulation data were performed over a wide range of Hatta numbers. For each Hatta number, a regression was made over five equidistant points in the range $Ha < x < Ha + 0.1$. From this, a_{GL}, k_L and their corresponding percentage errors with respect to the values in Table 1 were calculated. The results for a bubble column are shown in Figure 4. The shaded area corresponds to the intermediate reaction regime, i.e., $0.3 < Ha < 3$, where the Danckwerts' plot method should hold. In this range of Ha, the interfacial area is underestimated by up to 20%, it then becomes more accurate for faster reactions, i.e., $Ha > 3$, and starts to deviate again for $Ha > 100$, or $Ha \sim 0.5Ei$. At this point, the reaction is too fast for the assumption of pseudo-first-order to be valid. The source of error in the intermediate reaction regime comes from the misprediction of the Enhancement factor by Equation (4), which was derived under the assumption that the concentration of reactant A in the liquid bulk equals zero, an assumption that holds well for fast reactions but not so well for intermediate and slow reactions relative to the mass-transfer rate. These results show that chemical absorption can lead to very accurate measurements of the interfacial area if a fast reaction is used while a loss of accuracy should be expected when using intermediate reactions. The mass-transfer coefficient, on the other hand, lies close—without fully converging—to the predetermined value of 1.0×10^{-4} $m_L{}^3 \cdot m_i{}^{-2} \cdot s^{-1}$ when it is measured within the intermediate reaction regime but is highly mispredicted in the fast reaction regime. In this sense, it does not seem possible to determine, simultaneously, both mass-transfer parameters in an accurate way.

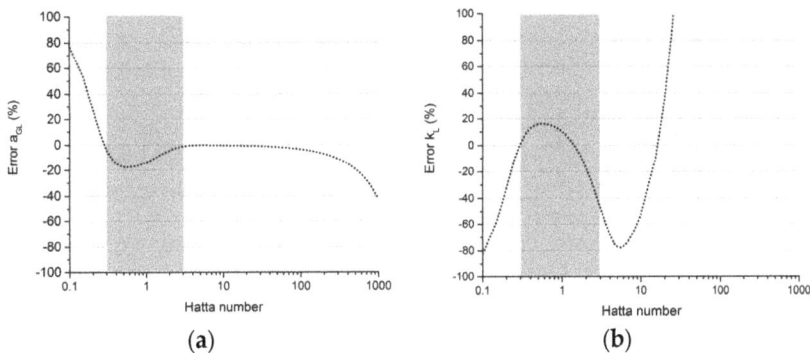

Figure 4. Percentage error in the estimation of (a) interfacial area (a_{GL}) and (b) mass-transfer coefficient (k_L) using the Danckwerts' plot over a wide range of Ha for a Bubble column, $\varepsilon_l/(a_{GL} \cdot \delta) = 1000$.

3.2. Bias in the Determination of k_L and Effect of the Ratio $\varepsilon_l/(a_{GL}\cdot\delta)$

Figure 5 shows similar plots to those in Figure 4 but for the intermediate reaction regime only. The different lines represent one of the three equipment types previously mentioned, with their characteristic hydrodynamic and mass-transfer parameters listed in Table 1. It can be seen in the figure that again there is always a bias in the determination of a_{GL} and k_L At higher Hatta numbers, a_{GL} can be determined more accurately, while the error in k_L is less, but never fully disappears, at lower Hatta numbers. Moreover, the error at lower Hatta numbers rapidly increases when $\varepsilon_l/(a_{GL}\cdot\delta)$ decreases.

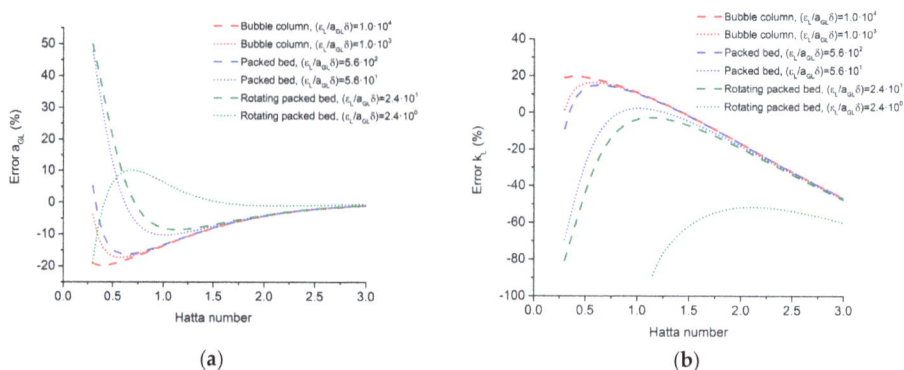

Figure 5. Percentage error in estimation of (**a**) interfacial area (a_{GL}) and (**b**) mass-transfer coefficient (k_L) using the Danckwerts' plot in the intermediate reaction regime for the different equipment.

These results show that the Danckwerts' plot method leads to acceptable estimations of the interfacial area but estimates of the mass-transfer coefficient are subject to a systematic error. This error becomes more significant as the ratio of liquid in the bulk over liquid in the film layers decreases, i.e., as $\varepsilon_l/(a_{GL}\cdot\delta)$ decreases. The method has been said to be valid for $0.3 < Ha < 3$; however, the exact range depends on the validity of the assumption that A^∞ is close to zero. This assumption can only hold when the rate of reaction in the bulk is large enough to prevent the presence of unreacted A in the bulk itself. The lower limit of this range is located at $Ha = 0.3$ for systems with large volumes of bulk, such as bubble columns. However, this lower limit shifts towards higher Ha numbers as $\varepsilon_l/(a_{GL}\cdot\delta)$ decreases. If this ratio becomes too small, which may occur with process-intensified equipment, at some point an accurate determination of k_L cannot be obtained any longer. This is true even without considering the possible accumulation in time of unreacted A for systems with limited bulk, a phenomenon previously described by Elk et al. [28]. A natural approach would then be to introduce a correction factor in terms of the ratio $\varepsilon_l/(a_{GL}\cdot\delta)$ to obtain a revised and more general expression for the range of validity of the method. In this way, it could be applied with more certainty to reactors with different relative amounts of liquid bulk. However, according to Equations (12) and (13) and boundary conditions (14)–(17), the range of validity depends not only on this ratio but also on the reaction rate constant, the stoichiometric coefficient, and the diffusivity of species A and B and their concentration at the interface and in the liquid bulk, respectively. Introducing such a revised range of validity for the method thus seems unfeasible and of little practical significance as it would require knowledge on the mass-transfer coefficient a priori.

Even inside the traditional limits of $0.3 < Ha < 3$, k_L is subject to a bias of up to 90% while the interfacial area is subject to a bias of up to 50%. In this perspective, it is interesting to note the findings of Cents et al. [29], who measured $k_L \cdot a_{GL}$ simultaneously by physical desorption and by chemical absorption, ending up with differences of about 64% between both measurements.

In intensified equipment, where there is a limited amount of liquid in the bulk and the mass-transfer coefficient is large, the interval in allowable Hatta numbers becomes too small to make accurate predictions with the Danckwerts' method. The method can therefore only be used to estimate the interfacial area accurately when considerable care is taken that the assumptions of zero bulk concentration of solute gas and large bulk volumes are satisfied. In addition, the method only allows for an order of magnitude estimation of k_L. A more accurate way to obtain the mass-transfer parameters from experimental data would be to fit k_L and a_{GL} in a numerical simulation of the equations of reaction and diffusion.

4. Conclusions

The Danckwerts' plot method results from a mathematical approximation for the enhancement factor for intermediate reactions. The derivation of the Danckwerts' plot equation has a solid mathematical basis and therefore it has been used often for the simultaneous determination of the mass-transfer coefficient and the effective interfacial area of gas–liquid absorbers and reactors. However, this method was developed more than 50 years ago, when computational capability was limited and process-intensified equipment types were not under development. At this time, there was thus no need and no computational possibility to study the effect that the underlying assumptions had on the accuracy of the method. This work shows that the Danckwerts' method can only lead to fair estimates of the mass-transfer coefficient over a very limited range of Hatta numbers, depending on the relative liquid volume in the bulk versus that in the film layers. Nonetheless, the accuracy in k_L comes at the cost of a loss in accuracy in the prediction of the interfacial area. On the other hand, if the method is applied to measure the interfacial area, it comes at a loss in accuracy in k_L. A systematic error appears then to exist in the method, since k_L cannot be determined accurately even using simulated absorption data. A bias of up to 90% in the mass-transfer coefficient and of up to 50% in the interfacial area was found for the equipment simulated. A preferred approach to isolate k_L from a_{GL} would be to use the chemical absorption method (either with a Danckwerts' plot or with single measurements) to find the interfacial area and to use physical absorption experiments to measure $k_L \cdot a_{GL}$. Moreover, with the current computational power, a more accurate method would be to fit k_L and a_{GL} in a numerical simulation of the equations of reaction and diffusion to experimental data that would typically be subjected to the graphical Danckwerts' method.

Acknowledgments: This research was carried out within the HighSinc program—a joint development between AkzoNobel and the Department of Chemical Engineering and Chemistry from Eindhoven University of Technology—where 12 PhD students will work on various aspects and applications of HiGee technologies.

Author Contributions: German E. Cortes Garcia: Conceived and designed the simulations, analyzed the data, and wrote the paper. Kevin M. P. van Eeten: Co-wrote the paper and helped interpret the results. Michiel M. de Beer: Co-wrote the paper and helped interpret the results. Jaap C. Schouten: Co-wrote the paper and helped interpret the results. John van der Schaaf: Initiated the research, co-wrote the paper, and helped interpret the results.

Conflicts of Interest: The authors declare no conflict of interest.

Nomenclature

a_{GL}	Effective gas–liquid interfacial area, $m_i^2 \cdot m_R^{-3}$
C_A	Concentration of solute gas A within stagnant liquid film, $mol \cdot m_L^{-3}$
C_A^*	Concentration of solute gas A at the gas–liquid interphase, $mol \cdot m_L^{-3}$
C_A^∞	Concentration of solute gas A in liquid bulk, $mol \cdot m_L^{-3}$
C_B	Concentration of liquid reactant B within stagnant liquid film, $mol \cdot m_L^{-3}$
C_B^∞	Concentration of liquid reactant B in liquid bulk, $mol \cdot m_L^{-3}$
D_A	Diffusion coefficient of component A in the liquid, $m_L^4 \cdot m_i^{-2} \cdot s^{-1}$
D_B	Diffusion coefficient of component B in the liquid, $m_L^4 \cdot m_i^{-2} \cdot s^{-1}$
E	Enhancement factor
E_i	Enhancement factor for an instantaneous reaction, defined by $E_i = 1 + \frac{D_{BL}}{D_A} \cdot \frac{C_B^\infty}{v_B \cdot C_A^*}$

ε_l	Liquid holdup
Ha	Hatta number
k_{app}	Apparent first-order rate constants, s^{-1}
k_G	Gas-phase mass-transfer coefficient, $m_G^3 \cdot m_i^{-2} \cdot s^{-1}$
$k_G \cdot a_{GL}$	Volumetric gas-phase mass-transfer coefficient, $m_G^3 \cdot m_R^{-3} \cdot s^{-1}$
k_L	Liquid-phase mass-transfer coefficient, $m_L^3 \cdot m_i^{-2} \cdot s^{-1}$
$k_G \cdot a_{GL}$	Volumetric liquid-phase mass-transfer coefficient, $m_L^3 \cdot m_R^{-3} \cdot s^{-1}$
k_2	Reaction rate constant for second-order reaction
$R \cdot a_{GL}$	Overall rate of absorption, $mol \cdot s^{-1}$
v_B	Stoichiometric coefficient of B
x	Position perpendicular to interface, m
y	Position parallel to interface, m

References

1. Van Eeten, K.M.P.; Verzicco, R.; Van Der Schaaf, J.; Van Heijst, G.J.F.; Schouten, J.C. A numerical study on gas-liquid mass transfer in the rotor-stator spinning disc reactor. *Chem. Eng. Sci.* **2015**, *129*, 14–24. [CrossRef]
2. De Beer, M.M.; Keurentjes, J.T.F.; Schouten, J.C.; Van Der Schaaf, J. Bubble formation in co-fed gas-liquid flows in a rotor-stator spinning disc reactor. *Int. J. Multiph. Flow* **2016**, *83*, 142–152. [CrossRef]
3. Haseidl, F.; Pottbäcker, J.; Hinrichsen, O. Gas–liquid mass transfer in a rotor–stator spinning disc reactor: Experimental study and correlation. *Chem. Eng. Process. Process Intensif.* **2016**, *104*, 181–189. [CrossRef]
4. Chen, Y.; Lin, C.; Liu, H. Mass transfer in a rotating packed bed with various radii of the bed. *Ind. Eng. Chem. Res.* **2005**, *44*, 7868–7875. [CrossRef]
5. Garcia, G.E.C.; Van Der Schaaf, J.; Kiss, A.A. A review on process intensification in higee distillation. *J. Chem. Technol. Biotechnol.* **2017**, *92*, 1136–1156. [CrossRef]
6. Danckwerts, P.V. *Gas-Liquid Reactions*; Mcgraw-Hill Book Co.: New York, NY, USA, 1970.
7. Rejl, J.F.; Linek, V.; Moucha, T.; Valenz, L. Methods standardization in the measurement of mass-transfer characteristics in packed absorption columns. *Chem. Eng. Res. Des.* **2009**, *87*, 695–704. [CrossRef]
8. Hegely, L.; Roesler, J.; Alix, P.; Solaize, D.; Rouzineau, D.; Meyer, M. Absorption methods for the determination of mass transfer parameters of packing internals: A literature review. *AIChE J.* **2017**, *63*. [CrossRef]
9. Last, W.; Stichlmair, J. Determination of mass transfer parameters by means of chemical absorption. *Chem. Eng. Technol.* **2002**, *25*, 385–391. [CrossRef]
10. Hoffmann, A.; Mackowiak, J.F.; Gorak, A.; Haas, M.; Loning, J.-M.; Runowski, T.; Hallenberger, K. Standardization of mass transfer measurements. a basis for the description of absorption processes. *Chem. Eng. Res. Des.* **2007**, *85*, 40–49. [CrossRef]
11. Danckwerts, P.V.; Kennedy, A.M.; Roberts, D. Kinetics of CO_2 absorption in alkaline solutions—II. Absorption in packed column and tests of surface-renewal models. *Chem. Eng. Sci.* **1963**, *18*, 63–72. [CrossRef]
12. Linek, V.; Kordac, M.; Moucha, T. Mechanism of mass transfer from bubbles in dispersions part II: Mass transfer coefficients in stirred gas–liquid reactor and bubble column. *Chem. Eng. Process.* **2005**, *44*, 121–130. [CrossRef]
13. Strumillo, C.; Kundra, T. Interfacial area in three-phase fluidized beds. *Chem. Eng. Sci.* **1976**, *32*, 229–232. [CrossRef]
14. Maalej, S.; Benadda, B.; Otterbein, M. Interfacial area and volumetric mass transfer coefficient in a bubble reactor at elevated pressures. *Chem. Eng. Sci.* **2003**, *58*, 2365–2376. [CrossRef]
15. Gourich, B.; Vial, C.; Soulami, M.B.; Zoualian, A.; Ziyad, M. Comparison of hydrodynamic and mass transfer performances of an emulsion loop-venturi reactor in cocurrent downflow and upflow configurations. *Chem. Eng. J.* **2008**, *140*, 439–447. [CrossRef]
16. Cents, A.H.G.; Brilman, D.W.F.; Versteeg, G.F. Gas absorption in an agitated gas-liquid-liquid system. *Chem. Eng. Sci.* **2001**, *56*, 1075–1083. [CrossRef]
17. Hatta, S. *Technological Reports of Tohoku University*; Tohoku University: Sendai, Japan, 1932; Volume 10, p. 119.
18. Whitman, W.G. The two-film theory of gas absorption. *Int. J. Heat Mass Transf.* **1962**, *5*, 429–433. [CrossRef]

19. Pohorecki, R.; Moniuk, W.; Zdrojkowski, A. Hydrodynamics of a bubble column under elevated pressure. *Chem. Eng. Sci.* **1999**, *54*, 5187–5193. [CrossRef]

20. Kulkarni, A.A.; Joshi, J.B.; Kumar, V.R.; Kulkarni, B.D. Simultaneous measurement of hold-up profiles and interfacial area using lda in bubble columns: predictions by multiresolution analysis and comparison with experiments. *Chem. Eng. Sci.* **2001**, *56*, 6437–6445. [CrossRef]

21. Bouaifi, M.; Hebrard, G.; Bastoul, D.; Roustan, M. A comparative study of gas hold-up, bubble size, interfacial area and mass transfer coefficients in stirred gas–liquid reactors and bubble columns. *Chem. Eng. Process.* **2001**, *40*, 97–111. [CrossRef]

22. Sahay, B.N.; Sharma, M.M. Effective interfacial area and liquid and gas side mass transfer coefficients in a packed column. *Chem. Eng. Sci.* **1973**, *28*, 41–47. [CrossRef]

23. Shulman, H.L.; Ulrich, C.F.; Wells, N. Performance of packed columns. *AIChE J.* **1955**, *1*, 247–253. [CrossRef]

24. Piché, S.; Grandjean, B.P.A.; Larachi, F. Reconciliation procedure for gas–liquid interfacial area and mass-transfer coefficient in randomly packed towers. *Ind. Eng. Chem. Res.* **2002**, *41*, 4911–4920. [CrossRef]

25. Zheng, X.; Chu, G.; Kong, D.; Luo, Y.; Zhang, J.; Zou, H. Mass transfer intensification in a rotating packed bed with surface-modified nickel foam packing. *Chem. Eng. J.* **2016**, *285*, 236–242. [CrossRef]

26. Luo, Y.; Chu, G.; Zou, H.; Zhao, Z.; Dudukovic, M.P.; Chen, J. Gas–liquid effective interfacial area in a rotating packed bed. *Ind. Eng. Chem. Res.* **2012**. [CrossRef]

27. Burns, J.R.; Jamil, J.N.; Ramshaw, C. Process intensification: Operating characteristics of rotating packed beds—Determination of liquid hold-up for a high-voidage structured packing. *Chem. Eng. Sci.* **2000**, *55*, 2401–2415. [CrossRef]

28. Van Elk, E.P.; Knaap, M.C.; Versteeg, G.F. Application of the penetration theory for gas-liquid mass transfer without liquid bulk. differences with systems with a bulk. *Chem. Eng. Res. Des.* **2007**, *85*, 516–524. [CrossRef]

29. Cents, A.H.G.; De Bruijn, F.T.; Brilman, D.W.F.; Versteeg, G.F. Validation of The Danckwerts-plot technique by simultaneous chemical absorption of CO_2 and physical desorption of O_2. *Chem. Eng. Sci.* **2005**, *60*, 5809–5818. [CrossRef]

MDPI

Article

Evaluation of Interfacial Heat Transfer Models for Flashing Flow with Two-Fluid CFD

Yixiang Liao * and Dirk Lucas

Institute of Fluid Dynamics, Helmholtz-Zentrum Dresden-Rossendorf, Bautzner Landstraße 400, 01328 Dresden, Germany; d.lucas@hzdr.de
* Correspondence: y.liao@hzdr.de; Tel.: +49-351-260-2389

Received: 4 May 2018; Accepted: 29 May 2018; Published: 1 June 2018

Abstract: The complexity of flashing flows is increased vastly by the interphase heat transfer as well as its coupling with mass and momentum transfers. A reliable heat transfer coefficient is the key in the modelling of such kinds of flows with the two-fluid model. An extensive literature survey on computational modelling of flashing flows has been given in previous work. The present work is aimed at giving a brief review on available theories and correlations for the estimation of interphase heat transfer coefficient, and evaluating them quantitatively based on computational fluid dynamics simulations of bubble growth in superheated liquid. The comparison of predictions for bubble growth rate obtained by using different correlations with the experimental as well as direct numerical simulation data reveals that the performance of the correlations is dependent on the Jakob number and Reynolds number. No generally applicable correlations are available. Both conduction and convection are important in cases of bubble rising and translating in stagnant liquid at high Jakob numbers. The correlations combining the analytical solution for heat diffusion and the theoretical relation for potential flow give the best agreement.

Keywords: flashing flow; interphase heat transfer coefficient; bubble growth in superheated liquid; two-fluid model; computational fluid dynamics

---

1. Introduction

Flash boiling is a vaporization process triggered by depressurization instead of heating, which is relevant to a number of industrial economic and safety concerns. For example, in the automobile industry, the atomization of fuel spray in a combustion chamber is affected significantly by its flashing characteristics inside the injector nozzle [1]. In the nuclear industry, during the hypothetical loss of coolant accident of pressurized water nuclear reactors, the rate of coolant loss is determined by the critical flashing flow through the crack [2]. In the chemical industry, the severity of failure of pressurized vessels or pipes containing liquefied chemical hazardous gases is characterized by the external flashing flow [3]. An additional flashing evaporation phenomenon was reported in [4], which refers to an aerospace application and concerns the leading edge cooling of a space vehicle. Another similar phenomenon often encountered in case of pressure variation is cavitation. In general, cavitation occurs at relatively low temperature levels, where bubble growth is controlled mainly by the pressure difference across the interface. In contrast, flashing of hot fluids is more like a boiling process, which is driven principally by the thermal non-equilibrium. The complexity of flashing flows is represented by gas-liquid mixture with rapid phase change and bubble dynamics [5], and numerical studies are directed towards the determination of vapour generation rate. Good reviews have been given by Pinhasi [3] and Liao & Lucas [6]. In general, two methods have been used for the evaluation of the interfacial mass transfer rate in flashing flows. One is based on the observation of non-equilibrium mechanical and thermal processes. The other treats the transition of the thermodynamic system from non-equilibrium to equilibrium as a relaxation process. The two states are bridged by means of an empirical coefficient, i.e., the relaxation time [7–10]. The present paper will focus on the former one, which is consistent with the two-fluid framework. Under this category there are again two alternatives having been adopted for the estimation of vapour generation rate. One is based directly on the interfacial heat transfer process

$$\dot{m} = \frac{\dot{q}}{L} , \tag{1}$$

where $\dot{m}$ is the mass flux, L the latent heat of vaporization, and $\dot{q}$ the heat flux transferring from the vapour and liquid to the phase interface. For vapour-liquid such as steam-water flows under most practical conditions, the interfacial heat transfer on the vapour phase is usually much smaller (less than 5%) than that on the liquid phase [11]. Therefore, it is usually neglected by assuming that the temperature is uniform inside the bubble and equal to that at the interface. This assumption is also made in the current work

$$\dot{q} = h^l T_{\text{sup}} , \tag{2}$$

where h^l is the overall heat transfer coefficient between the superheated liquid and the liquid-vapour interface, and T_{sup} is the superheat degree of the liquid.

An alternative approach is formulated in terms of the resultant bubble growth rate

$$\dot{m} = \rho_v \dot{R} , \tag{3}$$

where $\dot{R}$ is the growth rate of bubble radius given by an analytical solution, and ρ_v vapor density. Bubble growth in superheated liquid is known to be controlled successively by surface tension, liquid inertia and heat transfer [12]. The first stage is usually neglected in numerical analysis with the consideration of sufficient bubble size. The effect of liquid inertia is only important at the very early stage of depressurization [13] and for sufficiently small bubbles [14]. Therefore, the thermally controlled growth is of the greatest interest. In this domain the above two models are related to each other, and one gets

$$\dot{q} = h^l T_{\text{sup}} = L\rho_v \dot{R} , \tag{4}$$

$$h^l = \frac{L\rho_v \dot{R}}{T_{\text{sup}}} . \tag{5}$$

As a result, a primary concern of the numerical study on flashing flows turns out to be a reliable prediction of the interphase heat transfer coefficient or bubble growth rate. It is generally recognized that high uncertainty is present in choosing an appropriate heat transfer coefficient correlation for two-fluid computational fluid dynamics (CFD) simulations [6]. One major reason is that the insufficiency and limitation of the correlations is not completely identified, and a quantitative evaluation is missing. This work aims to present a thorough literature survey on existing theories and correlations, and evaluate their performance by carrying out CFD simulations and making comparisons with available experimental and Direct Numerical Simulation (DNS) data. Note that although the background of the present work is flashing flow, the results and discussions are not restricted to it. Certain similarities have been identified in the transfer scenarios of a liquid sphere exposure to blowing hot gas and a bubble rising in superheated liquid or dissolving in liquid. The correlations for heat (or mass) transfer in condensation, evaporation and dissolution are often exchangeable. A variety of correlations are available in the literature. They take into account the effect of conduction, convection and turbulence partially or totally, but mostly in a pure empirical or semiempirical way. A good review was given by Mathpati & Joshi [15]. An overview of the classical theories, analytical solutions and empirical correlations is given below.

2. Theories and Analytical Solutions

2.1. Conduction

Numerous analytical solutions are available for the heat transfer between spherical vapour bubbles and the surrounding liquid. Most of them account for the simplified heat conduction problem and neglected the momentum effect in the liquid and gas phases on the bubble growth and its shape. It states that the bubble growth problem is analogous to a one-dimensional, unsteady state heat diffusion process with moving boundary, which is described by

$$\frac{\partial T_l}{\partial t} = a_l \frac{\partial^2 T_l}{\partial x^2} , \tag{6}$$

where a_l is liquid thermal diffusivity, t the time coordinate, the direction x normal to the boundary surface, and T_l is the temperature of the surrounding liquid. Under certain initial and boundary conditions, the temperature field of the liquid around the bubble surface may be solved from Equation (6) analytically. The assumption of a thin "thermal boundary layer", i.e., the change of liquid temperature taking place only in a thin film adjacent to the interface, is often made in these solutions. The liquid temperature in the bulk, T_0, is uniform and constant. In addition, for constant pressure fields, the saturation temperature T_{sat} at the interface remains unchanged, and the vapor inside the bubble is often assumed to have the saturation temperature. The theory of thermal boundary layer is shown schematically in Figure 1.

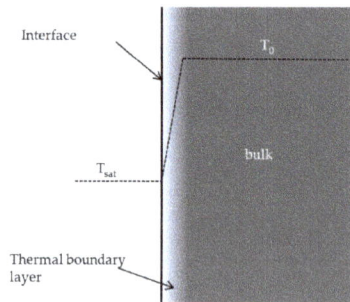

Figure 1. Temperature profile in the thermal boundary layer.

Knowing the temperature distribution, the heat flux transferring from the bulk to the bubble surface can be determined from the Fourier's Law

$$\dot{q} = \lambda_l \left. \frac{\partial T_l}{\partial x} \right|_{x=R} , \tag{7}$$

where λ_l is liquid thermal conductivity and R the bubble radius. Finally, the bubble growth rate $\dot{R}$ is obtained by substituting $\dot{q}$ into Equation (4). Fritz and Ende [16] solved the heat conduction across a semi-infinite plane slab under constant temperature boundary conditions. They derived the asymptotic bubble radius

$$R(t) = \frac{2}{\sqrt{\pi}} \cdot Ja_T \cdot (a_l t)^{1/2} , \tag{8}$$

where the Jakob number, Ja_T, is defined as

$$Ja_T = \frac{\rho_l c_{p,l} T_{sup}}{\rho_v L} . \tag{9}$$

Combining with Equation (5) the heat transfer coefficient for conduction is obtained

$$h_{cond}^l = \frac{\lambda_l}{\sqrt{\pi a_l t}} , \tag{10}$$

wherein the time t is related to Ja_T number and bubble radius via Equation (8). In terms of the dimensionless number, the Nusselt number Nu, above equation is expressed as

$$Nu_{cond} = \frac{4}{\pi} Ja_T . \tag{11}$$

Unsteady heat conduction across spherical bubble surfaces were studied by Plesset & Zwick [17] and Forster & Zuber [13] independently. Both solutions are in the same form as Equation (8) with the exception of a so-called "spherical factor", K_s, i.e.,

$$R(t) = K_s \frac{2}{\sqrt{\pi}} \cdot Ja_T \cdot (a_l t)^{1/2} . \tag{12}$$

The numerical constant K_s is greater than 1, which means that under the same temperature difference, heat flux across a spherical bubble is larger than a planar surface because the temperature gradient in the "thermal boundary layer" is increased by the curvature. In [17] $K_s = \sqrt{3}$ while in [13] $K_s = \pi/2$. Olek et al. [18] derived an alternative expression for the heat flux at the boundary of a sphere by using the hyperbolic heat conduction equation. For long times, the asymptotic solution approaches those obtained by using the Fourier heat conduction, but with a correction factor like

$$K_s = \frac{1}{2} \left[1 + \left(1 + \frac{2\pi}{Ja_T} \right)^{1/2} \right] . \tag{13}$$

With consideration of the correction factor K_s, the Nusellt number in Equation (11) turns into

$$Nu_{cond} = K_s^2 \frac{4}{\pi} Ja_T . \tag{14}$$

2.2. Convection

In principle, the above solutions without considering the effect of slip velocity are applicable for low void fraction and high superheat degrees. These conditions are expected to be satisfied only in a short time interval during a rapid depressurization, and at the transition of the flow changing from one-phase to two-phase [19]. For large bubbles one may expect a significant under-prediction by using these models. As observed and discussed in [20,21], the influence of slip velocity on the transfer rate is noticeable even in the case of bubbles rising in stagnant superheated liquid under normal gravity. The effect of slip velocity on interfacial heat transfer was firstly studied by Ruckenstein [22] and Sideman [23]. For the transfer between spherical independent vapour bubbles (influence of other bubbles and turbulence negligible) and the boiling liquid in motion, they suggested in a potential flow

$$h_{conv}^l = \frac{\lambda_l}{R} \cdot \frac{1}{\sqrt{\pi}} \cdot Pe^{1/2} , \tag{15}$$

or in terms of Nusselt number

$$Nu_{conv} = \frac{2}{\sqrt{\pi}} \cdot Pe^{1/2} , \tag{16}$$

where the Péclet number is defined by

$$Pe = \frac{d |\tilde{U}_{rel}|}{a_l} = Re_p Pr_l , \tag{17}$$

with the particle Reynolds number $Re_p = d|\vec{U}_{rel}|/v_l$ and the liquid Prandtl number $Pr_l = v_l/a_l$. Equation (15) or (16) is often interpreted by the so-called "penetration theory" [24]. It states that the liquid molecules in contact with the bubble surface are replaced at a constant time interval, which can be expressed as a ratio of bubble diameter to relative velocity

$$\tau_{conv} \approx \frac{d}{|\vec{U}_{rel}|}. \tag{18}$$

The schematic representation of the penetration theory is shown in Figure 2.

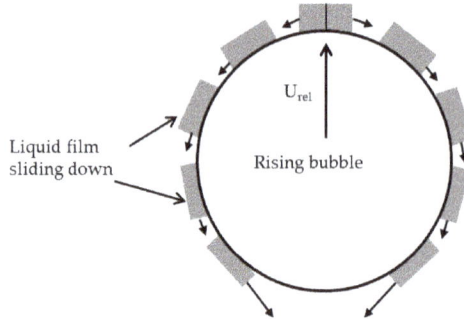

Figure 2. Schematic representation of the penetration theory.

Similar expressions can be obtained from the "penetration" theory and the "thermal boundary layer" theory [23] discussed above. The heat transfer takes place in a thin laminar sublayer, where a constant velocity can be assumed regardless of the hydrodynamics in the bulk of liquid. By substituting Equations (17) and (18) to Equation (15), one gets

$$h^l_{conv} = 2 \cdot \frac{\lambda_l}{\sqrt{\pi a_l \tau_{conv}}}, \tag{19}$$

which has the same form as the conduction transfer coefficient given in Equation (10) except the factor 2 and the characteristic time scale.

2.3. Effect of Turbulence

The effect of wake and freestream turbulence on the transfer from spheres has received relatively less attention, and is still not well understood. Conflicting arguments and observations exist. It is commonly believed that the transfer is enhanced in the presence of wake interactions and freestream turbulence. The experimental investigation on heat transfer from solid spheres reported by Lavender & Pei [25] and Raithby & Eckert [26] showed that the Nusselt number increased with increasing turbulence intensity in the ambient flow. Yearling & Gould [27] measured the convective heat and mass transfer rates from liquid droplets in turbulent air flow and also observed that the Nusselt number increased with increasing turbulence intensity. However, the augmentation was not duplicated by the experiment on the evaporative heat and mass transfer of suspended heptane droplets performed by Buchanan [28]. Theoretical interpretation of the turbulence effect is mainly based on the so-called "surface renewal theory", which is a modification of the "penetration theory" discussed above. As described in [22], if intense turbulent motions appear in the liquid, the contact surface between the liquid and the bubbles is continuously renewed by turbulence eddies, which brings the liquid from the bulk to the interface at average intervals of τ_{turb}, see Figure 3.

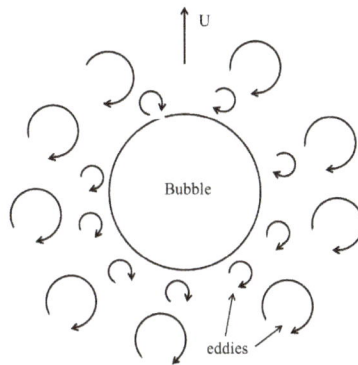

Figure 3. Schematic representation of the surface renewal theory.

The characteristic time is determined by the turbulent fluctuation velocity instead of the slip velocity used by the "penetration theory". Dackwerts [29] suggested that turbulence renews the volume elements at the interface, but the turbulence dies out as it approaches the interface. As a result, the liquid element itself has a non-turbulent structure, and the heat or mass transfer at the interface still has a molecular character. The heat transfer coefficient may be expressed as

$$h_{turb}^l = 2\lambda_l \cdot \frac{1}{\sqrt{\pi a_l \tau_{turb}}} .$$

(20)

It has the same form as Equation (19) except the time scale τ_{turb}, which is often estimated using the ratio between Kolmogorov length and velocity scales [30–32], i.e., $\tau_{turb} = \sqrt{\frac{\nu}{\varepsilon}}$. That means that the boundary layer adjacent to the interface is renewed by near-surface small scale eddies. In contrast, a large eddy model uses the scales of energy containing eddies [33]. Some researchers suggested that the small eddy model is more valid at higher Reynolds numbers while the large eddies dominate the surface renewal at lower Reynolds numbers, and thus proposed a two-regime model. However, there is no consistent definition of the Reynolds number and the criterion for transition. On the other hand, Sideman [34] approximated it as a ratio of the bubble diameter d to the fluctuation velocity at a distance of d

$$\overline{U'} \sim \begin{cases} (\varepsilon d)^{1/3} & \text{for inertial regime } (\eta < d \leq l) \\ d \left(\frac{\varepsilon}{\nu}\right)^{1/2} & \text{for viscous regime } (d \leq \eta) \end{cases},$$

(21)

where η, l is the Kolmogorov and integral length scale, respectively. Owing to advances in numerical algorithms and high performance computing, the transfer process occurring at the interface becomes amenable with the aid of DNS. Figueroa-Espinoza and Legendre [35] investigated the effect of bubble aspect ratio and bubble wake on the mass transfer from oblate spheroids by DNS solving the Navier-Stokes equations. They found that most of the transfer occurs on the front part of the bubble. The contribution in the wake region increases as the aspect ratio increases. The local transfer rate at the bubble surface as a function of the azimuthal angle deviates significantly from Equation (15) due to unsteady effects from vorticity production and wake destabilization. However, the total transfer rate expressed as the Nusselt number is shown to satisfy well the potential flow theory if the equivalent diameter is used as the characteristic length scale, see Figure 4.

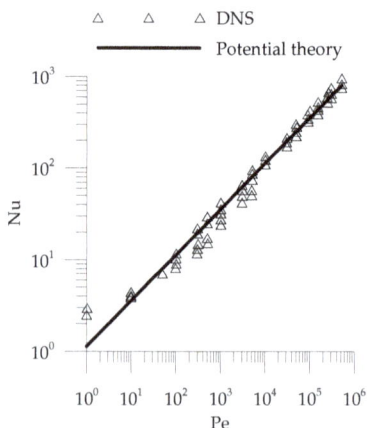

Figure 4. Applicability of the potential theory for calculating average transfer rate around deformed bubbles with wake interaction (DNS data from [35]). DNS: Direct Numerical Simulation.

Similar results about the wake effect on the transfer rate were obtained by Bagchi & Kottam [36] in their DNS simulation of heat transfer from a sphere in a turbulent flow. In addition, the freestream turbulence was shown to have a clear influence on the instantaneous and local Nusselt number. However, the time and surface averaged Nusselt number was found insensitive to the ambient turbulence and it can be predicted by correlations for steady and uniform flow.

3. Empirical Correlations

Besides analytical solutions, some researchers have proposed useful empirical correlations for calculating the heat transfer coefficient in flashing conditions. For conduction, a widely used expression in terms of the Jakob number was presented in [37]

$$Nu = \left[2 + \left(\frac{6\,Ja_T}{\pi} \right)^{1/3} + \frac{12}{\pi}\,Ja_T \right] \tag{22}$$

The correlation was validated for bubble growth in uniformly heated liquid, and adopted in [38–40] for the modelling of various flashing flows. A slightly modified expression was presented later on in [41].

$$Nu = \left[2 + (2\,Ja_T)^{1/3} + \frac{12}{\pi}\,Ja_T \right] . \tag{23}$$

The Ranz-Marshall correlation [42], which was proposed based on experimental data for spherical water drops evaporating in blowing hot dry air

$$Nu = \left(2 + C \cdot Re_p^A \, Pr_l^B \right), \tag{24}$$

has been often used for estimating interphase heat transfer rates also in the case of flashing flows. Among many others examples are one-dimensional simulations presented by Richter [43], Bird et al. [44] and Dobran [45] and three-dimensional CFD simulations by Giese [46], Laurien [47] and Frank [48]. The empirical constants in Equation (24) are $A = 1/2$, $B = 1/3$ and $C = 0.6$. Hughmark [49] suggested that these constants are valid for the range $Re_p < 450$ and $Pr_l < 250$. The exponents A and B may increase with the Reynolds number Re_p and the Prandtl number Pr_l, respectively. Actually, correlations with slightly different constants have been widely used for convective transfer, e.g., in [50] $C = 0.15$, while in [51,52] C was replaced by 0.46 and 0.55, respectively. At the same time, Lee and Ryley [53] found that the observations of water drops evaporating in superheated steam instead of air conform closely to Equation (24) but with $C = 0.738$. For a vapour bubble freely oscillating in liquid, $A = B = 0.5$, and $C = 1.0$ according to Nigmatulin et al. [54] and Mahulkar et al. [55]. The expression of Aleksandrov et al. [56] was modified slightly by Saha et al. [57] for the calculation of heat transfer rate in flashing nozzle flows. They combine the conduction and convection transfer in the way

$$\text{Nu} = \left(\frac{12^2}{\pi^2} \text{Ja}_T^2 + \frac{4}{\pi} \text{Pe} \right)^{1/2}. \tag{25}$$

A similar expression was used by Wolfert [58] to simulate the rapid depressurization processes of high pressure pipes

$$\text{Nu} = \left(\frac{12}{\pi} \text{Ja}_T + \frac{2}{\sqrt{\pi}} \text{Pe}^{1/2} \right). \tag{26}$$

The above two correlations can be reformulated as

$$\text{Nu} = \sqrt{(\text{Nu}_{cond})^2 + (\text{Nu}_{conv})^2}, \tag{27}$$

and

$$\text{Nu} = \text{Nu}_{cond} + \text{Nu}_{conv}. \tag{28}$$

The conduction and convection Nusselt number is evaluated by Equations (14) and (16), respectively. To account for the effect of turbulence, Wolfert et al. [59] introduced a so-called eddy conductivity, λ_t. The apparent thermal conductivity of liquid is given by

$$\lambda_l' = \lambda_l + \lambda_t. \tag{29}$$

And the overall heat transfer coefficient is computed through

$$\text{Nu} = \left(\frac{12}{\pi} \cdot \text{Ja}_T + \frac{2}{\sqrt{\pi}} \left(1 + \frac{\lambda_t}{\lambda_l} \right) \cdot \text{Pe}^{1/2} \right). \tag{30}$$

The eddy conductivity λ_t was assumed to be dependent on the liquid velocity. In case of one velocity component W_l, λ_t is expressed as

$$\lambda_t = \lambda_l \cdot \chi_t \cdot W_l, \tag{31}$$

where χ_t is an empirical constant. Based on a pressure release experiment on a test vessel and two blowdown experiments on the vessel and straight pipe, Wolfert et al. [59] found that $\chi_t = 0.8\,\text{sm}^{-1}$ gives the best agreement between calculated and experimental results. Whitaker [60] interpreted that the enhancement in transfer rates from a sphere due to the presence of turbulence comes purely from the wake contribution, while the transfer process at the front surface can be described by the law for potential flow, i.e., the Nusselt number $\text{Nu} \propto \text{Re}_p^{1/2}$, and the freestream turbulence has no effect. In the wake region, the functional dependence for the Reynolds number is $\text{Re}_p^{2/3}$, and it is cumulative to the laminar part

$$\text{Nu} = \left(2 + \left(0.4 \cdot \text{Re}_p^{1/2} + 0.06 \cdot \text{Re}_p^{2/3} \right) \text{Pr}_l^{0.4} \right). \tag{32}$$

Another empirical method often used to account for the turbulence enhancement is to increase the constant A in Equation (24). Issa et al. [61] compared the Nusselt number correlations for the case of saturated steam bubbles condensing in subcooled water, and found that the dependency upon Reynolds number grows as it increases, e.g., for Re_p up to 800, $\text{Nu} \propto \text{Re}_p^{0.5}$, and for Re_p up to 10^4, $\text{Nu} \propto \text{Re}_p^{0.7}$. They derived the following correlation for highly deformed large bubbles condensing in turbulent pipe flow,

$$\text{Nu} = 0.0609\,\text{Re}_p^{0.89}\,\text{Pr}_l^{0.33}. \tag{33}$$

Based on a numerical study on the transient heat transfer from a sphere, Feng & Michaelides [62] obtained a simple correlation for the Nusselt number, which describes the dependence on Reynolds and Peclet numbers as follows

$$\text{Nu} = \left(0.922 + \text{Pe}^{1/3} + 0.1 \cdot \text{Re}_p^{1/3}\,\text{Pe}^{1/3}\right). \tag{34}$$

The DNS performed by Dani et al. [63] showed that the average mass transfer (heat transfer) from spherical bubble is affected not only by the Reynolds and Schmidt (Prandtl) numbers but also by the surface mobility and contamination. As shown in Figure 5a, for solid spheres or fully contaminated bubbles in a creeping flow (low Re_p numbers) following expression proposed by Clift [64] reproduces the exact numerical solution with very high accuracy

$$\text{Nu} = 1 + (1 + \text{Pe})^{1/3}. \tag{35}$$

However, as the Re_p number increases, the influence of Re_p and Pr_l are no more similar and have to be considered separately. In these cases, the Ranz & Marshall [42] correlation was shown to be able to give perfect agreements, see Figure 5b.

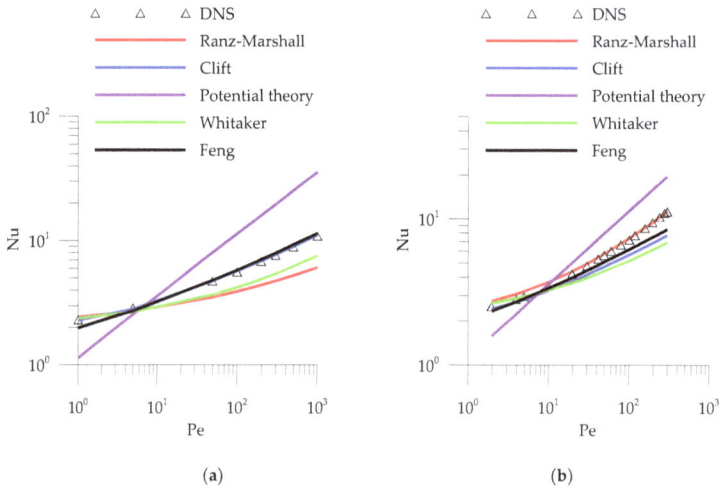

(a) (b)

Figure 5. Applicability of correlations for predicting the average Nusselt number of fully contaminated spherical bubbles (DNS data from [63]): (**a**) $\text{Re}_p = 0.1$, $\text{Pr}_l = 1 \sim 10^4$. (**b**) $\text{Re}_p = 1 \sim 150$, $\text{Pr}_l = 2$. DNS: Direct Numerical Simulation.

For clean spherical bubbles, Clift [64] suggested

$$\text{Nu} = 1 + (1 + 0.564\,\text{Pe}^{2/3})^{1/3}, \tag{36}$$

which achieves excellent agreement with the DNS data at low Reynolds numbers, e.g., $\text{Re}_p \leq 0.1$ (see Figure 6a). As shown in Figure 6b the aymptotic solution for both $\text{Pr}_l \to \infty$ and $\text{Re}_p \to \infty$ agrees well with the potential theory, i.e., $\text{Nu} \sim \text{Pe}^{1/2}$ given by Equation (15). In this case, the Ranz & Marshall [42] correlation under-predicts the growth rate significantly.

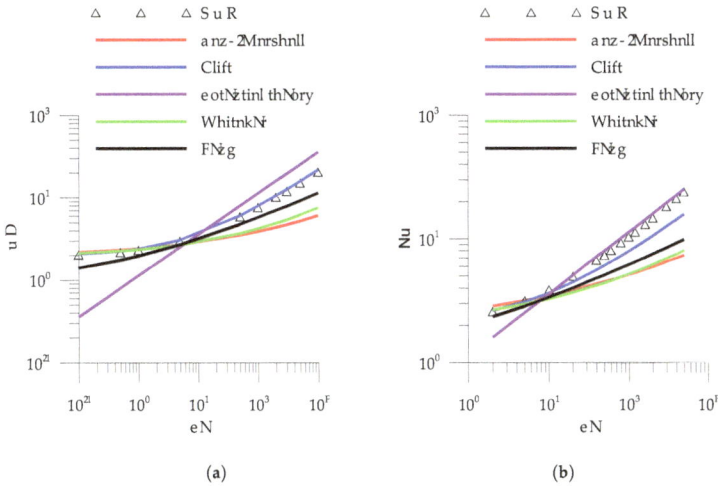

Figure 6. Applicability of correlations for predicting the average Nusselt number of clean spherical bubbles (DNS data from [63]): (**a**) $Re_p = 0.1$, $Pr = 1 \sim 10^4$. (**b**) $Re_p = 1 \sim 250$, $Pr_l = 2$. DNS: Direct Numerical Simulation.

4. CFD Simulation of Bubble Growth in Superheated Liquid

Bubble growing in stagnant liquid of uniform superheats has been adequately studied with high speed photography around the middle of the last century [20,65–68]. Dergarabedian [65] obtained bubble formation within the body of the liquid by heating a beaker of water slowly. In order to avoid wall nucleation, the Pyrex beaker surface was annealed carefully to be very smooth and free of pits. Nevertheless, thermal gradients were found to exist in a narrow boundary adjacent to the bottom of the beaker, and most of the bubbles were formed in this thermal layer. Hooper [66] and Florschuetz et al. [20] achieved uniform superheats by suddenly depressurizing heated pressurized water down to the atmospheric pressure within 5 ms. The bubble growth rates observed by Hooper [66] were found to be lower than the theoretical ones, and moreover, the deviation increased as the superheat was increased. Florschuetz et al. [20] investigated the effect of translational motion on the bubble growth rate at small superheat degrees under near zero- and normal-gravity conditions. The zero-gravity data was found to be suitable for the validation of theoretical solutions for heat-conduction controlled growth. In other words, the convective effects are negligible in these cases. On the other hand, the data taken at normal gravity indicate clearly that the enhancement of growth rates due to bubble translation becomes significant at later stages. Kosky [67] heated a tube of water uniformly in a silicone oil bath. The pressure in the test chamber was regulated with a vacuum pump, and recorded with a pressure transducer simultaneously.

In this section, the analytical and empirical correlations discussed above (see Table 1) are tested for the early stage of bubble growth in superheated liquid with the two-fluid CFD. Three configurations are considered, namely stationary bubble growth, translating bubble growth in stagnant and flowing liquid. These cases are chosen for validation considering the fact that the effect of heat conduction, convection and turbulence can be investigated separately to a certain extent, and many other uncertainties such as swarm effect and interphase momentum transfer can be excluded. Separate conservation equations are solved for the vapour and liquid phases. Further, the vapour is assumed to be saturated as done by many other researchers like Maksic [39] and Giese [46]. The particle model is applied for the mimic of interfacial morphology and computation of interfacial area density, where the vapour is modelled as spherical bubbles. A constant number concentration is assumed, which is consistent with the experimental observation by Florschuetz et al. [20] and valid for the early-stage of flashing.

Microscopic or nearly microscopic bubbles are present naturally in the domain at the instant of pressure release, and provide the nuclei for subsequent phase growth. For further details about the numerical setup, the reader is referred to [69,70]. The presumed bubble concentration is found to have a negligible effect on the results, and a value of 10^4 m^{-3} is adopted in all the cases.

Table 1. Correlations for estimating interphase heat transfer coefficient.

	Conduction	
Reference	**Correlation**	**Note**
[16]	$\text{Nu} = \frac{4}{\sqrt{12}}\text{Ja}_T$	analytical
[17]	$\text{Nu} = \frac{12}{\pi}\text{Ja}_T$	analytical
[13]	$\text{Nu} = \pi\,\text{Ja}_T$	analytical
[18]	$\text{Nu} = \frac{\text{Ja}_T}{\pi}[1 + (1 + \frac{2\pi}{\text{Ja}_T})^{1/2}]^2$	analytical
[37]	$\text{Nu} = 2 + (\frac{6\text{Ja}_T}{\pi})^{1/3} + \frac{12}{\pi}\text{Ja}_T$	empirical
	Convection	
[22]	$\text{Nu} = (\frac{2}{\sqrt{\pi}})\text{Pe}^{1/2}$	potential theory
[56]	$\text{Nu} = \left(\frac{12^2}{\pi^2}\text{Ja}_T^2 + \frac{4}{\pi}\text{Pe}\right)^{1/2}$	heuristic
[58]	$\text{Nu} = \left(\frac{12}{\pi}\text{Ja}_T + \frac{2}{\sqrt{\pi}}\text{Pe}^{1/2}\right)$	heuristic
[42]	$\text{Nu} = 2 + 0.6\,\text{Re}_p^{1/2}\,\text{Pr}_l^{1/3}$	empirical
	Turbulence	
[29]	$\text{Nu} = \dfrac{2}{\sqrt{\pi a_l \tau_{turb}}}$	surface renew theory
[60]	$\text{Nu} = 2 + (0.4\,\text{Re}_p^{1/2} + 0.06\,\text{Re}_p^{2/3})\,\text{Pr}_l^{0.4}$	empirical
[59]	$\text{Nu} = \frac{12}{\pi}\text{Ja}_T + \frac{2}{\sqrt{\pi}}\left(1 + \frac{\lambda_t}{\lambda_l}\right)\text{Pe}^{1/2}$	empirical
[62]	$\text{Nu} = 0.922 + \text{Pe}^{1/3} + 0.1 \cdot \text{Re}_p^{1/3}\,\text{Pe}^{1/3}$	empirical
[61]	$\text{Nu} = 0.0609\,\text{Re}_p^{0.89}\,\text{Pr}_l^{0.33}$	empirical

4.1. Stationary Bubble Growth

Two experimental test cases under atmospheric conditions at zero gravity are simulated with the CFD software ANSYS CFX (Version 18.0, ANSYS Inc., Canonsburg, PA, USA), where the buoyancy is deactivated and therefore there are no relative motion and convection effects. The experimental data are taken from the work of Florschuetz et al. [20] for steam-water systems. The water has an initial superheat degree of 2.9 K and 3.2 K, respectively. The simulation domain is a cube as shown in Figure 7. It is worth noting that the domain for the simulation of translating bubble growth in stagnant and flowing liquid is elongated in the stream direction.

No slip wall boundary conditions are applied to the bottom of the box. The top is set as opening, which allows both inflow and outflow of the gas phase. The other four sides (left, right, front, back) are treated as symmetrical planes. The results show that in the first ∼500 ms the liquid temperature remains nearly constant. Meanwhile, the phase distribution and flow parameters are uniform in the domain with the exception of the region adjacent to the bottom wall. Therefore, the simulation condition conforms with the experiment that a bubble grows in stagnant uniform superheated liquid, for which an analytical solution of the growth rate is possible. As shown in Figure 8 the numerical results for the case $T_{\text{sup}} = 2.9 \text{ K}$, which is averaged over the midplane of the domain(plane 1 in Figure 7), coincide with the analytical ones. This proves that the applied model is capable of simulating bubble growth in superheated liquid.

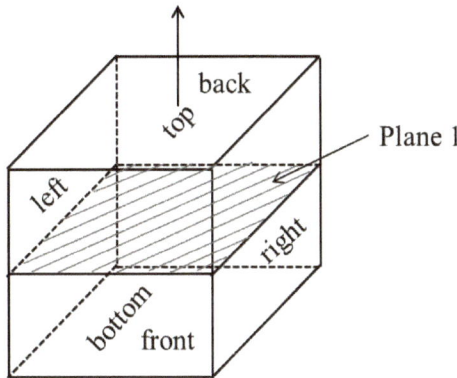

Figure 7. Simulation domain for cases at zero gravity.

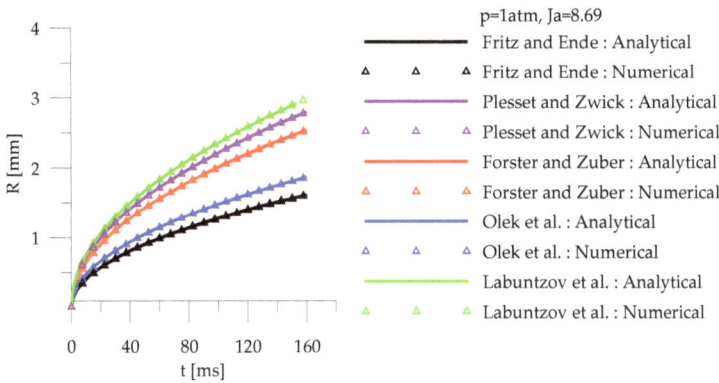

Figure 8. Numerical and analytical solution of bubble growth at zero gravity ($T_{sup} = 2.9$ K).

The comparison of the simulated and measured transient bubble size for the two cases is depicted in Figure 9a,b, respectively.

Note that in a log-log plot, the bubble growth exhibits a slope of 1/2 according to both theory and experiment. Nevertheless, linear-linear plots are used here in order to show the difference more clearly. For the case of $T_{sup} = 2.9$ K the results obtained by using the correlations of Labuntzov et al. [37], Plesset & Zwick [17] and Forster & Zuber [13] agree well with the measurement, while the Olek et al. [18] and Fritz & Ende [16] correlations under-predict the bubble growth rate substantially. As the liquid superheat increases from 2.9 K to 3.2 K, at the early stage the experimental data are more consistent with the predictions of the former three correlations. However, asymptotically they approach to the latter two, see Figure 5b. It is interesting to note that the overall uncertainty of bulk liquid temperature values and equivalent bubble radii was estimated to be about ±0.2 K and ±0.05 mm [20]. Another DNS case of bubble growth in superheated under the zero-gravity condition presented in Ye [71] is simulated. It is based on the thermal properties of water under the atmospheric conditions. A liquid superheat of 1 K is considered, and the Jakob number is estimated as 3.0. The predicted bubble growth rate is shown in Figure 10. The DNS results evidence that the asymptotic bubble growth follows the theoretical relation, $R(t) \propto t^{1/2}$. However, in the initial stage when the thermal boundary layer around the bubble is developing, the growth rate is evidently larger than the 1/2 law (see Figure 6a). In contrast, the two-fluid simulation results obey the theoretical law well. The correlation of Fritz & Ende [16] is shown to be able to reproduce the DNS asymptotic results satisfactorily, while all the others are prone to over-predict the bubble size.

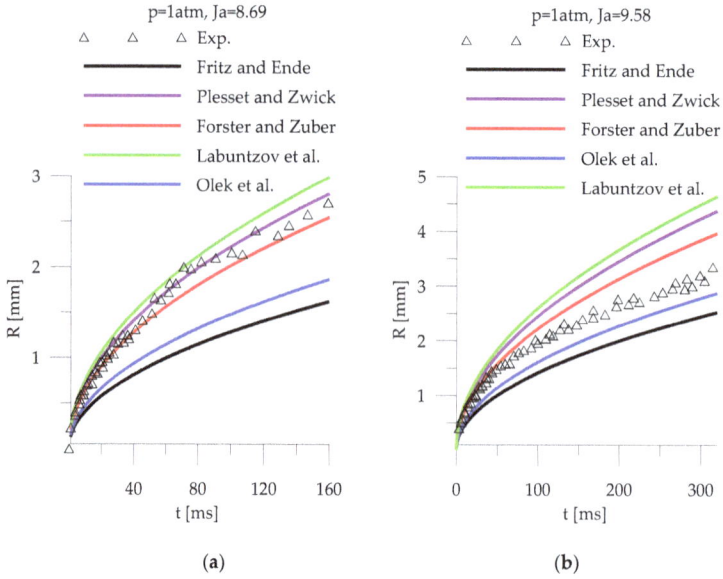

(a)

(b)

Figure 9. Simulated and measured bubble growth at zero gravity (experimental data from [20]): (a) $T_{sup} = 2.9$ K. (b) $T_{sup} = 3.2$ K.

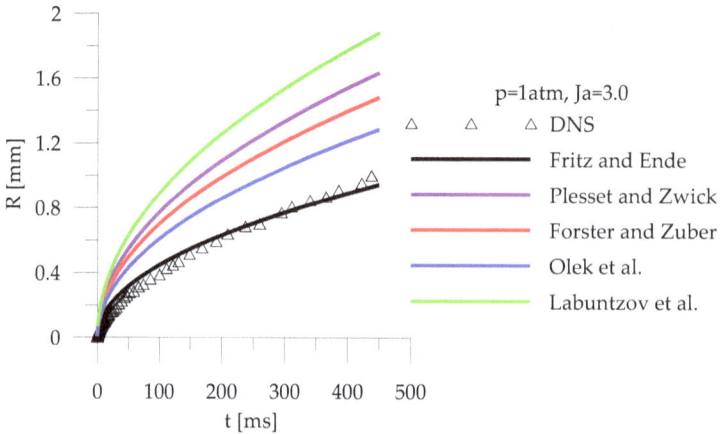

Figure 10. Two-fluid CFD and DNS simulated bubble growth at zero gravity (DNS data from [71], $T_{sup} = 1.0$ K). CFD: Computational Fluid Dynamics, DNS: Direct Numerical Simulation.

4.2. Translating Bubble Growth

Under the normal gravity condition, bubbles will be rising and translating simultaneously through the initially stagnant liquid, and the bubble growth rate is obviously larger than the diffusion limit 1/2. In the simulation, the buoyancy model is activated, which results in a relative motion between the bubbles and the liquid. Both the conduction and convection play a role in the interphase heat transfer, but the turbulence effect is negligible. The momentum interaction is modelled by interphase drag force, and the drag coefficient is calculated according to the Ishii & Zuber correlation [72].

4.2.1. Florschuetz et al. Cases

Two cases from Florschuetz et al. [20] with $T = 3.0\,K$ and $T = 3.9\,K$ respectively are simulated by the two-fluid CFD with different correlations for the heat transfer coefficient. The results are presented in Figure 11. As expected, the relative motion accelerates bubble growing and results in a steeper slope of growth line than that under zero-gravity conditions. At the early stage ($t < 5$ ms) the predictions given by Wolfert [58] and Aleksandrov et al. [56] agree well with the experimental data. Later on, the potential theory without consideration of heat conduction deliver the best agreement with the experimental data. It implies that heat conduction plays an important role at the initial stage, while convection becomes the dominant mechanism later. The results obtained by the correlation of Ranz & Marshall [42] under-predict the growth rate significantly in both cases. In addition, the asymptotic value of 2 is found to be insufficient to describe the heat transfer rate under the asymptotic condition of zero slip velocity, which is consistent with the findings of Walton [73]. In his experimental study on the evaporation of water droplets in hot air, Walton [73] measured the average Nusselt number of 3.8 for natural convection and an asymptotic value of 6 for forced convection.

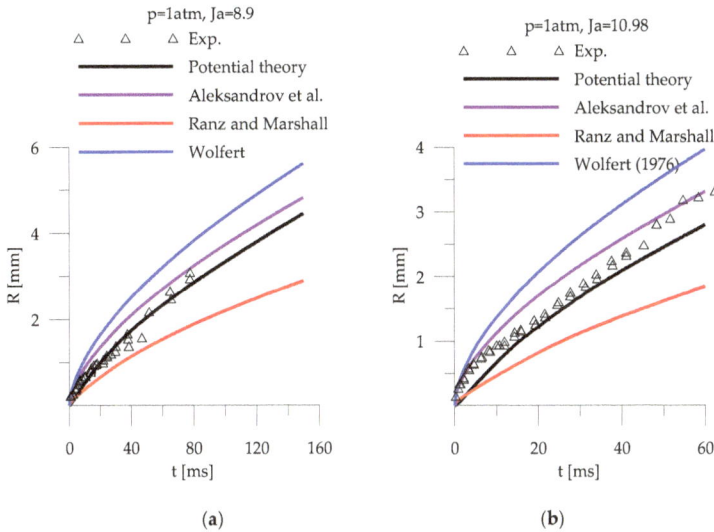

Figure 11. Simulated and measured bubble radius at normal gravity (experimental data from [20]): (a) $T_{sup} = 3.0$ K. (b) $T_{sup} = 3.9$ K.

4.2.2. Kosky Cases

Four cases from Kosky [67] with higher superheat degrees ($T = 10.5 \sim 23.2$ K) are also investigated. The results are shown in Figure 12. They have pressures other than 1 *atm* and relatively high superheats in comparison with the last two cases. Here, a generally good agreement with the experimental data is demonstrated by the heuristic correlations of Aleksandrov et al. [56] and Wolfert [58]. On the other hand, the potential theory and the empirical correlation of Ranz & Marshall [42] under-predict obviously the transient bubble size.

The thermodynamic conditions of all above cases are summarized in Table 2.

Table 2. Summary of test cases used for validation.

Case No.	$p\,[atm]$	$T_{sup}\,[K]$	Ja_T	Pr_l
1	1.0	2.9	8.69	1.70
2	1.0	3.2	9.58	1.69
3	1.0	1.0	3.0	1.72
4	1.0	3.0	8.9	1.69
5	1.0	3.9	10.98	1.68
6	1.19	10.5	26.5	1.50
7	0.642	16.0	71.0	1.68
8	0.477	19.5	113.25	1.75
9	0.613	23.2	107.0	1.58

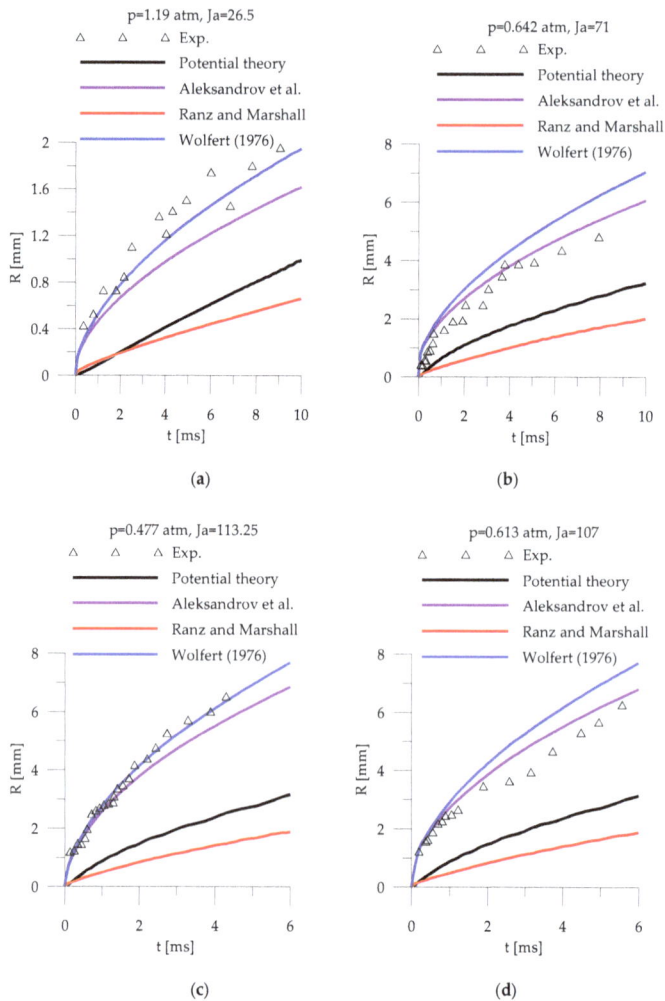

Figure 12. Simulated and measured bubble radius at normal gravity (experimental data from [67]): (a) $T_{sup} = 10.5$ K. (b) $T_{sup} = 16.0$ K. (c) $T_{sup} = 19.5$ K. (d) $T_{sup} = 23.2$ K.

4.3. Bubble Growth in a Flowing Liquid

Up to now the formulas obtained for stagnant liquid are commonly used for estimating the growth rate of vapor bubbles in high-velocity flowing liquid without checking the applicability. In his book, Avdeev [74] showed that these extrapolations are not justified and that they conflict with available measurements. Unfortunately, data on bubble growth rate in turbulent flows are quite limited because of the great difficulty of experiments. A set of experimental data regarding bubble growth in turbulent flow (Re = $1.8 \times 10^5 \sim 1.9 \times 10^6$) obtained by Kol'chugin et al. [74] and Lutovinov [75] was presented in [74]. The pressure of these experiments was in the range of $0.3 \sim 4.0$ MPa, and the liquid superheat varied from 0.7 to 2.5 K. It was found that bubbles were always deformed and had an irregular shape. The bubble velocity and growth rate were measured by high-speed filming at different positions along the flow direction and the double-exposure method. The comparison between the bubble size predicted by the empirical correlations and the measured one is shown in Figure 13a. An under-prediction is given by the correlations of Whitaker [60], Feng & Michaelides [62] and Issa et al. [61]. In contrast, the Wolfert et al. [59] correlation given by Equation (30) accounts for the intensification due to flowing velocity and turbulence adequately by introducing an eddy conductivity. The prediction of bubble growth rate is in a good agreement with the measurement. Finally, the surface renewal theory is tested by combining the theoretical parts of conduction, convection and turbulence cumulatively, i.e.,

$$\text{Nu} = \left(\frac{12}{\pi} \text{Ja}_T + \frac{2}{\sqrt{\pi}} \text{Pe}^{1/2} + \frac{2d}{\sqrt{\pi a_l \tau_{turb}}} \right). \tag{37}$$

The results are shown in Figure 13b. The difference between the small and large eddy model lies in the time scale τ_{turb} for surface renewal as discussed above. It is evident that the prediction by the small eddy model is closer to the measurement. It may indicate that small eddies instead of large eddies are responsible for the surface renewal and interphase transfer. However, further data are required for a reliable evaluation of the correlations and theories for turbulent flows.

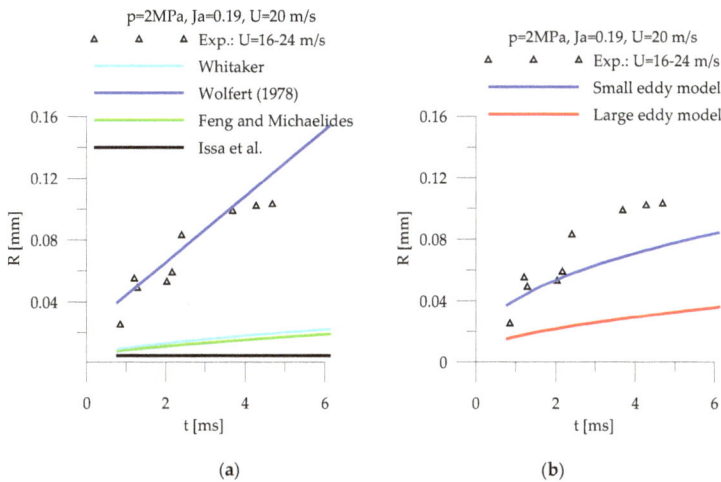

Figure 13. Simulated and measured bubble radius at turbulent conditions (experimental data from [74]): (a) Empirical correlations. (b) Theoretical models.

5. Conclusions

The interphase heat transfer in flashing flows is commonly believed to be a joint effect of heat conduction, convection and turbulence. The theory of bubble growth driven by each of the three mechanisms has been studied intensively. Whereas some general agreement exists on the contribution

Fluids **2018**, *3*, 38

of conduction and convection, the effect of turbulence is still under debate. In spite of the large uncertainty, correlations obtained under stagnant or potential conditions are often extrapolated to turbulent cases. An evaluation of the applicability of these correlations in the two-fluid modelling of practical flashing flows is difficult, since there is a great deal of uncertainty related to other closures such as momentum interactions and turbulence modulation by bubbles. In the present work, heat transfer models are evaluated for the early stage of bubble growth in stagnant and high-velocity liquid with CFD simulations. The cases are ideal for the purpose of evaluation, since the effect of conduction, convection and turbulence can be investigated separately, and other uncertainties are reduced to a minimum. From the comparison with available DNS and experimental data, the following conclusions can be drawn:

- For creeping flows ($\text{Re}_p \ll 1$) the two correlations presented by Clift [64] provide excellent agreement for immobile (contaminated bubbles) and fully mobile (clean bubbles) interfaces, respectively.
- For high Reynolds number, the Ranz & Marshall [42] correlation reproduces well the transfer rate from (to) solid spheres, droplets and contaminated bubbles, while it gives under-predictions in the case of clean bubbles, for which the potential theory is more suitable at least for Pe > 10.
- Stationary bubble growth follows the theoretical relation for heat conduction, i.e., $R(t) \propto t^{1/2}$. The numerical results are consistent with the analytical ones. The performance of the correlations is found to be dependent on the Jakob number. The correlation of Fritz & Ende [16] reproduces the bubble growth rate very well at low Jakob numbers, while those of Plesset & Zwick [17] and Forster & Zuber [13] give better predictions at moderate Jakob numbers. As the Jakob number increases further, the results of Olek et al. [18] get closer to the experimental data.
- For a reliable prediction of translating bubble growth, it is important to account for both heat conduction and convection. The conduction effect is evident in the initial stage even at moderate Jakob numbers. In cases with high Jakob numbers, the potential theory and the Ranz & Marshall [42] correlation under-predict the bubble size significantly, while the Wolfert [58] and Aleksandrov et al. [56] correlations, which account for both conduction and convection, deliver satisfying results.
- Wolfert et al. [59] is capable of reproducing the bubble growth rate in turbulent high-velocity flows by introducing an eddy conductivity, while significant under-prediction is given by other empirical correlations. The situation is improved by using the cumulative model proposed by Wolfert [58] supplemented with the surface renewal theory for turbulence. The time scale of small eddies is found to more suitable for the characterization of interfacial transfer than that of the large eddies. Nevertheless, acquisition of more detailed data is necessary for the quantitation of the turbulence effect.

Author Contributions: Investigation & draft preparation, Y.L.; Review & Supervision, D.L.

Conflicts of Interest: The authors declare no conflict of interest.

Abbreviations

The following abbreviations are used in this manuscript:

CFD   Computational fluid dynamics
DNS   Direct numerical simulation

References

1. Gopalakrishnan, S. Modeling of Thermal Non-Equilibrium in Superheated Injector Flows. Ph.D. Thesis, University of Massachusetts Amherst, Amherst, MA, USA, 2010.
2. Edwards, A.R.; O'Brien, T.P. Studies of phenomena connected with the depressurization of water reactors. *J. Br. Nucl. Energy Soc.* **1970**, *9*, 125–135.

3.   Pinhasi, G. Modeling of flashing two-phase flow. *Rev. Chem. Eng.* **2005**, *21*, 133–264. [CrossRef]

4.   De Luca, L.; Mongibello, L. Critical discharge in actively cooled wing leading edge of a reentry vehicle. *J. Thermophys. Heat Transf.* **2008**, *22*, 677–684. [CrossRef]

5.   Zhou, T.; Duan, J.; Hong, D.; Liu, P.; Sheng, C.; Huang, Y. Characteristics of a single bubble in subcooled boiling region of a narrow rectangular channel under natural circulation. *Ann. Nucl. Energy* **2013**, *57*, 22–31. [CrossRef]

6.   Liao, Y.; Lucas, D. Computational modelling of flash boiling flows: A literature survey. *Int. J. Heat Mass Transf.* **2017**, *111*, 246–265. [CrossRef]

7.   Bilicki, Z.; Kestin, J. Physical Aspects of the Relaxation Model in Two-Phase Flow. *Proc. R. Soc. A Math. Phys. Eng. Sci.* **1990**, *428*, 379–397. [CrossRef]

8.   Downar-Zapolski, P.; Bilicki, Z.; Bolle, L.; Franco, J. The non-equilibrium relaxation model for one-dimensional flashing liquid flow. *Int. J. Multiph. Flow* **1996**, *22*, 473–483. [CrossRef]

9.   Gopalakrishnan, S. Multidimensional simulation of flash-boiling fuels in injector nozzles. In Proceedings of the 21st Annual Conference on Liquid Atomization and Spray Systems, Orlando, FL, USA, 18–21 May 2008.

10. Neroorkar, K.D. Modeling of Flash Boiling Flows in Injectors with Gasoline-Ethanol Fuel Blends. Ph.D. Thesis, University of Massachusetts Amherst, Amherst, MA, USA, 2011.

11. Banerjee, S. A surface renewal model for interfacial heat and mass transfer in transient two-phase flow. *Int. J. Multiph. Flow* **1978**, *4*, 571–573. [CrossRef]

12. Miyatake, O.; Tanaka, I.; Lior, N. A simple universal equation for bubble growth in pure liquids and binary solutions with a nonvolatile solute. *Int. J. Heat Mass Transf.* **1997**, *40*, 1577–1584. [CrossRef]

13. Forster, H.K.; Zuber, N. Growth of a Vapor Bubble in a Superheated Liquid. *J. Appl. Phys.* **1954**, *25*, 474–478. [CrossRef]

14. Birkhoff, G.; Margulies, R.S.; Horning, W.A. Spherical Bubble Growth. *Phys. Fluids* **1958**, *1*, 201–204. [CrossRef]

15. Mathpati, C.S.; Joshi, J.B. Insight into Theories of Heat and Mass Transfer at the Solid-Fluid Interface Using Direct Numerical Simulation and Large Eddy Simulation. *Ind. Eng. Chem. Res.* **2007**, *46*, 8525–8557. [CrossRef]

16. Fritz, W.; Ende, W. Über den Verdampfungsvorgang nach kinematographischen Aufnahmen an Dampfblasen. *Phys. Z.* **1936**, *XXXVII*, 391–401.

17. Plesset, M.S.; Zwick, S.A. The Growth of Vapor Bubbles in Superheated Liquids. *J. Appl. Phys.* **1954**, *25*, 493–500. [CrossRef]

18. Olek, S.; Zvirin, Y.; Elias, E. Beschreibung des Blasenwachstums durch Wärmeleitungs-Gleichungen von hyperbolischer und parabolischer Form. *Wärme- und Stoffübertragung* **1990**, *25*, 17–26. [CrossRef]

19. Plesset, M.S.; Zwick, S.A. A Nonsteady Heat Diffusion Problem with Spherical Symmetry. *J. Appl. Phys.* **1952**, *23*, 95–98. [CrossRef]

20. Florschuetz, L.W.; Henry, C.L.; Khan, A.R. Growth rates of free vapor bubbles in liquids at uniform superheats under normal and zero gravity conditions. *Int. J. Heat Mass Transf.* **1969**, *12*, 1465–1489. [CrossRef]

21. Ivashnyov, O.E.; Smirnov, N.N. Thermal growth of a vapor bubble moving in superheated liquid. *Phys. Fluids* **2004**, *16*, 809–823. [CrossRef]

22. Ruckenstein, E. On heat transfer between vapour bubbles in motion and the boiling liquid from which they are generated. *Chem. Eng. Sci.* **1959**, *10*, 22–30. [CrossRef]

23. Sideman, S. The Equivalence of the Penetration and Potential Flow Theories. *Ind. Eng. Chem.* **1966**, *58*, 54–58. [CrossRef]

24. Higbie, R. The rate of absorption of a pure gas into a still liquid during short periods of exposure. *Trans. Am. Inst. Chem. Eng.* **1935**, *31*, 365–389.

25. Lavender, W.J.; Pei, D.C.T. The effect of fluid turbulence on the rate of heat transfer from spheres. *Int. J. Heat Mass Transf.* **1967**, *10*, 529–539. [CrossRef]

26. Raithby, G.D.; Eckert, E.R.G. The effect of turbulence parameters and support position on the heat transfer from spheres. *Int. J. Heat Mass Transf.* **1968**, *11*, 1233–1252. [CrossRef]

27. Yearling, P.R.; Gould, R.D. Convective heat and mass transfer from a single evaporating water, methanol, and ethanol droplet. *ASME FED* **1995**, *233*, 33–38.

28. Buchanan, C.D. The measurement of droplet evaporation rates at low Reynolds number in a turbulent flow. In Proceedings of the NHTC'00 34th National Heat Transfer Conference, Pittsburgh, PA, USA, 20–22 August 2000; pp. 1–8.

29. Danckwerts, P.V. Significance of Liquid-Film Coefficients in Gas Absorption. *Ind. Eng. Chem.* **1951**, 43, 1460–1467. [CrossRef]

30. Banerjee, S.; Scott, D.S.; Rhodes, E. Mass Transfer to Falling Wavy Liquid Films in Turbulent Flow. *Ind. Eng. Chem. Fundam.* **1968**, 7, 22–27. [CrossRef]

31. Deckwer, W.D. On the mechanism of heat transfer in bubble column reactors. *Chem. Eng. Sci.* **1980**, 35, 1341–1346. [CrossRef]

32. Garcia-Ochoa, F.; Gomez, E. Theoretical prediction of gas–liquid mass transfer coefficient, specific area and hold-up in sparged stirred tanks. *Chem. Eng. Sci.* **2004**, 59, 2489–2501. [CrossRef]

33. Fortescue, G.E.; Pearson, J.R.A. On gas absorption into a turbulent liquid. *Chem. Eng. Sci.* **1967**, 22, 1163–1176. [CrossRef]

34. Sideman, S.; Barsky, Z. Turbulence effect on direct-contact heat transfer with change of phase: Effect of mixing on heat transfer between an evaporating volatile liquid in direct contact with an immiscible liquid medium. *AIChE J.* **1965**, 11, 539–545. [CrossRef]

35. Figueroa-Espinoza, B.; Legendre, D. Mass or heat transfer from spheroidal gas bubbles rising through a stationary liquid. *Chem. Eng. Sci.* **2010**, 65. [CrossRef]

36. Bagchi, P.; Kottam, K. Effect of freestream isotropic turbulence on heat transfer from a sphere. *Phys. Fluids* **2008**, 20, 073305. [CrossRef]

37. Labuntzov, D.A.; Kolchugin, B.A.; Zakharova, E.A.; Vladimirova, L.N. High speed camera investigation of bubble growth for saturated water boiling in a wide range of pressure variations. *Thermophys. High Temp.* **1964**, 2, 446–453.

38. Blinkov, V.N.; Jones, O.C.; Nigmatulin, B.I. Nucleation and flashing in nozzles—2. Comparison with experiments using a five-equation model for vapor void development. *Int. J. Multiph. Flow* **1993**, 19, 965–986. [CrossRef]

39. Maksic, S. CFD-calculation of the flashing flow in pipes and nozzles. In Proceedings of the 2002 ASME Joint U.S.-European Fluids Engineering Conference, Montreal, QC, Canada, 14–18 July 2002.

40. Marsh, C. Three-dimensional modelling of industrial flashing flows. *Prog. Comput. Fluid Dyn.* **2009**, 9, 393–398. [CrossRef]

41. Valero, E.; Parra, I.E. The role of thermal disequilibrium in critical two-phase flow. *Int. J. Multiph. Flow* **2002**, 28, 21–50. [CrossRef]

42. Ranz, W.E.; Marshall, W.R.J. Evaporation from drops Part I. *Chem. Eng. Prog.* **1952**, 48, 141–146.

43. Richter, H.J. Separated two-phase flow model: Application to critical two-phase flow. *Int. J. Multiph. Flow* **1983**, 9, 511–530. [CrossRef]

44. Bird, R.B.; Stewart, W.E.; Lightfoot, E.N. *Transport Phenomena*, 1st ed.; John Wiley and Sons: New York, NY, USA, 1960.

45. Dobran, F. Nonequilibrium Modeling of Two-Phase Critical Flows in Tubes. *J. Heat Transf.* **1987**, 109, 731–738. [CrossRef]

46. Giese, T. Experimental and Numerical Investigation of Gravity-Driven Pipe Flow with Cavitation. In Proceedings of the 10th International Conference on Nuclear Engineering, Arlington, VA, USA, 14–18 April 2002.

47. Laurien, E. Influence of the model bubble diameter on three-dimensional numerical simulations of thermal cavitation in pipe elbows. In Proceedings of the 3rd International Symposium on Two-Phase Flow Modelling and Experimentation, Pisa, Italy, 22–25 September 2004.

48. Frank, T. Simulation of flashing and steam condensation in subcooled liquid using ANSYS CFX. In Proceedings of the 5th Joint FZR & ANSYS Workshop "Multiphase Flows: Simulation, Experiment and Application", Dresden, Germany, 26–27 April 2007.

49. Hughmark, G.A. Mass and Heat Transfer from Rigid Spheres. *AIChE J.* **1967**, 13, 1219. [CrossRef]

50. Schwellnus, C.F.; Shoukri, M. A two-fluid model for non-equilibrium two-phase critical discharge. *Can. J. Chem. Eng.* **1991**, 69, 188–197. [CrossRef]

51. AL-Sahan, M.A. On the Development of the Flow Regimes and the Formulation of a Mechanistic Non-Equilibrium Model for Critical Two-Phase Flow. Ph.D. Thesis, University of Toronto, Toronto, ON, Canada, 1988.

52. McAdams, W.H. *Heat Transmission*; McGraw-Hill: New York, NY, USA, 1954.

53. Lee, K.; Ryley, D.J. The Evaporation of Water Droplets in Superheated Steam. *J. Heat Transf.* **1968**, *90*, 445–451. [CrossRef]
54. Nigmatulin, R.I.; Khabeev, N.S.; Nagiev, F.B. Dynamics, heat and mass transfer of vapour-gas bubbles in a liquid. *Int. J. Heat Mass Transf.* **1981**, *24*, 1033–1044. [CrossRef]
55. Mahulkar, A.V.; Bapat, P.S.; Pandit, A.B.; Lewis, F.M. Steam bubble cavitation. *AIChE J.* **2008**, *54*, 1711–1724. [CrossRef]
56. Aleksandrov, Y.A.; Voronov, G.S.; Gorbunkov, V.M.; Delone, N.B.; Nechayev, Y.I. *Bubble Chambers*; Indiana University Press: Bloomington, IN, USA, 1967.
57. Saha, P.; Abuaf, N.; Wu, B.J.C. A Nonequilibrium Vapor Generation Model for Flashing Flows. *J. Heat Transf.* **1984**, *106*, 198–203. [CrossRef]
58. Wolfert, K. The simulation of blowdown processes with consideration of thermodynamic nonequilibrium phenomena. In Proceedings of the Specialists Meeting of Transient Two-Phase Flow, OECD/Nuclear Energy Agency, Toronto, ON, Canada, 3–4 August 1976.
59. Wolfert, K.; Burwell, M.J.; Enix, D. Non-equilibrium mass transfer between liquid and vapour phases during depressurization process in transient two-phase flow. In Proceedings of the 2nd CSNI Specialists Meeting, Paris, France, 12–14 June 1978.
60. Whitaker, S. Forced convection heat transfer correlations for flow in pipes, past flat plates, single cylinders, single spheres, and for flow in packed beds and tube bundles. *AIChE J.* **1972**, *18*, 361–371. [CrossRef]
61. Issa, S.A.; Weisensee, P.; Macian, R.J. Experimental investigation of steam bubble condensation in vertical large diameter geometry under atmospheric pressure and different flow conditions. *Int. J. Heat Mass Transf.* **2014**, *70*, 918–929. [CrossRef]
62. Feng, Z.G.; Michaelides, E.E. A numerical study on the transient heat transfer from a sphere at high Reynolds and Peclet numbers. *Int. J. Heat Mass Transf.* **2000**, *43*, 219–229. [CrossRef]
63. Dani, A.; Cockx, A.; Guiraud, P. Direct Numerical Simulation of Mass Transfer from Spherical Bubbles: The Effect of Interface Contamination at Low Reynolds Numbers. *Int. J. Chem. React. Eng.* **2006**, *4*. [CrossRef]
64. Clift, R. *Bubbles, Drops, and Particles*; Acadaenuc Press: New York, NY, USA, 1978.
65. Dergarabedian, P. The rate of growth of vapor bubbles in superheated water. *J. Appl. Mech.* **1953**, *20*, 537–545.
66. Hooper, F.C. The flashing of liquids at higher superheats. In Proceedings of the Third International Heat Transfer Conference, Chicago, IL, USA, 7–12 August 1966; pp. 44–50.
67. Kosky, P.G. Bubble growth measurements in uniformly superheated liquids. *Chem. Eng. Sci.* **1968**, *23*, 695–706. [CrossRef]
68. Prosperetti, A.; Plesset, M.S. Vapour-bubble growth in a superheated liquid. *J. Fluid Mech.* **1978**, *85*, 349–368. [CrossRef]
69. Liao, Y.; Lucas, D.; Krepper, E.; Rzehak, R. Flashing evaporation under different pressure levels. *Nucl. Eng. Des.* **2013**, *265*, 801–813. [CrossRef]
70. Liao, Y.; Lucas, D. 3D CFD simulation of flashing flows in a converging-diverging nozzle. *Nucl. Eng. Des.* **2015**, *292*, 149–163. [CrossRef]
71. Ye, T. Direct Numerical Simulation of a Translating Vapor Bubble with Phase Change. Ph.D. Thesis, University of Florida Digital Collections, Gainesville, FL, USA, 2001.
72. Ishii, M.; Zuber, N. Drag coefficient and relative velocity in bubbly, droplet or particulate flows. *AIChE J.* **1979**, *25*, 843–855. [CrossRef]
73. Walton, D.E. The evaporation of water droplets: A single droplet drying experiment. *Dry. Technol.* **2004**, *22*, 431–456. [CrossRef]
74. Avdeev, A.A. *Bubble Systems*; Springer: Berlin, Germany, 2016.
75. Lutovinov, S.Z. Investigation of Hot Water Discharge at Tube Rupture in Application to the Accident Sitation at Nuclear Power Plant. Ph.D. Thesis, Krzhizhanovsky Power Engineering Institute, Moscow, Russia, 1985.

fluids

MDPI

Article

Quality Measures of Mixing in Turbulent Flow and Effects of Molecular Diffusivity

Quoc Nguyen * and Dimitrios V. Papavassiliou

School of Chemical, Biological and Materials Engineering, The University of Oklahoma, Norman, OK 73019, USA; dvpapava@ou.edu
* Correspondence: quocnguyen@ou.edu; Tel.: +1-405-325-5811

Received: 27 June 2018; Accepted: 27 July 2018; Published: 30 July 2018

Abstract: Results from numerical simulations of the mixing of two puffs of scalars released in a turbulent flow channel are used to introduce a measure of mixing quality, and to investigate the effectiveness of turbulent mixing as a function of the location of the puff release and the molecular diffusivity of the puffs. The puffs are released from instantaneous line sources in the flow field with Schmidt numbers that range from 0.7 to 2400. The line sources are located at different distances from the channel wall, starting from the wall itself, the viscous wall layer, the logarithmic layer, and the channel center. The mixing effectiveness is quantified by following the trajectories of individual particles with a Lagrangian approach and carefully counting the number of particles from both puffs that arrive at different locations in the flow field as a function of time. A new measure, the mixing quality index Ø, is defined as the product of the normalized fraction of particles from the two puffs at a flow location. The mixing quality index can take values from 0, corresponding to no mixing, to 0.25, corresponding to full mixing. The mixing quality in the flow is found to depend on the Schmidt number of the puffs when the two puffs are released in the viscous wall region, while the Schmidt number is not important for the mixing of puffs released outside the logarithmic region.

Keywords: turbulent transport; turbulent mixing; Lagrangian modeling; turbulence simulations

---

1. Introduction

Turbulent flow is known to promote mixing in chemical reactors and other industrial processes. In fact, turbulent diffusion, which is one of the defining characteristics of turbulent flow [1], leads to enhancement in mixing [2]. A number of industrial applications depend on turbulent mixing [3–7], including combustion processes, and the design and operation of continuous stirred tank reactors (CSTRs) and plug flow reactors (PFRs), in addition to environmental processes that are important for weather changes and pollution dispersion [8,9]. While mixing because of molecular diffusion is a slow process [10], convective diffusion can significantly increase the mixing speed in turbulent flows. Yet, it has been noted that "the study of fluid mixing has very little scientific basis; processes and phenomena are analyzed on a case-by-case basis" (see J.M. Ottino's introduction in [10]). In fact, several definitions of mixing have been proposed in the literature, while finding one that can fully describe mixing for industrial applications is still a matter of investigation [11]. Often, the measure of mixing is defined by examining the segregation between two substances, instead of the mixing [11,12].

A rather fundamental mixing process occurs when two clouds of different substances are released from point sources in a flow field. A cloud of particles or scalar markers resulting from an instantaneous release from a point or a line is called a *puff* (for example, consider a puff of smoke that comes out of the exhaust pipe of a car or a puff of smoke exhaled by a smoker), while a cloud of particles that results from a continuous release of particles is called a *plume* [13,14]. The behavior of a puff is the most elementary dispersion in a turbulent flow field, and can be used to investigate and understand quite fundamental mechanisms of turbulent dispersion [15–17], even though it is not easy to study experimentally, where the study of plumes is easier [18–25].

Using computations, it is feasible to simulate the dispersion of particles that mark the trajectory of scalar quantities (like heat or mass concentration) in turbulent flow. These computations are based on the Lagrangian framework, and they consist of tracking individual scalar markers in a flow field obtained through a computational fluid dynamics model. A lot of work has been done with turbulent dispersion in homogeneous, isotropic turbulence [26–33], as well as in anisotropic turbulent flow, like channel flow [34–38], where the flow field was obtained through a direct numerical simulation (DNS) or a large eddy simulation (LES) approach. Lagrangian methods offer the advantage of simulating dispersion cases with a range of molecular diffusivities and Schmidt numbers (Sc) that may span a wide range. Combining a DNS with Lagrangian scalar tracking (LST), our laboratory has produced results for simulations with Sc that covers several orders of magnitude (from $Sc = 0.01$ to $Sc = 50,000$) [17]. The main disadvantage of Lagrangian methods is the need to use a large number of scalar markers to simulate turbulent transfer and the slow convergence of particle-based methods [39]. Hybrid Eulerian-Lagrangian methods have been utilized recently [34], but the highest Sc achieved is in the order of $Sc = 50$. The use of DNS for the generation of the velocity field has the advantage of providing a high fidelity representation of the flow in which mixing occurs, but has the disadvantage that the Reynolds number (Re) for the flow field is relatively small [40–43], while LES or hybrid methods can get to higher Re values but with sacrifice in fidelity.

In the present work, a DNS/LST approach is employed to investigate mixing from instantaneous line sources of scalars released at different locations of a turbulent channel flow. Such an approach has been implemented to investigate the changes of the shape of a puff or scalar markers released in the middle of a turbulent flow channel [44] and to investigate mixing [45]. The contributions of this work are distinct from the contributions of prior work in that they (a) describe the concept of mixing quality with a quantitative measure; (b) investigate the characteristics of turbulent mixing in anisotropic turbulent flow; and (c) explore the effects of different Sc on mixing, with the use of results for $Sc = 0.7, 6, 200$, and 2400.

2. Materials and Methods

2.1. Flow Field Simulation and Lagrangian Scalar Tracking Approach

The velocity field is obtained using a DNS for a Newtonian and incompressible fluid in a fully developed Poiseuille channel flow. The turbulent flow is anisotropic, so that the velocity fluctuations and the turbulence statistics depend on the distance from the wall. The details of the DNS methodology have been published elsewhere [46] and have been validated with experimental measurements [44,47], while the LST approach [48] has also been validated with comparisons of the computational results to both experimental and Eulerian computational results [14,49,50]. The dimensions of the DNS computational box are 16πh × 2h × 2πh in the x (streamwise), y (wall-normal), and z (spanwise) directions, respectively, where the half-channel height (h) is 300. The driving force for the flow is a constant mean pressure gradient, and the resolution of the simulation is a grid with 1024 × 129 × 256 points in the x, y, and z directions, respectively. The spacing of the mesh is uniform in the streamwise and spanwise directions with periodic boundary conditions, while Chebyshev polynomial collocation points are used in the direction normal to the channel walls with no-slip and no-penetration boundary conditions on the rigid channel walls. The DNS algorithm is based on a pseudospectral fractional step method with the pressure correction suggested by Marcus [51,52]. The friction Reynolds number is $Re_\tau = h = 300$, when the quantities are made dimensionless with the viscous wall parameters (i.e., the friction velocity, u^*, and the kinematic viscosity of the fluid, ν). The friction velocity is given as $u^* = (\tau_w/\rho)^{1/2}$, where τ_w is the shear stress at the wall and ρ is the fluid density. Based on the mean centerline velocity and the channel half height, the Reynolds number is $Re = 5760$. The time step was

0.1 in viscous wall units, and the iterations were carried out for 30,000 time steps for stationary channel flow in order to simulate 3000 viscous time units of flow after the puffs are released.

The trajectories of mass markers released from lines parallel to the z axis at different distances from the channel wall are calculated using the flow field created by the DNS. Data were generated for Sc covering four orders of magnitude (Sc = 0.7, 6, 200 and 2400) with markers released at a distance from the bottom channel wall equal to y_o = 0, 15, 75, and 300, as seen in Figure 1. This configuration allows the study of effects of turbulent convection and molecular diffusion on mixing at both small and relatively large times. The markers were released from 20 lines uniformly spaced along the x direction (at spaces of $\Delta x = 16\pi h/20$), with the first line starting at $x = 0$ for each distance from the wall. The released markers were also uniformly spaced in the z direction (spacing equal to $\Delta z = 2\pi h/5000$), so that 5000 markers were released on each one of the 20 lines. In this way, the total number of markers per distance from the wall, y_o, is 100,000 and the total number of markers per Sc is 400,000. Since the flow is homogeneous in the x and z directions, there is no statistically important difference in choosing the point of release, and by covering the x-y plane with particles released at each y_o, one can remove bias effects by velocity field peculiarities at the point of particle release. In order to obtain the statistics of the particle motion in the streamwise direction, the x coordinate of each particle was determined after subtracting the x coordinate at the marker point of release x_o, so that $(x-x_o)$ was used as the streamwise marker location in the data analysis. The mass markers are passive, and they do not affect the flow field. The flow field used for the Lagrangian scalar tracking of the markers is the same for all Sc cases, so that the effects of the Sc can be observed rather than the effects of the flow. The mass markers were released after the flow field reached a stationary state [48]. The motion of the scalar markers was decomposed into a convection part and a molecular diffusion part. The convective part can be calculated from the fluid velocity at the particle position, so that the Lagrangian velocity at time t of a marker released at location $\vec{X}_o$ at time zero, is assumed to be the same as the Eulerian velocity at that particle's location at the beginning of the convective step, i.e., $\vec{V}(\vec{X}_o, t) = \vec{U}[\vec{X}(\vec{X}_o, t), t]$, where $\vec{V}$ is the Lagrangian velocity and $\vec{U}$ is the Eulerian velocity vector. The equation of particle motion is then:

$$\vec{V}(\vec{X}_o, t) = \frac{\partial \vec{X}(\vec{X}_o, t)}{\partial t} \tag{1}$$

The particle velocity is found by using a mixed Lagrangian-Chebyshev interpolation between the Eulerian velocity field values at the surrounding mesh points. The particle position integration in time is approximated with a second order Adams-Bashforth scheme. The above approach has also been used in [44,45]. The effect of molecular diffusion follows from Einstein's theory for Brownian motion [53], in which the rate of molecular diffusion is related to the molecular diffusivity D as follows:

$$\frac{d\overline{X^2}}{dt} = 2D \tag{2}$$

for diffusion in the x dimension. The diffusion effect is simulated by adding a random movement on the particle motion at the end of each convective step. This random motion is a random jump that takes values from a Gaussian distribution with zero mean and standard deviation σ, which is found using Equation (2) to be $\sigma = \sqrt{2\Delta t/Sc}$, where Δt is the time step of the simulation in viscous wall units, (Δt = 0.1). This approach has also been applied in [14,16,17,54].

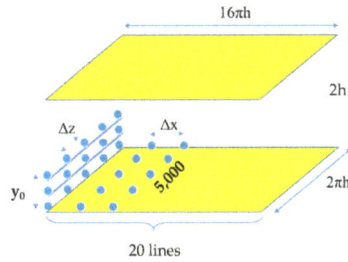

Figure 1. Channel geometry and points of release of markers at time $t = 0$.

2.2. Mixing Calculations

The mixing of scalar markers arriving at a location in the flow field was evaluated following the procedures detailed in [45]. Assuming that two puffs of particles (one of type A and one of type B) were released in the flow field, one needs to decide when they would be considered to be mixed. A length scale for mixing, r, can be defined as the maximum distance at which two mass markers can interact, so that when a particle A is close to a particle B at a distance smaller than r, the two particles are considered mixed. The length scale for mixing needs to be determined so that it is not too small (missing mixed particles) or too large (overestimating mixing) [12]. In this work, the mixing length scale was determined as the diameter of a sphere defined using the turbulent Schmidt number (Sc_t) and the Lagrangian time scale. The scale r is proportional to the length scale for particle turbulent diffusion, which is $\sigma = \sqrt{2\tau/Sc_t}$, in analogy to the standard deviation of the random molecular motion. This is the standard deviation of the distribution of particle motion for particles diffusing with a diffusivity related to the turbulent Schmidt number over a time scale τ that is defined based on a Lagrangian criterion. In analogy to the Lagrangian correlation coefficient for fluid particles, Saffman [55] defined a material correlation coefficient for the Lagrangian motion of scalar markers. The difference is that scalar markers can move off of fluid particles depending on their diffusivity. The integral of Saffman's material correlation coefficient over time provides a *material timescale* for scalar turbulent motion, and this is the time scale used herein to determine the value of σ. This time scale has been determined in prior results from our lab to be [56]:

$$\tau = 4.59 + 33.04 \left(\frac{y_0}{h}\right)^{0.87} \qquad (Sc_t > 3) \tag{3a}$$

$$\tau = (0.98 * Sc_t^{0.13})[4.59 + 33.04\left(\frac{y_0}{h}\right)^{0.87}] \qquad (Sc_t \leq 3) \tag{3b}$$

while the average Sc_t in the flow channel has been found to be $Sc_t = 1.0257$ (based on averaging the data in Figure 6 of ref. [35], where the Sc_t has been calculated as a function of distance from the channel wall). Since there is 95% probability for a Gaussian random variable to be within 2σ of its mean, we can assume that 95% of the motion of a scalar particle within time τ is going to be within a sphere of diameter 4σ, with σ determined as above. The average value of τ is 21.76, so that $\sigma = 6.5$ in viscous wall units. A circle with the diameter 4σ has an area of 533 square wall units. Therefore, the whole channel is divided into square bins of size 20×20 in the x and y directions, respectively, so that the bin area is in the same order of magnitude as the area of the particle motion when projected on the x-y plane (i.e., 400 square wall units is close to 533 square wall units). Bins of this size account for the mixing length scale used herein.

The total number of bins in the x and y directions were:

$$N_{binX} = \frac{4 * L_X}{\Delta bin_X}; \ N_{binY} = \frac{2h}{\Delta bin_Y} \tag{4}$$

where the bin size is $\Delta bin_X = \Delta bin_Y = 20$, L_X is the computational box length in the x direction, and $2h$ is the channel height. Each bin in the system can be denoted as bin (i,j), with $1 \leq i \leq N_{binX}$ and $1 \leq j \leq N_{binY}$. In the flow direction, there were enough bins to cover four times the channel length (this can be accomplished by taking advantage of periodicity in the streamwise direction and by allowing a particle that exits the box from one side to enter the box from the other side). This length was sufficient to track all the particles within 3000 viscous wall time units. Tests with bins starting at different positions showed negligible difference in the results regarding mixing [45].

3. Results

3.1. Development of Mixing Region with Time

Figure 2a–d are contour plots of the development of the mixing region from puffs released at the center line of the channel and at $y = 75$ as a function or time t after their release. In this case, particles of type A are particles with $Sc = 0.7$ released from the center of the channel (at $y_o = 300$). This puff is designated as 0.7y300. Particles of type B are particles of the same Sc ($Sc = 0.7$), but released at $y_o = 75$, designated as 0.7y75. This notation will be adopted herein to denote the Sc of a puff and the location of its release. The contour colors in Figure 2 denote the number of mixed particles in each location, defined as the minimum particle count of particles of type A or B found in a bin. The idea is that if there are $n_A(i,j)$ particles of type A in bin (i,j), and $n_B(i,j)$ particles of type B in the same bin, then the mixing will be characterized by the minimum of $n_A(i,j)$ and $n_B(i,j)$. When this number is high, it means that the intensity of mixing is high. Figure 2 is a presentation of the quantity $n_{AB}(i,j) = \min[n_A(i,j), n_B(i,j)]$, and in this respect, it represents the development in time of the shape of the mixing zone and of the mixing intensity in the flow field.

Figure 3a–d are plots of the mixing region for puffs 0.7y300 and 0.7y15. The Sc is the same for these two puffs, but the location of release of the second puff is at $y_o = 15$, so that differences in the development of mixing can be seen when compared with the case shown in Figure 1. Since past work has indicated that the Sc of the puff released in the center of the channel is not important for mixing, only figures for puffs with $Sc = 0.7$ released from the channel center are shown. However, the Sc of the puff released closer to the wall is important for the development of the mixing zone. This is why in Figure 4a–d, plots of the mixing region for puffs 0.7y300 and 2400y15 are presented.

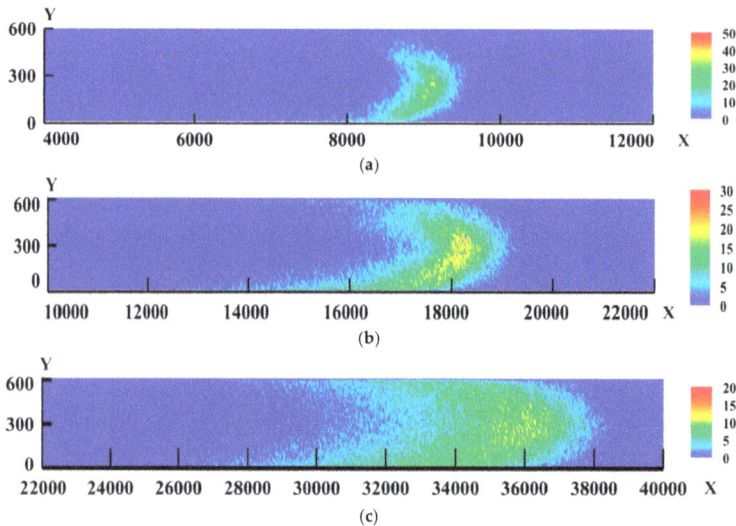

Figure 2. *Cont.*

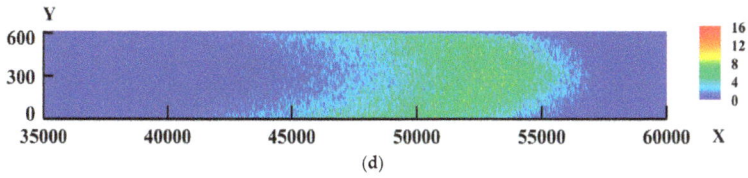

(d)

Figure 2. Development of the mixing region with time for two puffs, puff A: $Sc = 0.7$, $y_o = 300$, and puff B: $Sc = 0.7$, $y_o = 75$. (**a**) Contours of n_{AB} at $t = 500$; (**b**) Contours of n_{AB} at $t = 1000$; (**c**) Contours of n_{AB} at $t = 2000$; (**d**) Contours of n_{AB} at $t = 3000$.

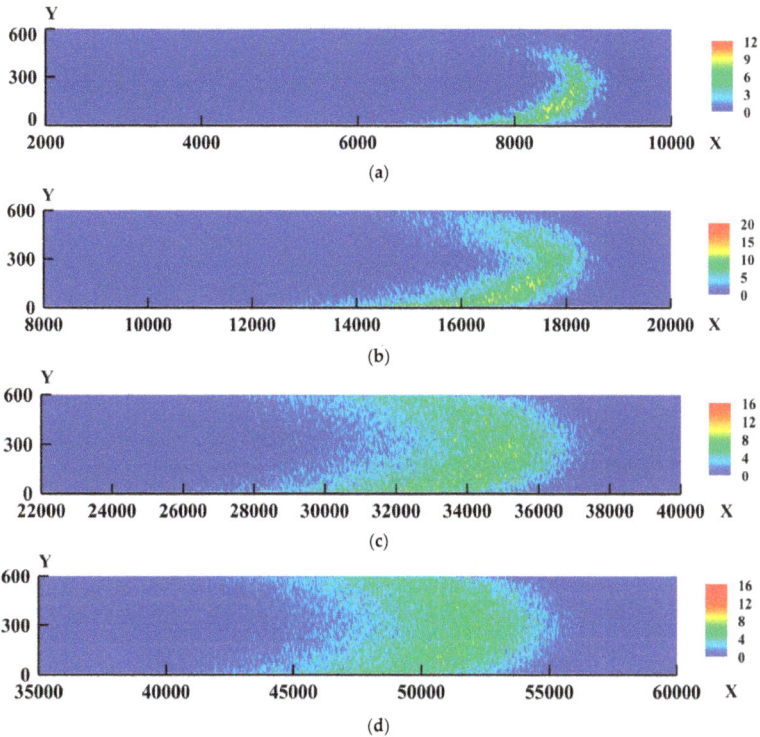

(a)

(b)

(c)

(d)

Figure 3. Development of the mixing region with time for two puffs, puff A: $Sc = 0.7$, $y_o = 300$, and puff B: $Sc = 0.7$, $y_o = 15$. (**a**) Contours of n_{AB} at $t = 500$; (**b**) Contours of n_{AB} at $t = 1000$; (**c**) Contours of n_{AB} at $t = 2000$; (**d**) Contours of n_{AB} at $t = 3000$.

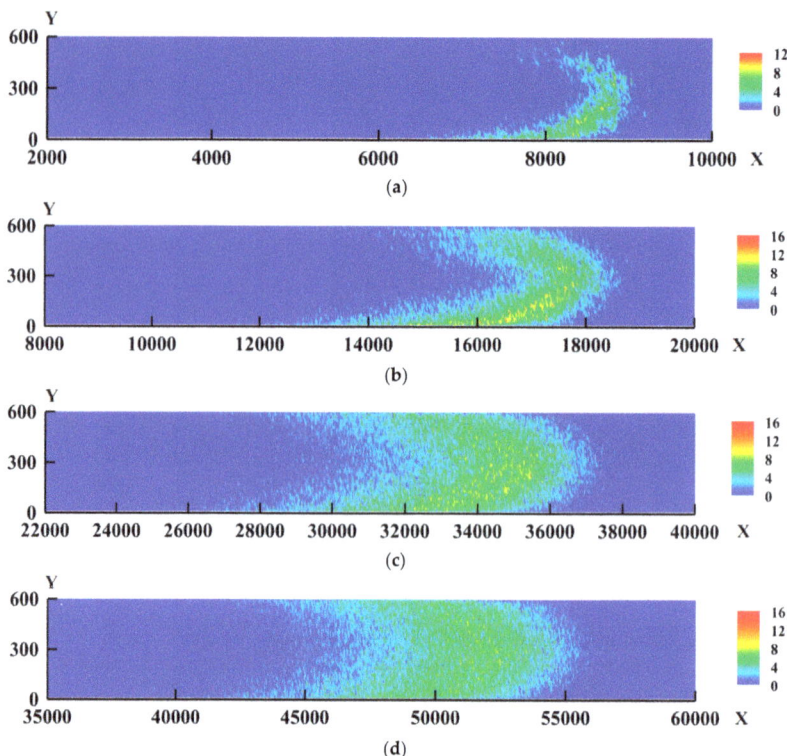

Figure 4. Development of the mixing region with time for two puffs, puff A: $Sc = 0.7$, $y_0 = 300$, and puff B: $Sc = 2400$, $y_0 = 15$. (**a**) Contours of n_{AB} at $t = 500$; (**b**) Contours of n_{AB} at $t = 1000$; (**c**) Contours of n_{AB} at $t = 2000$; (**d**) Contours of n_{AB} at $t = 3000$.

3.2. Quality of Mixing

Figures 2–4 are depictions of the mixing intensity, as expressed by n_{AB}, but there is a chance that particles are mixed, but not efficiently. This means that there might be many particles of one type in a bin, but there might be multiples of this number of particles of the other type in the same bin. To make this point clearer, when the number of particles of each type in a bin is the same, i.e., $n_A(i,j) = n_B(i,j)$, then the quality of mixing is considered to be best, as opposed to a case where $n_A(i,j) = C\, n_B(i,j)$, with C being a large coefficient. In order to quantify the *quality of mixing* in a location in the flow field, a new measure is introduced in this study, the mixing quality index, ϕ, defined as the product of the particle fraction in each bin (i,j), as follows:

$$\varphi(i,j) = \frac{n_A(i,j) \times n_B(i,j)}{[n_A(i,j) + n_B(i,j)]^2} \tag{5}$$

The above quantity has the advantage that it is monotonically increasing with the quality of mixing. When only particles from puff A or from puff B are present in a bin, the value of $\phi(i,j)$ is zero. The maximum value of ϕ occurs when half of the particles in a bin are from puff A and half from puff B—at that point the value of the mixing quality index is $\phi_{max} = 0.25$. Figures 5–7 are plots of the development of the mixing quality index for puffs 0.7y300 and 0.7y75, puffs 0.7y300 and 0.7y15, and 0.7y300 and 2400y15.

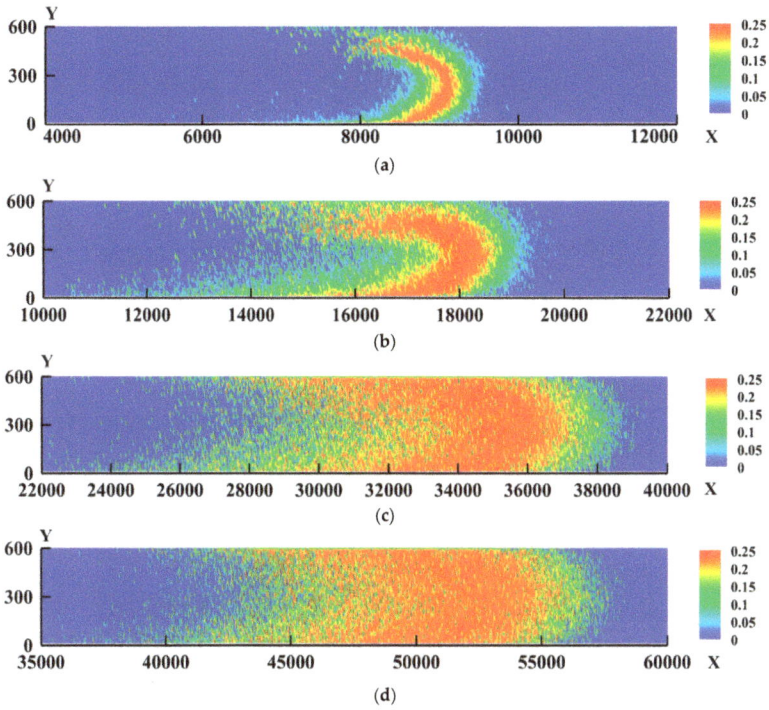

Figure 5. Mixing efficiency in the flow channel based on the contours of the mixing quality index, ϕ, for two puffs, puff A: $Sc = 0.7$, $y_o = 300$, and puff B: $Sc = 0.7$, $y_o = 75$. (**a**) Contours of ϕ at $t = 500$; (**b**) at $t = 1000$; (**c**) at $t = 2000$; and (**d**) at $t = 3000$.

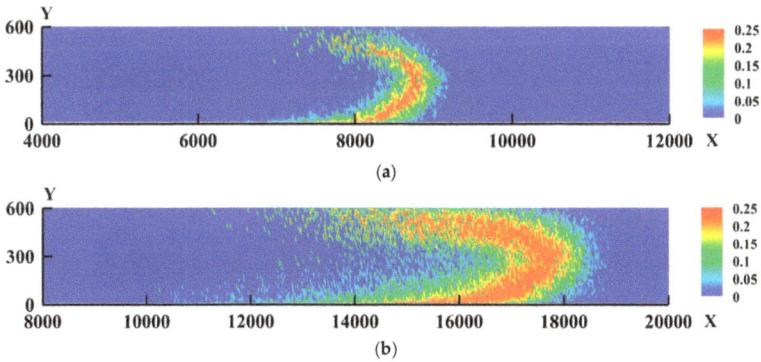

Figure 6. *Cont.*

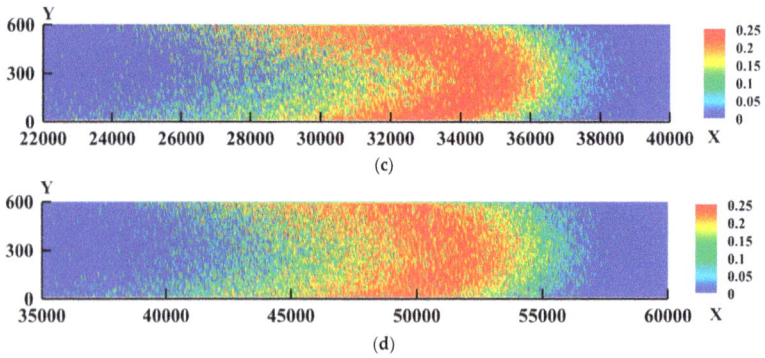

Figure 6. Mixing efficiency in the flow channel based on the contours of the mixing quality index, ϕ, for two puffs, puff A: $Sc = 0.7$, $y_o = 300$, and puff B: $Sc = 0.7$, $y_o = 15$. (**a**) Contours of ϕ at $t = 500$; (**b**) at $t = 1000$; (**c**) at $t = 2000$; and (**d**) at $t = 3000$.

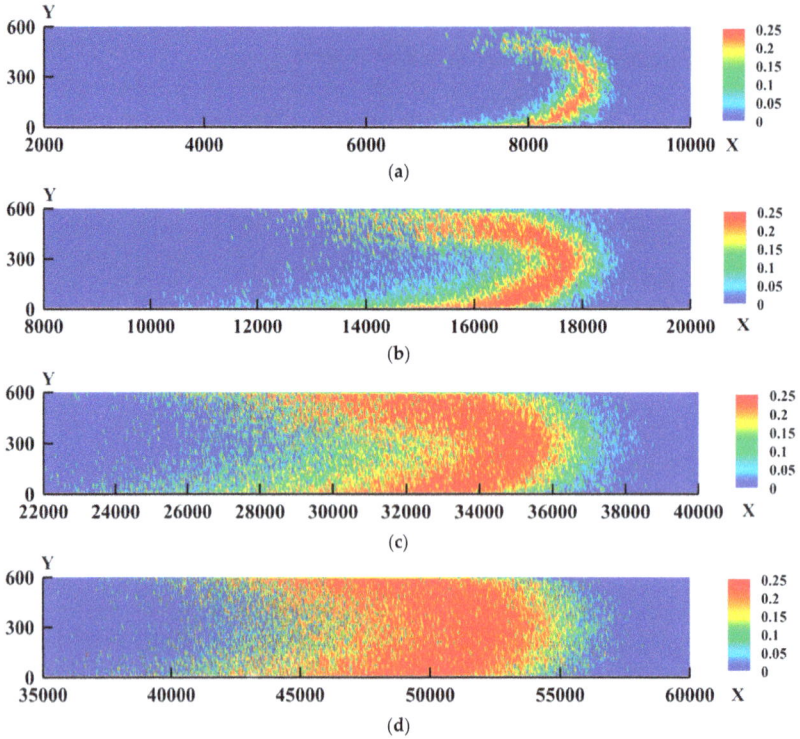

Figure 7. Mixing efficiency in the flow channel based on the contours of the mixing quality index, ϕ, for two puffs, puff A: $Sc = 0.7$, $y_o = 300$, and puff B: $Sc = 2400$, $y_o = 15$. (**a**) Contours of ϕ at $t = 500$; (**b**) at $t = 1000$; (**c**) at $t = 2000$; and (**d**) at $t = 3000$.

One can also define an overall mixing efficiency that can combine the intensity of mixing with the quality of mixing across the channel in the following manner:

$$m = \frac{\sum_{i=1}^{N_{binX}} \sum_{j=1}^{N_{binY}} n_{AB}(i,j)\, \varphi(i,j)}{N_t\, \varphi_{max}} \tag{6}$$

The measure m can take values between zero and one (N_t is the total number of particles released in each puff, in this case, $N_t = 100{,}000$). When it approaches zero, there is either no mixing or the quality of mixing is very poor, and when it approaches one, the particles are half of type A and half of type B in each bin. Figure 8 is a plot of the change of m with time for the cases of puff A being released from the center of the channel (0.7y300) and puff B released at different distances from the wall of the channel.

Figure 9 is a plot of the change of m for two puffs that have different Sc values. Puff A is the same as above, but puff B is a puff of drastically higher Sc, with a value of $Sc = 2400$. Figure 10 is a plot of the change of m for two puffs released at the same time at the wall of the channel. While the Sc of puffs released in the center region of the channel might not be that important in determining the dynamics of the mixing region, the Sc effects for puffs released in the viscous region close to the channel walls are important. In this region, the dynamics of the puff are affected by convection and by molecular diffusion. Figure 11 is a plot of m for two puffs that were released either with the high Sc puff released at the wall or the low Sc released at $y_o = 15$ (puff A is 0.7y15 and puff B is 2400y0) and for two puffs with the Sc reversed (puff is 2400y15 and puff B is 0.7y0).

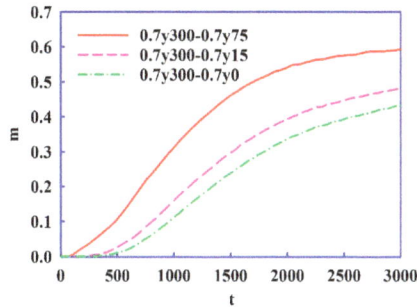

Figure 8. Normalized mixing efficiency m for mixing between a puff released at the channel center (puff A: $Sc = 0.7$, $y_o = 300$) and puffs of the same Sc released at distances $y_o = 75$, 15, and 0.

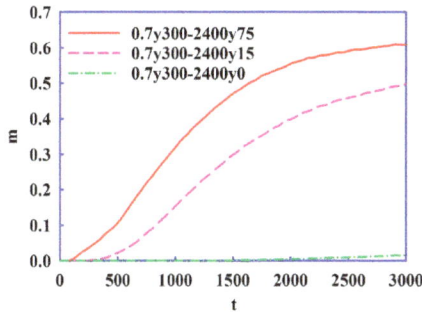

Figure 9. Normalized mixing efficiency m for mixing between a puff released at the channel center (puff A: $Sc = 0.7$, $y_o = 300$) and puffs of drastically different Sc values ($Sc = 2400$) released at distances $y_o = 75$, 15, and 0.

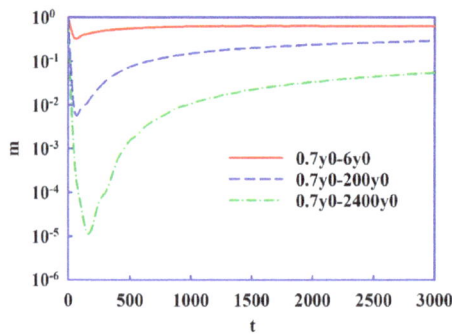

Figure 10. Normalized mixing efficiency *m* for mixing between two puffs released at the channel wall, but with different Schmidt numbers (here we have all 4 *Sc*).

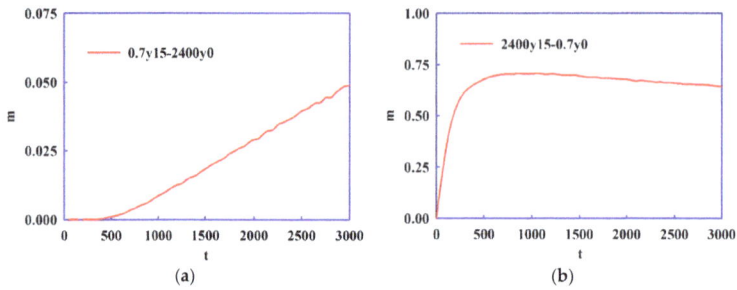

(a)   (b)

Figure 11. Normalized mixing efficiency m for mixing between two puffs released at the channel wall, and at $y = 15$, but with Schmidt numbers switching between the two configurations ((**a**): mixing between puff 0.7y15 and puff 2400y0, (**b**): mixing between puff 2400y15 and puff 0.7y0).

4. Discussion

For the mixing of puffs of the same *Sc* released at different distances from the wall, it is seen in Figures 2 and 3 that the mixing region starts from a crescent shape and slowly extends to cover the whole channel width. The crescent-shaped mixing region occurs because it represents the overlap region between puff A (released at the center of the channel and convecting downstream with a high velocity) and puff B (which is released at the logarithmic layer, where the mean flow velocity is less than at the channel centerline). The crescent shape is more pronounced in Figure 3, when one of the two mixing puffs is released closer to the wall (at $y_o = 15$). In Figure 12, we plot the individual puffs that resulted from the mixing shown in Figure 3a. One can see that the shape of the mixing region is the result of the shape of the individual puffs, and that the relative velocity of the puffs to each other and the shape of each puff (which is never a circle or an ellipse) play a very significant role in the dynamics of the mixing region. In anisotropic turbulence, such as the case of puffs released in a channel flow, mixing is affected by viscous effects and molecular diffusion, since the puff shape is the result of the interplay between molecular and convective transfer. At longer times, times larger than $t = 2000$ viscous wall units, the mixing region looks more homogenous, and the extent of it is quite large (see, for example, in Figure 3d, that the mixing region extends from roughly $x = 50,000$ to $x = 54,000$).

In Figure 4, it is seen that the *Sc* for the puff released in the viscous layer plays a role in the development of the mixing region, since the puff for higher *Sc* stays close to the channel wall for a longer time (note that the molecular diffusion for a high *Sc* puff is small, so that it takes longer for particles released close to the wall to get to the outer region of the flow field and mix with a puff released from the channel center). In Figure 4, it is seen that the crescent shape is narrower and the development of a mixing region that extends across the channel takes longer than seen in Figure 3.

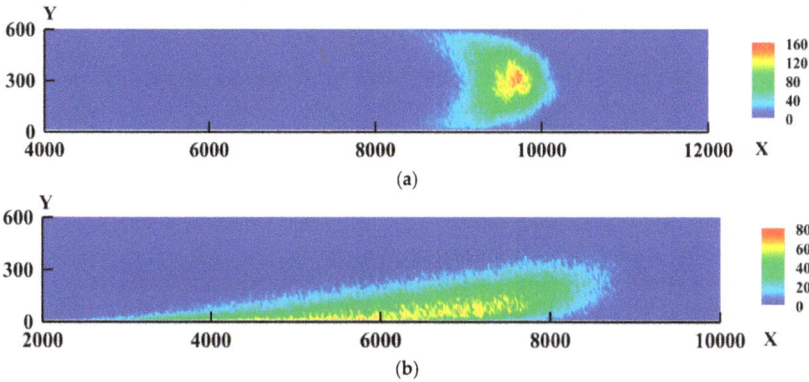

Figure 12. Contours of the location of the two puffs mixing at $t = 500$, corresponding to the conditions of Figure 3a. (a) Puff A ($Sc = 0.7$, $y_o = 300$), and (b) Puff B ($Sc = 0.7$, $y_o = 15$). Note that the shape of puff B is elongated in the streamwise direction and a large part of it is close to the wall. The shape of puff A looks like a jellyfish, because of the high mean velocity at the channel center. Mixing occurs at the overlap of these two puffs.

The index for mixing quality ϕ, as introduced in Equation (5), at $t = 500$, offers more insights to the mixing process. While the mixing zone, which is mostly located in the lower half of the channel at this time, seems to be homogeneous in terms of mixing intensity (see Figure 1a), the quality of mixing is not the same across this area, with peak mixing quality at the semicircular region clearly seen in Figure 4a. Similar results are seen for the other cases considered. The intensity of mixing and the mixing location do not necessarily imply a high mixing quality, when one considers Figures 2–4 along with their counterparts, Figures 5–7. In theory, if the mixing intensity reaches a maximum value of 100%, then based on our previous definitions, the mixing quality will also reach a value of 0.25 at every mixing bin, implying that mixing is homogenous across the channel. However, mixing intensity cannot necessarily reach 100% (as in this study, the measurements showed that up to 70% of the total number of particles can be mixed), and a rather extended amount of time may be required for it to happen, assuming that there is no reaction between different types of particles during this period of time. Under this circumstance, the mixing quality index allows us to evaluate the mixing zone in terms of its homogeneity. From the observed values of this index, it may not be accurate to assume that perfect mixing always occurs in turbulent flow. One case where such effects could be more pronounced is in the case of a fast chemical reaction happening, where one reactant may be fully depleted. For example, if puffs A and B represent two reactants that react with one molecule of A reacting with one molecule of B, then the reaction will occur in the pattern seen in Figures 5–7, where mixing is most homogeneous. If the reaction is exothermic, heat will be released in locations that follow the same pattern. This information might be important for the design and performance of chemical reactors in engineering, where controlling the reaction rate/location and heat transfer during the process is of high priority.

In a previous study [57], a mixing parameter was defined to quantify mixing that happens due to a Richtmyer-Meshkov instability, which involves the motion of two fluids with different densities driven by an impulsive acceleration. This parameter was said to be a measure for a fast reaction rate where one reactant is fully depleted, and the amount of mixing was quantified by measuring volume fractions along the x direction. The normalized mixing efficiency that is introduced here, depicted in Figures 8–11, is a measure that considers both mixing intensity and quality of mixing in both x and *y* directions. It is seen that given the conditions prescribed in the simulations, there appears to be a maximum *m* for each pair of puffs. This maximum is approached asymptotically, and it indicates that the best mixing (*m* = 1) is not achieved. It is seen in Figures 8 and 9 that the mixing is better when the two puffs are released in locations that are closer (see that *m* asymptotically goes to larger values when puff B is closer to the channel center). However, less mixing occurs when one puff is released at the center of the channel and the other at the channel wall, because the puff released at the wall is more elongated in the *x* direction. In Figure 9, where there is a large difference between the *Sc* of the two puffs, there is hardly any mixing when the second puff is released at the channel wall (see the green curve in Figure 9). When the *Sc* of puff B changes from 0.7 to 2400, there are no significant changes in the maximum *m* obtained when puff B is released at y_o = 75. The same is true when puff B is released at y_o = 15 (see Figures 8 and 9). The values of *m* exhibit differences when puff B is released from the channel wall. Furthermore, when two puffs of significantly different *Sc* values are released from the wall of the channel, even when they are released at the same time and at the same location on the wall, the value of *m* starts at its maximum value of 1, then decreases to a minimum, and then increases again. The local minimum of *m* indicates a decrease in mixing–in other words, particles released together separate first and mix again later in the channel. This is a behavior that has been observed and analyzed recently [58,59] for small times, while here, it is clearly observed during a larger time interval.

The mixing behavior observed for the puffs examined in Figure 11 is quite interesting and might have implications for the design of reactors. The same two puffs are released in the flow field, but the *Sc* of the two puffs is switched. When the low *Sc* puff is released at the channel wall, large molecular diffusion helps to disperse the puff in the flow field and to reach the regions where the y_o = 15 puff reaches. Mixing occurs in that case, as seen by the value of *m* that is relatively high at long times (*m* goes to 0.64 asymptotically). However, when the high *Sc* puff is released from the channel wall, there is hardly any mixing between the two puffs. One would expect to observe this behavior given the details of the development of puffs released from the channel wall, and given the transition of the puff from molecular diffusion-dominated development to convection-dominated dispersion [58]. However, if one were to predict mixing behavior without a deeper appreciation of the molecular effects on turbulent dispersion in anisotropic turbulence, the predictions might not have been correct.

5. Conclusions

Mixing in an anisotropic turbulent velocity field generated by channel flow has been studied using Lagrangian methods (direct numerical simulation of turbulent flow and Lagrangian scalar tracking techniques). Measures that quantify the efficiency of mixing in terms of both quality and intensity have been introduced and utilized to explore the simulation results. It is found that the mixing of puffs released in the flow field can depend strongly on the Schmidt number and on the location of puff release. It is also found that only calculating how many particles are mixed is not enough to obtain a full picture of the process, which is completed when the quality of mixing is taken into account. While the *Sc* of the puff released at the center of the flow field is not critical to the mixing process (as has previously been shown) it is now found that the *Sc* of puffs released at the wall or in the viscous wall region is important. Finally, it appears that one could control particle mixing and particle separation as a function of time after particle release by determining the position of puff injection in the flow field. This realization could be taken advantage of in the design of equipment for applications where the control of mixing is important. While the results presented here are obtained for passive scalar and cannot be extended to non-passive particles, one could change the equation of motion of the particles and include effects like drag, lift etc., to apply a similar analysis for non-passive particles.

Author Contributions: Both D.V.P. and Q.N. conceived and designed the experiments; Q.N. performed the simulations; and both D.V.P. and Q.N. analyzed the data and wrote the paper.

Funding: This research was partially funded by the National Science Foundation of the US, grant number CBET-18-0260.

Acknowledgments: The use of computing facilities at the University of Oklahoma Supercomputing Center for Education and Research (OSCER) and at XSEDE (under allocation CTS-090025) is gratefully acknowledged.

Conflicts of Interest: The authors declare no conflict of interest. The funders had no role in the design of the study; in the collection, analyses, or interpretation of data; in the writing of the manuscript, and in the decision to publish the results.

References

1. Tennekes, H.; Lumley, J.L. *A First Course in Turbulence*; MIT Press: Boston, MA, USA, 1972.
2. Dimotakis, P.E. Turbulent Mixing. *Annu. Rev. Fluid Mech.* **2005**, *37*, 329–356. [CrossRef]
3. Fox, R.O. *Computational Models for Turbulent Reacting Flows*; Cambridge University Press: Cambridge, UK, 2003.
4. Curl, R.L. Dispersed phase mixing: I. Theory and effects in simple reactors. *AIChE J.* **1963**, *9*, 175–181. [CrossRef]
5. Subramaniam, S.; Pope, S.B. A mixing model for turbulent reactive flows based on Euclidean minimum spanning trees. *Combust. Flame* **1998**, *115*, 487–514. [CrossRef]
6. Oldshue, J.W. *Fluid Mixing Techology*; McGraw Hill: New York, NY, USA, 1983.
7. Paul, E.L.; Atiemo-Obeng, V.A.; Kresta, M.S. *Handbook of Industrial Mixing: Science and Practice*; John Wiley & Sons: Hoboken, NJ, USA, 2004.
8. Mylne, K.R.; Mason, P.J. Concentration fluctuation measurements in a dispersing plume at a range of up to 1000 m. *QJRMS* **1991**, *117*, 177–206. [CrossRef]
9. Pasquill, F.; Smith, F.B. *Atmospheric Diffusion*; Horwood, E., Ed.; John Wiley & Sons: Hoboken, NJ, USA, 1983; p. 437.
10. Ottino, J.M. *The Kinematics of Mixing: Stretching, Chaos, and Transport*; Cambridge University Press: Cambridge, UK, 1989.
11. Kukukova, A.; Aubin, J.; Kresta, S.M. A new definition of mixing and segregation: Three dimensions of a key process variable. *Chem. Eng. Res. Des.* **2009**, *87*, 633–647. [CrossRef]
12. Calzavarini, E.; Cencini, M.; Lohse, D.; Toschi, F. Quantifying Turbulence-Induced Segregation of Inertial Particles. *Phys. Rev. Lett.* **2008**, *101*, 084504. [CrossRef] [PubMed]
13. Green, D.W.; Perry, R.H. *Perry's Chemical Engineers' Handbook*, 8th ed.; McGraw Hill: New York, NY, USA, 2007.
14. Papavassiliou, D.V. Turbulent transport from continuous sources at the wall of a channel. *Int. J. Heat Mass Transf.* **2002**, *45*, 3571–3583. [CrossRef]
15. Hanratty, T.J. Heat transfer through a homogeneous isotropic turbulent field. *AIChE J.* **1956**, *2*, 42–45. [CrossRef]
16. Papavassiliou, D.V.; Hanratty, T.J. Transport of a passive scalar in a turbulent channel flow. *Int. J. Heat Mass Transf.* **1997**, *40*, 1303–1311. [CrossRef]
17. Mitrovic, B.M.; Le, P.M.; Papavassiliou, D.V. On the Prandtl or Schmidt number dependence of the turbulent heat or mass transfer coefficient. *Chem. Eng. Sci.* **2004**, *59*, 543–555. [CrossRef]
18. Incropera, F.P.; Kerby, J.S.; Moffatt, D.F.; Ramadhyani, S. Convection heat transfer from discrete heat sources in a rectangular channel. *Int. J. Heat Mass Transf.* **1986**, *29*, 1051–1058. [CrossRef]
19. Raupach, M.R.; Legg, B.J. Turbulent dispersion from an elevated line source—Measurements of wind concentration moments and budgets. *J. Fluid Mech.* **1983**, *136*, 111–137. [CrossRef]
20. Poreh, M.; Cermak, J.E. Study of diffusion from a line source in a turbulent boundary layer. *Int. J. Heat Mass Transf.* **1964**, *7*, 1083–1095. [CrossRef]
21. Fackrell, J.E.; Robins, A.G. Concentration fluctuations and fluxes in plumes from point sources in a turbulent boundary layer. *J. Fluid Mech.* **1982**, *117*, 1–26. [CrossRef]
22. Shlien, D.J.; Corrsin, S. Dispersion measurements in a turbulent boundary layer. *Int. J. Heat Mass Transf.* **1976**, *19*, 285–295.
23. Beguier, C.; Dekeyser, I.; Launder, B.E. Ratio of scalar and velocity dissipation time scales in shear flow turbulence. *Phys. Fluids* **1978**, *21*, 307–310. [CrossRef]
24. Ma, B.K.; Warhaft, Z. Some aspects of the thermal mixing layer in grid turbulence. *Phys. Fluids* **1986**, *29*, 3114–3120. [CrossRef]

25. Dahm, W.J.A.; Dimotakis, P.E. Mixing at large Schmidt number in the self-similar far field of turbulent jets. *J. Fluid Mech.* **1990**, *217*, 299–330. [CrossRef]

26. Yeung, P.K.; Xu, S. Schmidt number effects on turbulent transport with uniform mean scalar gradient. *Phys. Fluids* **2002**, *14*, 4178–4191. [CrossRef]

27. Borgas, M.S.; Sawford, B.L.; Xu, S.; Donzis, D.A.; Yeung, P.K. High Schmidt number scalars in turbulence: Structure functions and Lagrangian theory. *Phys. Fluids* **2004**, *16*, 3888–3899. [CrossRef]

28. Buaria, D.; Yeung, P.K.; Sawford, B.L. A Lagrangian study of turbulent mixing: Forward and backward dispersion of molecular trajectories in isotropic turbulence. *J. Fluid Mech.* **2016**, *799*, 352–382. [CrossRef]

29. Yeung, P.K.; Donzis, D.A. High-Reynolds-number simulation of turbulent mixing. *Phys. Fluids* **2005**, *17*, 081703. [CrossRef]

30. Sawford, B.L. Lagrangian modeling of scalar statistics in a double scalar mixing layer. *Phys. Fluids* **2006**, *18*, 085108. [CrossRef]

31. Sawford, B.L.; Kops, S.M.D.B. Direct numerical simulation and Lagrangian modeling of joint scalar statistics in ternary mixing. *Phys. Fluids* **2008**, *20*, 095106. [CrossRef]

32. Brethouwer, G.; Hunt, J.C.R.; Nieuwstadt, F.T.M. Micro-structure and Lagrangian statistics of the scalar field with a mean gradient in isotropic turbulence. *J. Fluid Mech.* **2003**, *474*, 193–225. [CrossRef]

33. Yeung, P.K.; Xu, S.; Donzis, D.A.; Sreenivasan, K.R. Simulations of three-dimensional turbulent mixing for Schmidt numbers of the order 1000. *Flow Turbul. Combust.* **2004**, *72*, 333–347. [CrossRef]

34. Srinivasan, C.; Papavassiliou, D.V. Heat Transfer Scaling Close to the Wall for Turbulent Channel Flows. *Appl. Mech. Rev.* **2013**, *65*. [CrossRef]

35. Srinivasan, C.; Papavassiliou, D.V. Prediction of the turbulent Prandtl number in wall flows with Lagrangian simulations. *Ind. Eng. Chem. Res.* **2011**, *50*, 8881–8891. [CrossRef]

36. Srinivasan, C.; Papavassiliou, D.V. Comparison of backwards and forwards scalar relative dispersion in turbulent shear flow. *Int. J. Heat Mass Transf.* **2012**, *55*, 5650–5664. [CrossRef]

37. Hasegawa, Y.; Kasagi, N. Low-pass filtering effects of viscous sublayer on high Schmidt number mass transfer close to a solid wall. *Int. J. Heat Fluid Flow* **2009**, *30*, 525–533. [CrossRef]

38. Mito, Y.; Hanratty, T.J. Lagrangian stochastic simulation of turbulent dispersion of heat markers in a channel flow. *Int. J. Heat Mass Transf.* **2003**, *46*, 1063–1073. [CrossRef]

39. Koumoutsakos, P. Multiscale flow simulations using particles. *Annu. Rev. Fluid Mech.* **2005**, *37*, 457–487. [CrossRef]

40. Moin, P.; Mahesh, K. Direct numerical simulation: A tool in turbulence research. *Annu. Rev. Fluid Mech.* **1998**, *30*, 539–578. [CrossRef]

41. Lee, M.; Moser, R.D. Direct numerical simulation of turbulent channel flow up to Re-tau approximate to 5200. *J. Fluid Mech.* **2015**, *774*, 395–415. [CrossRef]

42. Lyons, S.L.; Hanratty, T.J.; McLaughlin, J.B. Direct numerical simulation of passive heat transfer in a turbulent channel flow. *Int. J. Heat Mass Transf.* **1991**, *34*, 1149–1161. [CrossRef]

43. Alfonsi, G. On Direct Numerical Simulation of Turbulent Flows. *Appl. Mech. Rev.* **2011**, *64*, 64. [CrossRef]

44. Nguyen, Q.; Feher, S.E.; Papavassiliou, D.V. Lagrangian Modeling of Turbulent Dispersion from Instantaneous Point Sources at the Center of a Turbulent Flow Channel. *Fluids* **2017**, *2*. [CrossRef]

45. Nguyen, Q.; Papavassiliou, D.V. Scalar mixing in anisotropic turbulent flow. *AIChE J.* **2018**. [CrossRef]

46. Lyons, S.L.; Hanratty, T.J.; McLaughlin, J.B. Large-scale computer simulation of fully developed turbulent channel flow with heat transfer. *Int. J. Numer. Methods Fluids* **1991**, *13*, 999–1028. [CrossRef]

47. Gunther, A.; Papavassiliou, D.V.; Warholic, M.D.; Hanratty, T.J. Turbulent flow in a channel at a low Reynolds number. *Exp. Fluids* **1998**, *25*, 503–511. [CrossRef]

48. Kontomaris, K.; Hanratty, T.J.; McLaughlin, J.B. An algorithm for tracking fluid particles in a spectral simulation of turbulent channel flow. *J. Comput. Phys.* **1992**, *103*, 231–242. [CrossRef]

49. Mitrovic, B.M.; Papavassiliou, D.V. Transport properties for turbulent dispersion from wall sources. *AIChE J.* **2003**, *49*, 1095–1108. [CrossRef]

50. Na, Y.; Papavassiliou, D.V.; Hanratty, T.J. Use of direct numerical simulation to study the effect of Prandtl number on temperature fields. *Int. J. Heat Fluid Flow* **1999**, *20*, 187–195. [CrossRef]

51. Orszag, S.A.; Kells, L.C. Transition to turbulence in plane Poiseuille and plane Couette flow. *J. Fluid Mech.* **1980**, *96*, 159–205. [CrossRef]

52. Marcus, P.S. Simulation of Taylor-Couette flow. 1. Numerical methods and comparison with experiment. *J. Fluid Mech.* **1984**, *146*, 45–64. [CrossRef]

53. Einstein, A. Uber die von der molekular-kinetischen Theorie der Warme geforderte Bewegung von in ruhenden Flussigkeiten suspendierten Teilchen. *Ann. Phys.* **1905**, *17*, 549. [CrossRef]

54. Papavassiliou, D.V.; Hanratty, T.J. The use of Lagrangian-methods to describe turbulent transport of heat from a wall. *Ind. Eng. Chem. Res.* **1995**, *34*, 3359–3367. [CrossRef]

55. Saffman, P.G. On the effect of the molecular diffusivity in turbulent diffusion. *J. Fluid Mech.* **1960**, *8*, 273–283. [CrossRef]

56. Le, P.M.; Papavassiliou, D.V. Turbulent Dispersion from Elevated Line Sources in Channel and Couette Flow. *AIChE J.* **2005**, *51*, 2402–2414. [CrossRef]

57. Thornber, B.; Drikakis, D.; Youngs, D.L.; Williams, R.J.R. The influence of initial conditions on turbulent mixing due to Richtmyer-Meshkov instability. *J. Fluid Mech.* **2010**, *654*, 99–139. [CrossRef]

58. Nguyen, Q.; Srinivasan, C.; Papavassiliou, D.V. Flow-induced separation in wall turbulence. *Phys. Rev. E* **2015**, *91*, 033019. [CrossRef] [PubMed]

59. Nguyen, Q.; Papavassiliou, D.V. A statistical model to predict streamwise turbulent dispersion from the wall at small times. *Phys. Fluids* **2016**, *28*, 125103. [CrossRef]

fluids

MDPI

Article

Coherent Vortical Structures and Their Relation to Hot/Cold Spots in a Thermal Turbulent Channel Flow

Suranga Dharmarathne *, Venkatesh Pulletikurthi and Luciano Castillo

School of Mechanical Engineering, Purdue University, 585 Purdue Mall, West Lafayette, IN 47907, USA;
vpulleti@purdue.edu (V.P.); Lcastillo@purdue.edu (L.C.)
* Correspondence: dharmara@purdue.edu; Tel.: +1-806-281-7638

Received: 22 December 2017; Accepted: 4 February 2018; Published: 8 February 2018

Abstract: Direct numerical simulations of a turbulent channel flow with a passive scalar at $Re_\tau = 394$ with blowing perturbations is carried out. The blowing is imposed through five spanwise jets located near the upstream end of the channel. Behind the blowing jets (about $1D$, where D is the jet diameter), we observe regions of reversed flow responsible for the high temperature region at the wall: hot spots that contribute to further heating of the wall. In between the jets, low pressure regions accelerate the flow, creating long, thin, streaky structures. These structures contribute to the high temperature region near the wall. At the far downstream of the jet (about $3D$), flow instabilities (high shear) created by the blowing generate coherent vortical structures. These structures move hot fluid near the wall to the outer region of the channel; thereby, these are responsible for cooling of the wall. Thus, for engineering applications where cooling of the wall is necessary, it is critical to promote the generation of coherent structures near the wall.

Keywords: coherent structures; wall heat flux; flow control; blowing; channel flow; direct numerical simulations (DNS)

1. Introduction

Jets in cross-flow, which are also known as transverse jets, are common in engineering applications, and the environment. Examples include: gas turbine film-cooling, dilution jets in gas turbine combustors, ash plumes from volcanic eruptions, etc. A comprehensive review of jets in cross-flow can be found in Mahesh [1]. Most of the previous studies focused on the parameters of the jets and cross-flow that affect the flow field. The generation of complex coherent vortical structures was studied in [2–4]. However, there is no general consensus about the genesis and evolution of these motions and their effects on passive scalars. The configuration of these vortical motions changes with the characteristic flow parameters like the blowing ratio (the ratio between the mass flow rates/velocities of the cross-flow and the jet) [1].

Blowing and suction have been tested for controlling turbulence, as well. Park and Choi [5] used direct numerical simulations (DNS) on a spatially-developing turbulent boundary layer to study the effect of small blowing and suction perturbations on wall skin friction. They imposed steady blowing and suction through a spanwise slot. Their results show that the skin friction coefficient significantly decreases near downstream and slightly increases far downstream of the slot. The number of coherent vortical structures is also increased downstream. Further, the increase of drag downstream of the slot was attributed to the stretching and tilting of vortices due to blowing. Kim and Sung [6] applied periodic and steady blowing through a spanwise slot to study their effects on the spatially-developing boundary layer. Their results showed that local steady blowing increased the number of vortical structures downstream. Araya et al. [7] studied the effects of steady and unsteady blowing on a turbulent channel flow. They found that forcing frequency, $f^+ = 0.044$ ($f^+ = f\nu/u_\tau^2$, where ν-kinematic viscosity and u_τ-friction velocity), is responsible for the local increase in skin friction coefficient. Their findings also include the generation of more vortex structures downstream of blowing. The effects of localized steady blowing on the thermal transport were studied by using DNS in a turbulent channel flow by [8,9]. They imposed the blowing through a spanwise slot. The results of their numerical experiment demonstrate that the critical streamwise length of the blowing slot is 30 wall units. At the critical streamwise length of the slot, they detected a strong enhancement of heat flux. Although the previous studies have found interesting phenomena and mechanisms related to both velocity and thermal fields with blowing perturbations, most studies imposed blowing through spanwise slots. However, the blowing through spanwise slots is rarely encountered in practice. In studies pertaining to gas turbine film-cooling, blowing is imposed through round jets [10]. However, those studies did not focus on local variations of velocity and temperature [11]. Although it is known that jets in cross-flow generate coherent vortical structures and reduce the wall temperature downstream, the influence of vortices on heat fluxes and thereby on wall temperature has not been investigated.

Therefore, we seek to demonstrate the importance of coherent vortical structures in cooling or heating of the wall in a thermal turbulent channel flow. The study uses a turbulent channel flow with small steady blowing perturbations at the bottom wall. Both walls are kept at constant temperatures, and perturbations are set through five holes located at 25% of the channel length from the upstream end of the channel and distributed in the spanwise direction. The results show that the generation of coherent vortical structures increases the wall-normal heat flux compared to the streamwise heat flux. The article is organized as follows. Section 2 describes the numerical procedure. The results are laid out in Section 3, and the conclusions are given in Section 4.

2. Numerical Procedure

In this section, a brief description of the numerical procedure of the current simulations is given. Continuity, momentum and passive scalar transport equations are shown in their non-dimensionalized form by Equations (1)–(3), respectively. The non-dimensional form of the equations was obtained by using velocity scale U_c and length sale h, where U_c is the unitary mean laminar centerline velocity and h is the channel half height. If the dimensional form of the variables is denoted by the superscript *, $u_i = u^*/U_c$, $x_i = x^*/h$, $p = p^*/(\rho U_c)$, and $t = t^*U_c/h$ represent non-dimensional forms of instantaneous velocity, spatial coordinates, instantaneous pressure and the time coordinate. The non-dimensional temperature is given by $\theta = 1 - 2[(\Theta_{bw} - \theta^*)/(\Theta_{bw} - \Theta_{tw})]$, where Θ_{bw} is the constant temperature at the bottom wall and Θ_{tw} is the temperature at the top wall. Here, $\rho\, u_i$, p and θ represent the density of the fluid, instantaneous velocity components, instantaneous pressure and instantaneous temperature, respectively. In Equation (2), $\pi\delta_{1i}$ is the instantaneously-changing pressure gradient to maintain a constant flow rate in the channel. The governing equations were discretized in a staggered grid using a second order central differencing scheme. The Reynolds number of the flow is given by $Re_h = U_c h/\nu$, and Pr ($=\alpha/\nu$) is the molecular Prandtl number of the fluid, where α and ν stand for the thermal diffusivity and the kinematic viscosity of the fluid. The fractional step method was used in which viscous terms and advective terms are respectively treated implicitly and explicitly. An approximate factorization method was used to invert the large sparse matrix resulting from explicit treatment of viscous terms. Details of the numerical procedure can be found in [12]. The equations are solved for the baseline case (unperturbed) and the perturbed case.

$$\frac{\partial u_i}{\partial x_i} = 0 \tag{1}$$

$$\frac{\partial u_i}{\partial t} + u_j \frac{\partial u_i}{\partial x_j} = -\frac{1}{\rho}\frac{\partial p}{\partial x_i} + \frac{1}{Re_h}\frac{\partial^2 u_i}{\partial x_j \partial x_j} + \pi\delta_{1i} \tag{2}$$

$$\frac{\partial \theta}{\partial t} + u_j \frac{\partial \theta}{\partial x_j} = \frac{1}{Re_h\,Pr}\frac{\partial^2 \theta}{\partial x_j \partial x_j} \tag{3}$$

Figure 1 shows the physical domain of the flow, together with the profiles of perturbation velocity. The dimensions of the computational box are: $L_z = \pi h$, $L_y = 2h$ and $L_x = 8\pi h$. A grid-independence test was performed as detailed in [7]. The number of grid points in the streamwise, spanwise and wall-normal directions is 1153, 193, and 193 respectively. The mesh resolution in the unperturbed case is: $\Delta z^+ = 6.4$, $\Delta y^+_{min} = 0.095$, $\Delta y^+_{max} = 11.3$ and $\Delta x^+ = 8.6$. Note that we use the friction velocity u_τ of the unperturbed case in the scaling of the velocity field and both thermal and velocity fields were normalized using wall variables. The validation of the first and second order statistics for the unperturbed case is shown in Araya et al. [7] and Dharmarathne et al. [13].

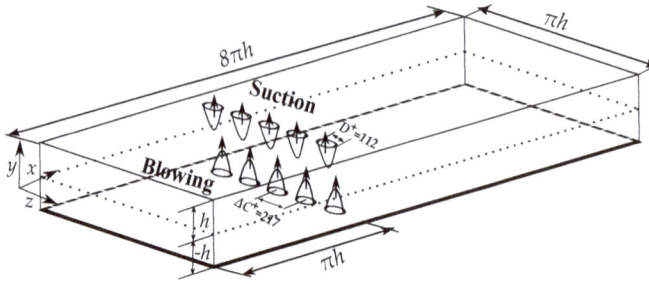

Figure 1. Schematic of the channel with spanwise local perturbations.

Periodic conditions are prescribed along the streamwise and spanwise directions. The no-slip boundary condition is imposed at both walls, except at locations where the jets are placed. The temperature boundary condition at the bottom wall is 1, and that at the top wall is -1. The fluid that comes through jets at the bottom wall has a non-dimensional temperature of 0.8, and at the top wall, the jet temperature is -1.2. The Courant–Friedrichs–Lewy (CFL) parameter remains constant during simulations and the time step $\Delta t^+ \approx 0.121$–0.159 for all cases.

A mean parabolic velocity profile with random fluctuations was used as an initial condition in the entire domain. The molecular Prandtl number, Pr, is 0.71, and the friction Reynolds number ($Re_\tau = hu_\tau/\nu$, where $u_\tau = \sqrt{\tau_w/\rho}$ is the friction velocity of the unperturbed channel) is 394. Here, τ_w is the shear stress at the wall. The local forcing, V_p, which creates the vertical perturbation, is modeled as follows:

$$V_p = A \sin(\alpha) \sin((R - x'_c)\pi/2R) \sin((R - z'_c)\pi/2R) \qquad (4)$$

where R is the radius of a blowing/suction jet and $x'_c = z'_c = 0$ is the center of a perturbing jet. Therefore, the range of jets is within $-R < x'_c < R$, $-R < z'_c < R$. The circular jets are approximated to the Cartesian grid with a percentage error $\approx 7\%$. The parameter A represents the ratio of the jet centerline velocity to channel centerline velocity (blowing ratio), and $A = 0.2$ in this study. This value of A complies with the widely-used blowing ratios in gas turbine film cooling [14]. Five equally-spaced jets were imposed at both walls in the spanwise direction; they are located at $L_x/4$ downstream from $x = 0$ as shown in Figure 1. The spanwise separation between the centers of two adjacent jets, ΔC^+, is approximately 217 in wall units, which accommodates the average separation of near-wall streaks in terms of spanwise wavelength, $\lambda_z^+ = 100 \pm 20$ [15], in between two adjacent jets. The diameter of the jets in wall units, D^+, is 112. The value of D^+ is approximately equal to the thickness of near-wall high and low speed streaks. In order to ensure the conservation of the mass flow rate inside the computational box, spatially-sinusoidal blowing at the bottom wall was synchronized with sinusoidal suction at the top wall, as shown in Figure 1.

3. Results

First, the effect of blowing on the mean velocity and temperature fields will be demonstrated. The next subsection describes the changes of the velocity and temperature fluctuations due to blowing perturbations. Then, we show the existence of hot and cold spots in the near-wall region by using two-dimensional contours on the xz plane. We then direct our focus to the heat fluxes, $\overline{u'\theta'}$ and $\overline{v'\theta'}$, in the following section. Subsequently, we show the generation of coherent vortical structures downstream of the jets.

3.1. Mean Velocity and Temperature Field

Figure 2 shows the variation of mean streamwise velocity (Figure 2a) and mean temperature (Figure 2b) along the centerline of the blowing jets (behind the jets, hereafter) at different downstream locations. The mean streamwise velocity at $1D$ downstream of blowing becomes negative in the near-wall region, as seen in Figure 2a. The negative streamwise mean velocity implies a reversed flow region just behind the blowing jets. The presence of jets obstructs the incoming boundary layer, which in turn creates a reversed flow region behind the jets. This region may significantly change the exchange processes between the wall and the boundary layer. The reversed flow region exists until $y^+ \approx 30$. The velocity quickly recovers from $y^+ = 30$ to $y^+ = 100$. This sudden change of velocity causes steep gradients of velocity. Steep velocity gradients generate more turbulence and increase turbulent momentum and heat transport in that region. The influence of blowing in the near-wall region is considerable even at $3D$ downstream of the jets. Although the deceleration of the flow in the near-wall region at $3D$ downstream is not as notable as at $1D$ downstream, it is noticeable at the outer-layer even at $10D$ downstream. Interestingly, we can observe a slight flow acceleration in the near-wall region at $5D$ and $10D$ downstream. This can be ascribed to the entrainment of accelerated flow in between two jets. The acceleration is a clear manifestation of the three-dimensionality of the perturbations, which may not be seen in slot blowing cases studied previously. The velocity deficit created by the presence of blowing jets gradually moves to the outer-layer of the channel, as suggested by the outward movement of the mean velocity deficit.

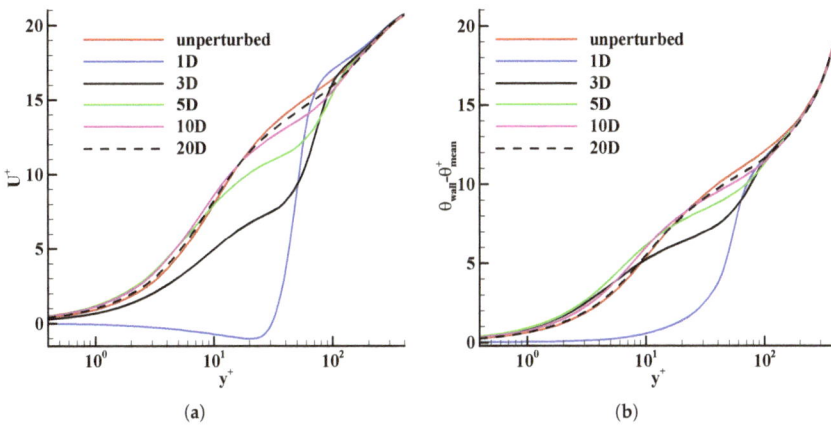

Figure 2. (a) Mean streamwise velocity variation and (b) mean temperature variation in the wall normal direction behind the jets.

We can observe simultaneous changes of the mean temperature profiles as Figure 2b shows. Due to the reversed flow region in the near-wall region at $1D$ downstream of blowing, velocity fluctuations might have been reduced. The reduction of velocity fluctuations reduces turbulent heat transport in the near-wall region; therefore, the temperature of the flow close to the wall is nearly similar to that of the wall. At the end of the recirculation region (around $y^+ = 30$), the mean temperature starts rising rapidly, creating steep temperature gradients. Steep temperature gradients generate temperature fluctuations.

It is peculiar to see that the temperature in the near-wall region at $3D$ downstream is lower than that of unperturbed flow. Because the mean velocity at $3D$ downstream is lower than that of the unperturbed flow, one would expect the temperature to be higher than the unperturbed flow. This inconsistency in the near-wall region at $3D$ downstream becomes more striking since the temperature profile becomes consistent with the velocity profile above the buffer region. Further, the mean temperature profiles at $5D$ and $10D$ downstream of the jets are also in compliance with the velocity profiles. The unpredictable behavior of the mean temperature profile (extra cooling effect) at $3D$ downstream may be attributed to the fluctuations of velocity and temperature fields.

3.2. Fluctuations of Velocity and Temperature

In this section, we discuss the variations of velocity and temperature fluctuations. Figure 3a depicts the root mean square (RMS) value of streamwise velocity fluctuations, u_{rms}^+, in wall coordinates. At the near-wall region (below $y^+ = 10$), the profiles of u_{rms}^+ at all downstream locations, except $1D$ downstream, are not significantly different from the unperturbed flow. At $1D$ downstream of blowing, streamwise velocity fluctuations are small compared to the unperturbed case. The obstruction of the cross-flow in the presence of blowing jet sets a wake at the immediate downstream of the jets. Since the wake is filled in with slowly-moving fluid, the recirculation region near the downstream of blowing jets attenuates turbulence. Above the reversed flow region ($y^+ \approx 30$), fluctuations increase rapidly due to the turbulence production caused by steep velocity gradients, and the maximum u_{rms}^+ occurs around $y^+ = 70$. Furthermore, for $3D$, $5D$ and $10D$ downstream locations, peak values of u_{rms}^+ are in the log-layer. The movement of the peak values of u_{rms}^+ towards the outer-layer could be attributed to the high v fluctuations due to blowing.

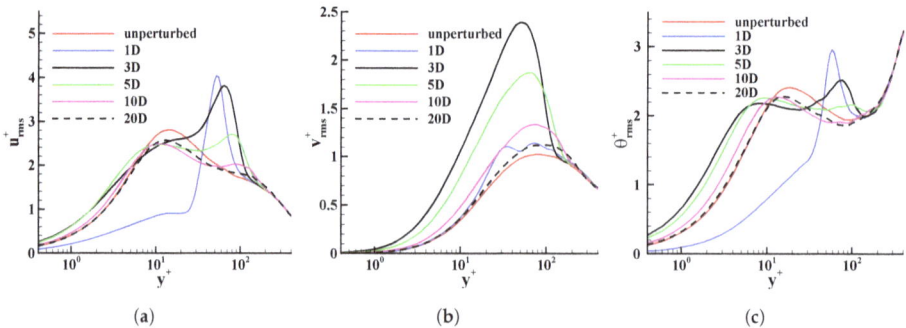

Figure 3. Variation of (**a**) streamwise velocity fluctuations; (**b**) wall normal velocity fluctuations; and (**c**) temperature fluctuations downstream, behind the jets.

RMS values of wall-normal velocity fluctuations at different downstream locations are shown in Figure 3b. It is interesting to see that v_{rms}^+ at $1D$ downstream is similar to that of the unperturbed flow particularly in the near-wall region. This indicates that the mechanism that creates wall-normal velocity fluctuations does not change due to the flow reversal just downstream of jets. However, wall-normal velocity fluctuations have considerably increased at $3D$ downstream. In fact, this location shows the highest increase in v_{rms}^+ from its unperturbed case out of all the observed locations. We speculate that wall-normal velocity fluctuations intensify due to the generation of more coherent vortical structures in the same region. The high fluctuations of wall-normal velocity move warm near-wall fluid upwards. Therefore, the amplification of wall-normal velocity fluctuations at the downstream of blowing may closely correspond to the low wall temperatures observed at $3D$ downstream. The increase of fluctuations is also significant at $5D$ and $10D$ downstream. However, the magnitude of fluctuations gradually reduces as the location moves away from the blowing jet.

Figure 3c depicts the variation of RMS values of temperature fluctuations, θ^+_{rms}. At $1D$ downstream, temperature fluctuations in the near-wall region are mitigated due to the reversed flow. It seems that temperature fluctuations correspond to streamwise velocity fluctuations at $1D$ downstream. As the location moves downstream, θ^+_{rms} behaves similar to v^+_{rms}. Temperature fluctuations in the near-wall region are amplified at the $3D$, $5D$ and $10D$ downstream locations. The observations of the temperature fluctuations suggest that turbulent mixing is intensified in the near-wall region after $1D$ downstream. The amplified wall-normal velocity fluctuations and temperature fluctuations relate to the abrupt change in mean temperature at $3D$ downstream. Moreover, we can predict significant differences between turbulent heat fluxes in the streamwise direction, $\overline{u'\theta'}$, and that of the wall-normal direction, $\overline{v'\theta'}$.

3.3. Hot Spots Near the Wall

To elucidate the changes of instantaneous and mean temperature in the near-wall region, we observe the temperature field around $y^+ = 5$. Figure 4a illustrates instantaneous high temperature spots (red color) in between jets where the flow is susceptible to acceleration. Right behind the jets where a reverse flow region is observed, high wall temperatures can also be detected. These instantaneous hot spots in between and right behind the jets contribute to the regions of high mean temperature in those regions, as we see in Figure 4b. According to Figure 4a, at $1D$ behind the jets, the high temperature spots exist. They are also observed on the mean temperature contours. When we move slightly downstream, around $3D$, we can see that cold (blue) spots are emerging right behind the jets. These sporadic events are directly coupled with sudden changes of flow phenomena like coherent vortical structure proliferation. The difference of instantaneous temperature between the hot and cold spots around $3D$ downstream is clearly noticeable. However, the changes that we see in mean temperature contours at the same region is not very distinguishable. Therefore, these instantaneous changes of thermal field are absolutely necessary in the design process. This highlights the importance of understanding the underlying flow physics of the changes in the instantaneous temperature field.

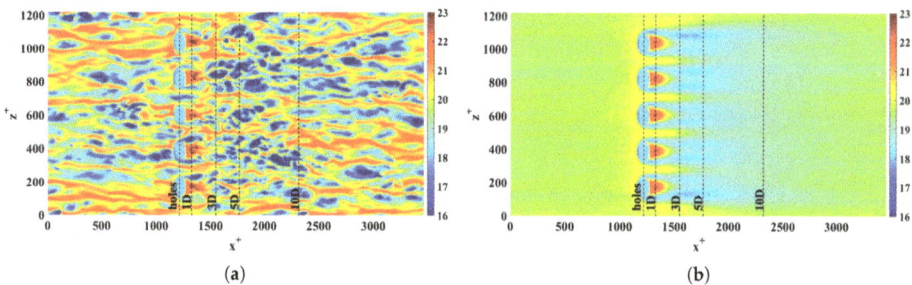

Figure 4. (a) Instantaneous and (b) mean temperature contours at $y^+ = 5$ on the xz plane. Both instantaneous and mean temperatures are normalized by Θ_{bw} and Θ_{tw}, as shown in Section 2.

Comparing the thermal field behind the jets with the unperturbed incoming flow, one can notice a significant reduction of hot spots behind the jets, i.e., $x^+ \sim 1300$–2400. This further confirms the importance of blowing jets in the proliferation of coherent vortical structures, which in turn, promotes the cooling of the wall, particularly beyond $3D$. These observations are clearly noticed in both the instantaneous and mean temperature contours, which extend far downstream in the channel, as shown in Figure 4.

3.4. Turbulent Heat Fluxes

This section discusses the changes of turbulent heat fluxes (streamwise heat flux, $\overline{u'\theta'}$, and wall-normal heat flux, $\overline{v'\theta'}$) due to blowing. Here, we are particularly interested in heat fluxes in the near-wall region to see which one of $\overline{u'\theta'}$ and $\overline{v'\theta'}$ influences temperature more.

In Figure 5, we observe that both $\overline{u'\theta'}$ and $\overline{v'\theta'}$ are negligibly small in the near-wall region at $1D$ downstream. The flow reversal just downstream of blowing jets diminishes the fluctuations of streamwise velocity and temperature. The reduction of fluctuations decreases turbulent transport near the downstream of the blowing jets. This observation complies with the high mean temperature at that location. A clear distinction between $\overline{u'\theta'}$ and $\overline{v'\theta'}$ can be seen at the $3D, 5D$ and $10D$ downstream locations particularly in the near-wall region. At the $3D, 5D$ and $10D$ locations, both $\overline{u'\theta'}$ and $\overline{v'\theta'}$ are higher than the respective values of the unperturbed case in the near-wall region until $y^+ \approx 8$. In between $y^+ = 10$ and $y^+ = 70$, one can clearly see that $\overline{u'\theta'}$ downstream of the perturbations is lower than that of the unperturbed flow.

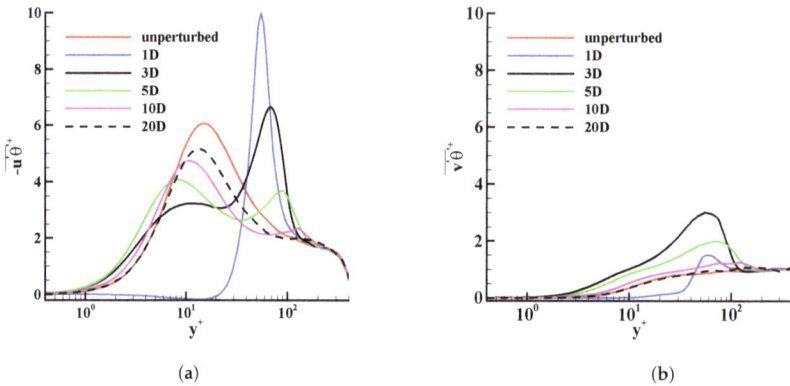

Figure 5. The variation of turbulent heat fluxes on the xy plane. (a) $\overline{u'\theta'}$ along the jets and (b) $\overline{v'\theta'}$ along the jets.

$\overline{u'\theta'}$ at downstream locations surpasses unperturbed conditions beyond $y^+ = 70$. On the other hand, $\overline{v'\theta'}$ is higher than its unperturbed counterpart for most of the boundary layer at all downstream locations, except at $1D$ downstream. The intensification of $\overline{v'\theta'}$ at downstream locations suggests a change in the mechanism that generates turbulent heat fluxes. We examine the changes of vortex structures downstream of blowing to see whether the vortex structure generation influences turbulent heat fluxes.

3.5. Generation of Vortex Structures

Figure 6a,b illustrates two-point correlations of streamwise velocity fluctuations, ρ_{uu}, in the near-wall region at $1D$ downstream of blowing behind the jets and in between jets, respectively. The streamwise length of ρ_{uu} demonstrates the flow acceleration in between jets (Figure 6b) in comparison to the deceleration of the flow behind the jets (Figure 6a). It is clear that the streamwise length scale of ρ_{uu} in between jets is much larger than ρ_{uu} behind the jets. This suggests that streaks in the near-wall region are stabilized in between jets, while they are destabilized behind the jets. In other words, the flow acceleration in between jets stabilizes low and high speed streaks. The flow acceleration is a result of the favorable pressure gradient that occurs between the jets. Conversely, there exists an adverse pressure gradient behind the jets. This spanwise pressure heterogeneity leads to a difference in coherent vortical structures' generation in between and behind the jets.

Figure 7 shows coherent vortical structures identified by λ_2 [16] iso-contours. The color spectrum of the contours depicts instantaneous temperature on the surfaces of the vortices. The number of coherent vortical structures downstream is higher right behind the blowing jets than that in between blowing jets because of the adverse pressure gradient. The rapid generation of coherent vortical structures has recently also been observed in reverse flow regions, even in high Reynolds number adverse pressure gradient flows, by Vinuesa et al. [17]. In fact, between the jets, the flow is accelerated (this is evident from the streamwise length of ρ_{uu} shown in Figure 6b), and this shows the evidence of less coherent vortical structures than behind the jets.

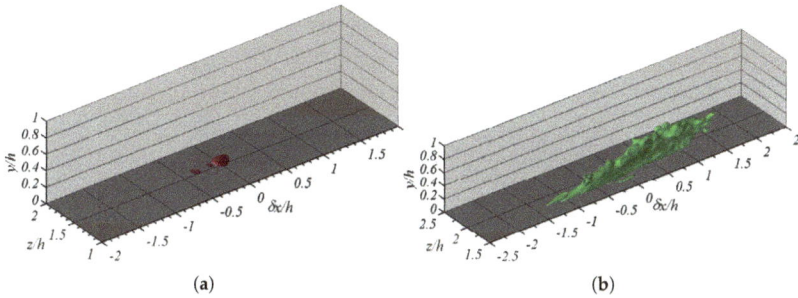

(a)   (b)

Figure 6. The two-point correlation of streamwise velocity fluctuations, ρ_{uu} (a) behind the jets and (b) between the jets.

The interaction of the jet and the incoming flow sets Kelvin–Helmholtz instability at the interface of the jet and the cross-flow. The process creates strong spanwise vortices. These spanwise vortices connect with quasi-streamwise vortices that are generated by destabilized low-speed streaks [18] behind the jet due to blowing perturbations. This leads to proliferation of hairpin vortices (as shown in Figure 7) downstream of blowing perturbations. In between jets, flow accelerates due to wall pressure gradient effects induced by blowing perturbations. The accelerated flow stabilizes low-speed streaks and these stabilized streaks are less vulnerable to break up; therefore, suppressing the generation of streamwise vortices downstream. This leads to a significant reduction in the number of vortices between two jets (leading to hot spots). The quasi-heterogeneity of vortex structure generation in the spanwise direction of the flow creates spanwise heterogeneity of turbulent thermal transport, as well. The red color of the temperature contours of Figure 7 directly downstream of blowing jets implies that vortical motions efficiently lift up hot fluid from the near-wall region to the outer region, thus promoting cooling. However, in between jets, heat fluxes are seen to be significantly lower than at $1D$ from the blowing jets; thus, they do not effectively remove hot fluid from the wall to the outer region. This is contrary to what we observed behind the jets in which high proliferation of coherent vortical structures led to effective cooling of the wall and, thus, generated extensive cold spot regions.

The previous section demonstrated that the wall-normal heat flux is highest around $3D$ downstream. To see whether this increase of turbulent heat transport has any relation to the generation of coherent vortical structures downstream of blowing, λ_2 structures are taken on two cross-sectional views at the $1D$ and $3D$ downstream locations, as shown in Figure 8a,b, respectively. It can be seen that the number of coherent vortical structures is greater at $3D$ downstream of blowing jets than at $1D$ downstream. The generation of more coherent vortical structures amplifies wall-normal velocity fluctuations in the near-wall region.

The wall-normal fluctuations in the near-wall region move low-speed fluid away from the wall. This action increases streamwise velocity fluctuations further from the wall. Figure 9 clearly shows the generation of wall-normal and streamwise velocity fluctuations with respect to the generation of vortex structures downstream of blowing. As the figure depicts, wall-normal velocity fluctuations increase at the near-wall region due to the proliferation of vortices. This is different than what we see in the unperturbed case in general, where wall-normal fluctuations are lower in the near-wall region due to boundary conditions. The increase of v_{rms}^+ in the near-wall region due to the proliferation of vortices moves the peak of u_{rms}^+ away from the wall. The steep mean velocity gradient in the streamwise direction is due to the wake recovery as observed previously in Figure 2a. The velocity gradients in the streamwise direction stretch vortices, and vortex stretching strengthens them. Since vortex stretching and the generation of vortices intensify near $3D$ downstream, wall-normal turbulent heat fluxes also amplify at the same region. This phenomena transports more heat flux from the wall, which in turn, reduces the temperature at $3D$ downstream. Importantly, streamwise heat fluxes in the near-wall region do not amplify due to the generation of coherent vortical structures. This is why one could see reduced $\overline{u'\theta'}$ at $3D$, $5D$ and $10D$ downstream than that of unperturbed flow (see in Figure 5a), particularly in the near-wall region. However, Figure 5b clearly shows that the generation and stretching of vortices directly affect the wall-normal heat flux throughout the boundary layer.

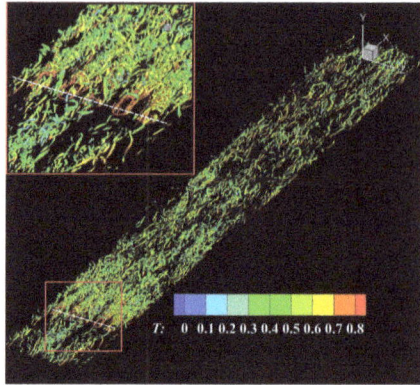

Figure 7. λ_2 contours colored with instantaneous temperature. The inset is a zoomed-in view of the flow field near the jets. The dashed white line shows the centerline of the jets. The iso-surfaces are drawn for $\lambda_2 = -3$.

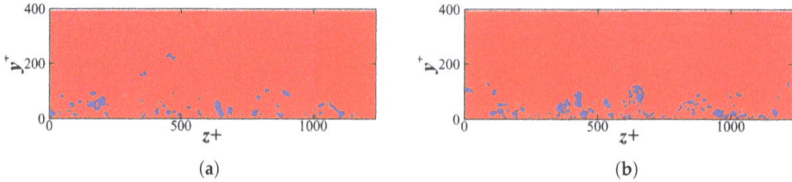

(a)   (b)

Figure 8. Iso-contours of λ_2 vortices at (**a**) $1D$ and (**b**) $3D$ downstream of the jets. The iso-contours are drawn for $\lambda_2 = -3$. The instantaneous realization corresponds to $t = 2200$.

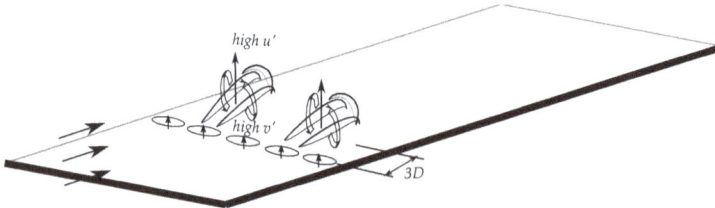

Figure 9. A schematic of hairpin vortices generated downstream of blowing. Wall-normal and streamwise velocity fluctuations are shown with respect to the vortex structure generation.

Fluids **2018**, *3*, 14

4. Conclusions

DNS of turbulent channel flow with small blowing perturbations was performed at a friction Reynolds number (Re_τ) of 394. Due to the obstruction of the flow by the presence of blowing jets, a recirculation region is created downstream of the blowing. The reverse flow region attenuates the intensity of turbulence, which in turn, reduces the turbulent transport of heat fluxes. Thus, it creates a high temperature region at the wall just behind the jets. We observed a peculiar change in mean temperature at $3D$ downstream at which we noticed low wall temperature even though the mean velocity of the flow is considerably lower than the unperturbed flow. The results show that $\overline{u'\theta'}$ is lower at $3D$ downstream than that of the unperturbed flow particularly from $y^+ = 10$ to $y^+ = 70$. On the other hand, $\overline{v'\theta'}$ is higher than that of unperturbed flow throughout the channel. We found that the generation of coherent vortical structures increases near $3D$ downstream, and they are intensified by vortex stretching due to steep velocity gradients. The results clearly indicate that the proliferation of coherent vortical structures downstream of the jets contributes to the removal of hot fluid from the wall to the outer region. However, in between the jets, the flow is accelerated mainly due to low pressure regions, which prevents the proliferation of coherent vortical structures, leading to high temperature regions.

Acknowledgments: This research was funded by NSF-CBET-ONR:1512393: International Collaboration: The role of initial conditions on LSMs/VLSMsin turbulent boundary layers. We acknowledge Dr. Stefano Leonardi at UT-Dallas for generously allowing us to use channel flow code for this study.

Author Contributions: Suranga Dharmarathne performed the simulations, coded all the post-processing programs and contributed to data analysis and interpretation. Venkatesh Pulletikurthi generated most of the figures, performed the literature review and contributed to data analysis. Luciano Castillo conceptualized the study, performed the data analysis and provided insights and guidance throughout the study.

Conflicts of Interest: The authors declare no conflict of interest.

References

1. Mahesh, K. The Interaction of Jets with Crossflow. *Annu. Rev. Fluid Mech.* **2013**, *45*, 379–407.
2. Keffer, J.; Baines, W. The round turbulent jet in a cross-wind. *J. Fluid Mech.* **1963**, *15*, 481–496.
3. Kamotani, Y.; Greber, I. Experiments on a turbulent jet in a cross flow. *AIAA J.* **1972**, *10*, 1425–1429.
4. Yuan, L.L.; Street, R.L.; Ferziger, J.H. Large-eddy simulations of a round jet in cross-flow. *J. Fluid Mech.* **1999**, *379*, 71–104.
5. Park, J.; Choi, H. Effects of uniform blowing or suction from a spanwise slot on a turbulent boundary layer flow. *Phys. Fluids* **1999**, *11*, 3095–3105.
6. Kim, K.; Sung, H.J. Effects of unsteady blowing through a spanwise slot on a turbulent boundary layer. *J. Fluid Mech.* **2006**, *557*, 423–450.
7. Araya, G.; Leonardi, S.; Castillo, L. Numerical assessment of local forcing on the heat transfer in a turbulent channel flow. *Phys. Fluids* **2008**, *20*, 085105.
8. Tardu, S.; Doche, O. Turbulent passive scalar transport under localized blowing. *J. Vis.* **2008**, *11*, 285–298.
9. Doche, O.; Tardu, S. Mechanism of wall transfer under steady localized blowing. *Int. J. Heat Mass Transf.* **2012**, *55*, 1574–1581.
10. Bogard, D.; Thole, K. Gas turbine film cooling. *J. Propul. Power* **2006**, *22*, 249–270.
11. Coletti, F.; Elkins, C.J.; Eaton, J.K. An inclined jet in cross-flow under the effect of streamwise pressure gradients. *Exp. Fluids* **2013**, *54*, 1589.
12. Orlandi, P. *Fluid Flow Phenomena: A Numerical Toolkit*; Fluid Mechanics and Its Applications; Kluwer Academic Publishers: Dordrecht, The Netherlands, 2000; Volume 55.
13. Dharmarathne, S.; Tutkun, M.; Araya, G.; Castillo, L. Structures of scalar transport in a turbulent channel. *Eur. J. Mech. B/Fluids* **2016**, *55 Pt 2*, 259–271.
14. Bunker, R.S. A review of shaped hole turbine film-cooling technology. *J. Heat Transf.* **2005**, *127*, 441–453.
15. Moin, P.; Moser, R.D. Characteristic-eddy decomposition of turbulence in a channel. *J. Fluid Mech.* **1989**, *200*, 471–509.

16. Jeong, J.; Hussain, F. On the identification of a vortex. *J. Fluid Mech.* **1995**, *285*, 69–94.
17. Vinuesa, R.; Örlü, R.; Schlatter, P. Characterisation of backflow events over a wing section. *J. Turbul.* **2017**, *18*, 170–185.
18. Schoppa, W.; Hussain, F. Coherent structure generation in near-wall turbulence. *J. Fluid Mech.* **2002**, *453*, 57–108.

MDPI

St. Alban-Anlage 66

4052 Basel

Switzerland

Tel. +41 61 683 77 34

Fax +41 61 302 89 18

www.mdpi.com

Fluids Editorial Office

E-mail: fluids@mdpi.com

www.mdpi.com/journal/fluids

www.ingramcontent.com/pod-product-compliance
Lightning Source LLC
Chambersburg PA
CBHW051846210326
41597CB00033B/5792